D1783332

METHODOLOGY, METAPHYSICS AND THE

HISTORY OF SCIENCE

BOSTON STUDIES IN THE PHILOSOPHY OF SCIENCE

EDITED BY ROBERT S. COHEN AND MARX W. WARTOFSKY

VOLUME 84

BENJAMIN NELSON

Courtesy of the Benjamin Nelson Foundation

METHODOLOGY, METAPHYSICS AND THE HISTORY OF SCIENCE

IN MEMORY OF BENJAMIN NELSON

Edited by

ROBERT S. COHEN

and

MARX W. WARTOFSKY

D. REIDEL PUBLISHING COMPANY

A MEMBER OF THE KLUWER ACADEMIC PUBLISHERS GROUP

DORDRECHT / BOSTON / LANCASTER

Library of Congress Cataloging in Publication Data

Main entry under title:

Methodology, metaphysics, and the history of science.

(Boston studies in the philosophy of science ; v. 84)
Bibliography: p.
Includes index.
  1. Science—Philosophy—Addresses, essays, lectures. 2. Science—Methodology—Addresses, essays, lectures. 3. Metaphysics—Addresses, essays, lectures. 4 Nelson, Benjamin, 1911–1977. I. Cohen, Robert Sonné. II. Wartofsky, Marx W. III. Nelson, Benjamin, 1911–1977. IV. Series.
Q174.B67 vol. 84   [Q175.3]   501s   [501]   84-15140
ISBN 90-277-1711-7

Published by D. Reidel Publishing Company,
P.O. Box 17, 3300 AA Dordrecht, Holland.

Sold and distributed in the U.S.A. and Canada
by Kluwer Academic Publishers,
190 Old Derby Street, Hingham, MA 02043, U.S.A.

In all other countries, sold and distributed
by Kluwer Academic Publishers Group,
P.O. Box 322, 3300 AH Dordrecht, Holland.

All Rights Reserved
© 1984 by D. Reidel Publishing Company, Dordrecht, Holland
No part of the material protected by this copyright notice may be reproduced or
utilized in any form or by any means, electronic or mechanical,
including photocopying, recording or by any information storage and
retrieval system, without written permission from the copyright owner.

Printed in The Netherlands.

TABLE OF CONTENTS

PREFACE ix

RICHARD M. BURIAN / Scientific Realism and Incommensurability: Some Criticisms of Kuhn and Feyerabend 1

MICHAEL MARTIN / How To Be a Good Philosopher of Science: A Plea for Empiricism in Matters Methodological [Commentary on Burian] 33

ROGER J. FABER / Feedback, Selection, and Function: A Reductionistic Account of Goal-Orientation 43

PAUL K. FEYERABEND / Philosophy of Science 2001 137

JOST HALFMANN / The Dethroning of the Philosophy of Science: Ideological and Technical Functions of the Metasciences 149

ALLAN JANIK / Comments on Jost Halfmann's 'Dethroning of the Philosophy of Science: Ideological and Technical Functions of the Metasciences' 173

HYMAN HARTMAN / Philosophy of Science and the Origin of Life 183

ANTHONY LEEDS / Sociobiology, Anti-Sociobiology, Epistemology, and Human Nature 215

ELENA PANOVA / Substance and Its Logical Significance 235

ARMAND SIEGEL / Tracking Down the Misplaced Concreton in the Neurosciences 247

ELISABETH STRÖKER / Does Popper's Conventionalism Contradict his Critical Rationalism? Objections against Popper in German Philosophy and Some Metacritical Remarks 263

A. SZABO / How to Explore the History of Ancient Mathematics? 283

E. V. WALTER / Nature on Trial: The Case of the Rooster that Laid an Egg 295

NATHAN SIVIN / Reflections on 'Nature on Trial' 323

DONALD D. WEISS / Toward the Vindication of Friedrich Engels 331

BIBLIOGRAPHY OF THE WRITINGS OF BENJAMIN NELSON 359

NAME INDEX 367

PREFACE

This selection of papers that were presented (or nearly so!) to the Boston Colloquium for the Philosophy of Science during the seventies fairly represents some of the most disturbing issues of scientific knowledge in these years. To the distant observer, it may seem that the defense of rational standards, objective reference, methodical self-correction, even the distinguishing of the foolish from the sensible and the truth-seeking from the ideological, has nearly collapsed. In fact, the defense may be seen to have shifted; the knowledge business came under scrutiny decades ago and, indeed, from the time of Francis Bacon and even far earlier, the practicality of the discovery of knowledge was either hailed or lamented. So the defense may be founded on the premise that science may yet be liberating. In that case, the analysis of philosophical issues expands to embrace issues of social interest and social function, of instrumentality and arbitrary perspective, of biological constraints (upon knowledge as well as upon the species-wide behavior of human beings in other relationships too), of distortions due to explanatory metaphors and imposed categories, and of radical comparisons among the perspectives of different civilizations.

Some of our contributors are frankly programmatic, showing how problems must be formulated afresh, how evasions must be identified and omissions rectified, but they do not reach their own completion. It is enough to clarify a problem, we have often heard in the history of philosophy, and yet we look forward to the further works of Burian, Feyerabend, Halfmann, and Szabo. Some are research reports, complete in themselves while parts of larger works too, and we think of the monographic essay of Faber on feedback and goal-orientation, as well as Hartman's exciting analysis of the origin of life, and Walter's often hilarious and always incisive account of Nature on juridical trial. Some are critical and elucidating, the papers of Leeds, Panova, Siegel, Ströker, Szabo, and Weiss; and others are equally clarifying as critical commentary, i.e., those of Martin, Janik, and Sivin.

*   *   *

Nearly twenty years ago, Benjamin Nelson's classical essay on 'The Early Modern Revolution in Science and Philosophy' appeared in the third volume of these *Boston Studies*. The essay had the intriguing sub-title 'Fictionalism, Probabilism, Fideism, and Catholic *Prophetism*', and its author urged upon us his view of modern studies of the origins of science thus: "That a conventionalist theory of science principally promoted by a devout Catholic physicist, philosopher, and historian of science was to be taken up and given world currency by the proudly anti-theological and anti-metaphysical Vienna positivists is one of the many ironies of this development." But Nelson's concern to challenge Duhem was matched in importance by his concern to establish the differences among those civilizational perspectives we mentioned above, to investigate the reasons *why* new perspectives arise in one civilization and *why not* in another. As E. V. Walter reminds us (p. 300), Nelson called this 'Needham's Challenge', to understand why modern science developed in Western Europe and nowhere else, to understand, moreover, just exactly the significance of Christian universalist theology for that development.

Benjamin Nelson's work has been described elsewhere, and we commend readers to the unique collection of his essays on the social and intellectual foundations of European culture, scientific and otherwise, edited by Professor Toby Huff under the title *On the Roads to Modernity: Conscience, Science, and Civilizations* (Rowman and Littlefield, Totowa, New Jersey, 1981). We dedicate this volume, close in spirit to aspects of his thought, to the ever-lively memory of a dear friend, Benjamin Nelson.

*   *   *

Nelson's bibliography completes this book, in cooperation with Toby Huff. Editing of the entire volume was immensely aided by the care and intelligence of Carolyn Fawcett, who also prepared the index.

July 1984

Center for Philosophy and History of Science,	ROBERT S. COHEN
Boston University	
Dept. of Philosophy,	MARX W. WARTOFSKY
Baruch College, City University of New York	

RICHARD M. BURIAN

SCIENTIFIC REALISM AND INCOMMENSURABILITY: SOME CRITICISMS OF KUHN AND FEYERABEND*

> ... only those revolutions in science will prove fruitful and beneficial whose instigators try to change as little as possible and limit themselves to the solution of a particular and clearly defined problem. Any attempt to make a clean sweep of everything or to change things quite arbitrarily leads to utter confusion. . . . True, I don't know whether scientific revolutions can be compared with social revolutions, but I suspect that even historically the most durable and beneficial revolutions have been the ones designed to serve clearly defined problems and which left the rest strictly alone.
>
> Werner Heisenberg
> *Physics and Beyond* (p. 148)

Over twenty years ago, the Boston Colloquium for the Philosophy of Science heard a series of papers concerning Paul Feyerabend's philosophy of science. On that occasion, Feyerabend defended his views against a variety of criticisms brought forward by J. J. C. Smart, Hilary Putnam, and Wilfrid Sellars. Although I was not privileged to hear these papers, my study of the published version of this debate, appearing in the *Boston Studies for the Philosophy of Science*, volume II,[1] greatly influenced my own views and was one of the chief factors leading me to write my dissertation on Feyerabend's philosophy of science. It is, therefore, only appropriate for me to continue the debate over Feyerabend's work in the present paper.

As it turns out, my central interest is in Feyerabend's notion of the incommensurability of theories — a notion which he shares with Thomas Kuhn. In spite of the great differences in the ways Kuhn and Feyerabend use this notion, much of what I have to say pertains to both of them. I shall deal primarily with Feyerabend's views in the sequel, but shall occasionally stop to take note of the relevance of what I say to Kuhn's work.

I

In order to set the stage for the main body of this paper, let me indicate briefly the context in which I wish to criticize Kuhn and Feyerabend. Contemporary philosophy of science may well be viewed as being in the throes of a Kuhnian change of paradigm. The regnant logical empiricist orthodoxy of the 1950s and early 1960s has come under increasingly severe attack;

matters have progressed so far that a Kuhnian might well contend that philosophy of science faces the proliferation of paradigms which Kuhn considers typical of a period of crisis. It now appears that the new generation of philosophers of science will be more deeply influenced by the work of such diverse thinkers as Agassi, Bromberger, Feyerabend, Hesse, Hanson, Kuhn, Lakatos, G. Maxwell, some bits of Popper's work, Quine, Sellars, Toulmin, and Wartofsky than it will be by Carnap, Feigl, Frank, Goodman, Hempel, Nagel, and Reichenbach. Whether, as of yet, the old tradition can fairly be said to be dethroned I am not sure. How soon a new dominant tradition will emerge out the current situation is anyone's guess!

Kuhn and Feyerabend have been among the leaders of the attack on the old orthodoxy. It will be well to bear in mind a sketch — or rather a caricature — of the position they attack. What follows is a modest example of such a caricature.

Logical empiricism holds that the aim of philosophy of science is to offer a rational reconstruction or an explication of fully developed scientific theories 'in the context of justification'. All, or nearly all, of the philosophically interesting and philosophically useful tasks of philosophy of science can be accomplished by the use of such reconstructions. They can be used, for example, to explore the general structure of scientific theories, to reveal the logic of scientific explanation, and to provide the necessary apparatus for assessing the precise content and the epistemic merit of scientific knowledge claims. With the addition of appropriate information about particular theories and experiments, such rational reconstructions are supposed to enable one to carry through an analysis of the kinds of claims particular theories make about the world and to evaluate the worth of such claims.

The empiricist side of logical empiricism expressed itself in the treatment of all scientific knowledge as contingent and as grounded in sensory experience. Feyerabend to the contrary notwithstanding,[2] most logical empiricists did not conceive of sensory experience as incorrigible or epistemically certain; nonetheless, they did require that the distinction between observational and theoretical claims be conceptually clear even if the borderline between the two was difficult to draw in practice. The usual way of making this distinction was to note that theories typically have two sorts of subject matter. Thus an atomic theory of matter (call it T) has atoms of the kind it describes as its *internal subject matter* and it has ordinary matter (chairs, rocks, metals, trees, stars) as its *external subject matter*.

If we label the language in which the terms characteristic of T appear L_T, the internal subject matter is determined by L_T. But if we are to know what it is that T explains, to what it is that the terms of L_T apply, we must be able to describe the external subject matter of T independently of T and L_T. This means that there must be a language L_O (usually called the observation language of T) in which we can describe and identify chairs, rocks, etc., without any recourse to the terminology of L_T.[3] To be sure, the observation language for some theories may, in turn, be a theory, but then we need only reapply the same argument to establish a hierarchy of ever-less-theoretical observation languages. If this hierarchy has a first member, it will be a completely theory-neutral language in which certain parts of our world and/or our experience are described. I shall call such a language an absolute or an a-theoretical observation language. We shall return to consider such languages below.

It is now possible to describe the chief kind of model for the language of a science used by logical empiricists. In this model there are two languages, an axiomatically structured theoretical language L_T and an observation language L_O. These languages are meshed together by a set of 'correspondence rules' or 'coordinating definitions' which connect some (usually highly derivative) expressions of L_T with expressions of L_O.[4] L_O is at least relatively stable as compared with L_T, for it contains pre-theoretical descriptions of that which the theory explains, and it must therefore remain unchanged by a change of theory. This stability is connected with an epistemic priority; theoretical claims are generally thought to be more indirectly supported and less secure than observational claims. Indeed, the epistemic assessment of theoretical claims proceeds, in this model, by an analysis of the logical relations established by the correspondence rules between the more problematic theoretical claims and the less problematic observational claims. But I shall not pursue the difficulties of confirmation theory today.

The asymmetry of L_O and L_T is brought out further by a brief consideration of correspondence rules. These rules are not usually conceived as connecting two fully meaningful languages, but rather as supplying the experiential content of the theoretical language which, in their absence, would be unconnected with experience. It follows that it is an error (though one which has often been committed) to treat correspondence rules as rules of *translation* between L_T and L_O. The sentences of L_T do not have the same meaning as any or all of the observation sentences with which they are coordinated by the correspondence rules. Their logical form and the inferential connections they exhibit differ essentially from those of the corresponding observation sentences.[5]

II

I turn now to the logical empiricist treatment of a perennial issue in the philosophy of science, namely the controversy between realism and instrumentalism. Feyerabend's critique of the old orthodoxy will enter the discussion.

Roughly speaking, instrumentalism is "the view that scientific theories are instruments of prediction which do not possess any descriptive meaning" (Feyerabend [15], p. 143), while realism is the view that singular theoretical claims are capable of literally describing the world. Consider a sentence of some well-defined theory of elementary particles — for example, 'The proton in space-time region R is in state S'. An extreme realist (provided he does not consider the whole theory 'bracketed' as an acknowledged approximation to a better theory) would maintain that this sentence claims, truly or falsely, that there is an invisible, non-touchable, non-colored object of extremely small dimensions, an object with properties assigned to it by the theory, located in a certain sub-microscopic, and hence not directly examinable, domain. An extreme instrumentalist would deny that this sentence had any such meaning. He would unpack it, instead, as some sort of shorthand means of deriving a large, perhaps infinite, set of claims about observations, for example, such claims as 'If such and such an (observationally described) apparatus were introduced at R', the probability of observing a scintillation would be p'.

The point of immediate interest is that the instrumentalist takes the meaning of the theoretical claim to be such that — analytically — its truth or falsity (so far as these terms may be applied to it) is simply a function of the truth or falsity of purely observational claims; no descriptive claim beyond observational ones is included in theoretical statements. The realist, in contrast, treats such observational claims as supplying contingently appropriate evidence for the truth of the theoretical claim, but not, properly speaking, as its truth condition.

It should be clear by now that once the two-language model of scientific discourse is adopted, the focus of the controversy over realism and instrumentalism will be over the truth conditions of the sentences of L_T. An extreme instrumentalist will claim that, so far as truth conditions are pertinent, they can be expressed entirely in L_O. An extreme realist will claim that singular sentences of L_T (or straightforward translations of those sentences) are requisite for expressing the truth conditions of L_T. A number of intermediate positions are possible and can be found in the literature.

At this point, let me bring Feyerabend into the picture. He claims[6] that the two-language model is inherently instrumentalist in that it can supply only observational truth conditions for theoretical sentences. He holds, furthermore, that any attempt to assign independent 'descriptive meaning' to theoretical claims will break down the distinction between L_T and L_O since *any* descriptive content in the theoretical language would infect the observation language via the correspondence rules. Thus, if an atomic theory of matter has any descriptive content, acceptance of such a theory is tantamount to acceptance of a re̲description of what it is to be a chair or a rock, even if one studiously avoids using any other descriptive *terms* than those of L_O! It would follow that evidence in favor of such a theory is *ipso facto* evidence *against* the pre-theoretical descriptions of the behavior of rocks and chairs. This result is intolerable for a two-language theorist who wishes these pre-theoretical descriptions to be able to serve as evidence for an atomic theory of matter. As a result, the two-language theorist is forced to deny independent descriptive content to theoretical claims. As evidence that this *is* the result, Feyerabend points to the fact that, while there is a traditional problem concerning the existence of theoretical entities, very few people have noted that there is an equally pressing problem concerning the existence of observable entities *precisely because of the evidence in favor of the existence of theoretical entities*.

Now, Feyerabend wishes to treat the language of science as a unitary language. He believes that the distinction between L_O and L_T cannot be made conceptually clear. He believes that our ordinary concepts *ought* to fall victim to scientific advance. Therefore, he is quite willing to accept the consequences of his criticism of the independence of L_O and L_T. I shall turn to these matters shortly. For now, the central point to be aware of is his claim that any attempt to keep L_O independent of L_T must deprive L_T of descriptive content and therefore result in an instrumentalist interpretation of theories. He concludes that standard two-language models of science exclude *a priori* logically and physically cogent descriptions of the world — and hence that these models should be abandoned.

How fair is this to Feyerabend's opponents? We shall have to leave this question open to a degree. I shall point out certain moves that such philosophers as Carnap, Hempel, and Nagel have made in the direction of accommodating a realistic point of view, but I shall also enter some reservations about the degree of accommodation open to them. It is impossible to resolve this complex question, on which there is a large literature, with a few brief comments.

Consider first the matter of stating the truth conditions for theoretical statements. As Carnap and Hempel have pointed out explicitly, their views require these conditions to be stated in a semantic meta-language. Whether or not this language should treat theoretical statements as independently meaningful or not is an open question, not necessarily settled by reference to the structure of the object language. Hempel, for example, points out that, in order to state the relevant truth conditions, "the semantic meta-language must contain either the theoretical expressions themselves or their translations", and maintains on this basis that the theoretical expressions, "must be antecedently understood if the semantical criteria are to be intelligible" ([27], p. 696). But this means that the issue between instrumentalism and realism must be settled *before* applying the apparatus of the (Carnap—Tarski) semantic theory of truth.[7] This makes the orthodox position vis-a-vis the instrumentalism—realism controversy more opaque than it might at first appear.

Feyerabend does not seem to appreciate the importance to his opponents of this distinction between an object language (in which the *evidence conditions* for theoretical claims are stated) and a semantic meta-language (in which the *truth conditions* for theoretical claims are stated). Therefore he does not distinguish clearly between truth and evidence conditions and does not appreciate the flexibility of the orthodoxy.

It must be admitted that Feyerabend's opponents were often unclear about the distinction between truth conditions and evidence conditions. Given the two-language analysis of theories, it is certainly tempting to adopt the standard view and to limit the semantic meta-language to the observation language combined with a syntactic meta-language. Carnap especially leans in this direction. But since this would require the truth conditions of theoretical statements to be only partially interpreted, it is not clear that this choice of meta-languge is best. And, as the quotation from Hempel two paragraphs above demonstrates, it is not the only choice open to the two-language theorist.[8]

One of the ways in which such men as Carnap, Hempel, and Nagel tend to accommodate realistic views is their recognition that as a matter of fact scientific laws cannot be formulated adequately in any known observation language (c.f., e.g., [5], Section 24). Accordingly, scientific concepts are accepted as indispensable; their content must go beyond that of the concepts of any (known) observation language. Indeed, so far as the answer to the theoretician's dilemma is that theories are logically (and not just pragmatically) indispensable, it would seem impossible to apply a purely instrumentalist account to them. Nagel tries to show that, at this point, the whole controversy is bogus, being but "a conflict over preferred modes of speech" ([41], p. 152) while Hempel tries to show that, on his reading, "To assert that the terms of a theory have factual reference, that the entities they purport to refer to actually exist, is tantamount to asserting that what the theory tells us is true; and this in turn is tantamount to asserting the theory" ([28], Section 10, p. 219).

Unfortunately, these moves by no means suffice to show that the two-language model can be naturally allied with a realistic outlook. The central issue for the realist is the status of singular claims in theories agreed to be correct or, better, in theories which would be agreed to be correct in an ideal science. Should such claims be considered true in the same sense as correct observation claims are considered to be true? Now the two-language model (as developed here) leaves the singular sentences of theories without any independent interpretations as long as those sentences are only 'partially interpreted' in terms of observation sentences. This suggests that even a philosopher who, recognizing the indispensability of singular theoretical claims, wishes to extend the status of truth to them may still have to recognize that the theoretical truths are in important ways dependent ones and are not, as observational truths are supposed to be, independent of the truth of the claims of any other framework.

In spite of their arguments to the contrary, there is just such an asymmetry in the theories of Hempel and Nagel. To adopt a thoroughgoing realism it does not suffice, *pace* Hempel, to 'assert the theory' (which even an extreme instrumentalist may do), nor to assert the theory while granting its sentences 'the status of significant statements' ([28], p. 219). It is necessary, in addition, that the singular sentences of the theory be interpreted as singular statements. And this is precisely where the notion of 'partial interpretation' gets Hempel and Nagel into trouble, for both treat these singular theoretical sentences, when partially interpreted, as statement *forms* rather than statements.

The underlying reason for this appears to be their view that, like implicit definitions, partially interpreted theoretical sentences may be satisfied by many different entities; partially interpreted sentences are therefore like statement forms in that they have indeterminate factual reference.[9] But, insofar as they are statement forms, they are not properly true or false.

Hempel and Nagel do, however, offer analogues of the singular statements (which their view makes into statement forms), analogues which they claim may be true or false in the usual sense. In effect, these analogues are obtained by adding appropriate existential quantifiers in front of every theoretical sentence.[10] Thus the inscriptions which we have treated as singular theoretical sentences and which Hempel and Nagel have treated as sentential forms may be replaced by existentially quantified sentences which are then treated as true or false.

This, however, does *not* resolve the difficulty, for Nagel and Hempel still treat partially interpreted predicates as predicate *variables* rather than predicate *constants*. As a result there will be no descriptive constants available as substitution instances for the existential quantifiers governing theoretical terms. It follows that there will be no true singular sentences of the theory.[11] Furthermore, it may well be argued that the sense in which existentially quantified theoretical sentences (which can have no true substitution instances) are true must differ from that in which existentially quantified observation sentences (which are true only if they have true substitution instances) are true. At the very least, we may say with Sellars that:

it is logically odd, given that '*f*' is a *descriptive* variable, to suppose that the theoretical statement

$$(\exists f)(x)\, fx$$

might be true even though no substitution instance of the function '*fx*' *could* be true. ([51], p. 177.)

I conclude, then, that the two-language model as here described cannot accommodate the strongest versions of scientific realism, although it is by no means confined to a pure instrumentalism.

To close this section, it might be well to note that most of the issues we have been discussing may be seen to revolve around the status of correspondence rules. As long as some form of the observational–theoretical distinction is preserved, these rules will determine the interrelations of observational concepts and theoretical concepts. One way of explicating a philosopher's doctrines concerning these interrelations is by examining his treatment of correspondence rules. Accordingly, these rules will provide a convenient focus for some of our subsequent discussions.

III

Now to Kuhn and Feyerabend. I shall start by sketching certain of their disagreements with logical empiricism.

Both men are acutely aware of a *chorismos* between the history and the philosophy of science, and both seek to bridge the gap by dealing with fundamental philosophical questions which grow out of historical studies. They blame philosophers for ignoring these questions, and suggest that it is partly because of the extreme degree of idealization of actual science built into a program of rational reconstruction that philosophers have been able to ignore them.

A particularly important set of questions concerns major changes of theory — the kind of changes which Kuhn has labelled revolutionary. These questions eluded logical empiricists because their central concern was to elaborate the structure of single theories rather than to compare competing theories. Among the questions involved are these: are such major changes of theory subject to experiential control? To what extent are they justified by observations and experiments? Can they be accounted for primarily by the 'internal logic' of scientific development or are they chiefly a consequence of external factors? Can they be identified and pinpointed clearly enough for the historian to be able to determine whether they are sudden or whether they involve extended, piecemeal adjustments in the theoretical structure of science? To what extent are such changes discontinuous; what sorts of continuities do they exhibit? How does one determine whether revolutionary changes of theory are progressive? What is the extent of reconceptualization which such changes of theory bring in their wake?

Questions like these are not purely historical. They presuppose considerable philosophical and methodological apparatus, for example, in individuating theories, in determining what the internal logic of scientific development allows, in assessing the extent of reconceptualization involved in a given case, and so on. Kuhn and Feyerabend both argue that if one accepts the logical empiricist doctrine, most of these questions will either be answered falsely or will turn out to be unanswerable.

One example will suffice to illustrate this contention. Both Kuhn and Feyerabend argue that two-language theorists will be unable to justify the choice of one theory over another when they deal with serious historical cases of theory competition. That is, logical empiricists will have to treat each competing theory in such cases as being as good (or as bad) as the others. After all, any halfway decent theory will be able to account for a large range of experience, and can be made to conform to whatever structural constraints a rational reconstructionist comes up with. (There is no *structural* reason to prefer, say, the special theory of relativity to Newtonian mechanics.) On this score, logical empiricism will have to award competing theories equally high marks. On the other hand, every major theory is born false and is unable to eliminate completely the difficulties it faces. There are always systematic and experimental errors, mistaken applications of the theory due to mistaken auxiliary hypotheses, domains in which the application of the theory is uncertain, etc.[12] There is usually conceptual difficulty as well, resulting from the conflict of every new theory with the more-or-less well confirmed theory which it aspires to replace. In short, logical empiricists ought to treat all theories as being falsified and untrustworthy. They should hold that *no* theory is acceptable except as a tool of prediction, and they should admit that they can judge the value of the tool only after the prediction has succeeded or failed. But to say this would be to abdicate as philosophers of science.

Kuhn and Feyerabend account for these supposed shortcomings in logical empiricism in rather similar ways. Both men, for example, offer a mélange of historical, psychological, and philosophical arguments against the claim that what is observed is independent of the theories which are held by those who do the observing.[13] Both men argue that two-language models of the language of science are radically wrong because the whole of human language is thoroughly theoretical. Correspondingly, both men include far more under the notion of a theory than their predecessors; they treat theories as extremely comprehensive frameworks of presupposition and belief rather allied to metaphysical systems.[14] Both men insist that observational categories undergo radical alteration in cases of a major change of theory. And, finally, when both men were in Berkeley, they worked together in developing an account of revolutionary change-or-theory as a change from a given theory to an *incommensurable* theory.

I must now hasten to undercut the false impression which the preceding paragraphs may have created that Kuhn and Feyerabend are as similar as Tweedledum and Tweedledee. Nothing could be further from the truth.[15] In spite of the similarity of their diagnoses of the faults of orthodox philosophy of science, their prescriptions for philosophy and for scientific methodology are radically different. The central matter is their attitude toward scientific revolutions. Both insist on the disruptive character of these revolutions and both insist that they are at least occasionally necessary. But they disagree about the reasons for which they are necessary, about the way in which they are brought about, and about their desirability.

To Kuhn, scientific revolutions are necessary evils, brought about by the inability of a theory or a paradigm to remain empirically adequate. Normal science is a non-revolutionary puzzle-solving activity which fits the logical empiricist mold in many respects; it involves the progressive improvement of the fit between recalcitrant experience and theory. But the very puzzles involved are set by the theory which is being elaborated; the criteria of success and failure are intra-theoretic, and therefore essentially mono-theoretic. Now, it often happens that a particular puzzle proves intractable. In such cases, the puzzle may be set aside for a while in favor of more tractable ones, or it may be ignored, or it may be taken seriously as threatening the paradigmatic theory. When enough puzzles prove intractable, or when particularly deep puzzles prove intractable, or when a wide range of puzzles prove simultaneously intractable, the theory enters a period of crisis. The result is a cessation or attenuation of normal science in favor of philosophical and foundational debate. It is at this stage that a scientific revolution is most likely to occur; if it does, the world is reconceived in accordance with a new paradigm governed by a new, incommensurable theory which determines the new set of puzzles dominating the new normal science. Kuhn argues that the mono-theoretic periods of normal science are necessary in order to ensure the full and progressive development of each theory and in order to develop the empirical basis for a revolution. In order, therefore, to ensure that revolutionary new theories accomplish more than those they replace, revolutions must be resisted as long as possible, and old theories supported tenaciously. This tenacity should be great enough so that theories are *not* abandoned in the face of seeming falsification or even because they encounter a large battery of intractable problems — after all, such difficulties *may* be accommodated in the end precisely by the problem-solving techniques of normal science.

Feyerabend, in contrast, adopts the slogan "revolution in permanence!". He maintains that experience can be made to accord with *any* theory with sufficient tenacity and with liberal employment of auxiliary hypotheses, especially if the mono-theoretic authoritarianism of Kuhn's "normal science" is allowed to take hold.[16] He argues further that since every theory is born false, as Kuhn admits ([33], p. 146), it will be impossible to distinguish cases of deviational noise from cases in which a theory is radically in error *unless the theory is judged against a background of alternative theories which mark off certain deviations as significant*. As a typical example of this sort, Feyerabend discusses the notorious $43''$/century advance of the perihelion of Mercury beyond Newtonian predictions.[17] He suggests that no Newtonian should take this rather typical recalcitrant puzzle so seriously as to think that it could overthrow the whole Newtonian edifice *except for the presence of a radical alternative*, such as the general theory of relativity, which singles out this bit of deviational noise as being of fundamental importance.

Feyerabend therefore opposes Kuhn's mono-theoretic methodology with a call for theoretical pluralism. To Kuhn's principle of tenacity — retain and develop a theory even in the face of extreme counterevidence in the hope of improving the theory and accommodating the evidence — he adds a principle of proliferation: invent, elaborate, and defend theories incommensurable with those currently accepted, even if the latter are well confirmed. Feyerabend's primary justification for the principle of proliferation is that its acceptance is requisite to maximize the *testability* of whatever theories one accepts.

Because of the place they assign to scientific revolutions, both Kuhn and Feyerabend deny that scientific knowledge converges to an ideal. Kuhn nonetheless thinks that there is measurable progress in science, and offers as criteria of progress: "accuracy of prediction, particularly of quantitative prediction; the balance between esoteric and everyday subject matter; and the number of different problems solved" ([33], p. 206).[18] Feyerabend's more radical approach results in a new conception of knowledge as "an ever increasing ocean of alternatives, each of them forcing the others into greater articulation, all of them contributing via this process of competition, to the development of our mental faculties" ([23], pp. 224–25).

IV

I now turn to a brief examination of what it means for two theories to be incommensurable. Since Feyerabend is more explicit than Kuhn on this matter, and since his doctrine is more clear-cut and more radical, I shall deal primarily with his account of the matter.

The first point to note is the sweeping nature of the theories which are involved. Feyerabend says that only *universal* theories can be incommensurable. He characterizes such theories in various ways — e.g., as those which say something about everything in the world, as those which contain their own observation language, and as those which have no external subject matter. Whichever of these accounts one adopts, Feyerabend argues that the meaning of every term of a universal theory is determined by the theory itself. In particular, there is no ostensive component of meaning. (Kuhn disagrees about this.)[19] The fact that a term or sentence is uttered as a response to stimulation has no bearing on its meaning, for its meaning is determined by the inferential connections it has within the theory in question.[20]

It is now easy to understand why Feyerabend holds that communication between differing theorists is possible only if they hold some universal theory in common. His doctrine of incommensurability, which has undergone a number of revisions,[21] makes this explicit — as well as making it clear that Feyerabend sees no necessity for two people to hold *any* theories in common.

I shall discuss this doctrine in its penultimate form, the form it took shortly before Feyerabend wrote *Against Method*. I shall later argue that it is his mistaken acceptance of the doctrine of incommensurability in this form which leads him into the anarchistic excesses of which *Against Method* is the culmination.

Two theories are said to be incommensurable when, if they are interpreted realistically, "they do not share a single statement and cannot be brought into deductive relationships."[22] Feyerabend offers a number of arguments to show that incommensurable theories *do* exist, that they *can* be used to criticize each other, and that it is *desirable* so to use them. For the moment I shall not comment on his arguments for the existence of incommensurable theories except to say that they depend essentially on his position that theories without external subject matter are fully meaningful. We shall have to ask whether or not this is so in the sequel.

The problem concerning the possibility of using incommensurable theories to criticize each other can be stated very briefly. If two theories do not share any statements and cannot be brought into deductive relation with each other, then no claim, no prediction, made within one, can contradict any claim made within the other. In short, it is hard to see how the two theories could be made either to agree or to disagree with each other. And it is certain that there can be no crucial experiments between them, since to have a crucial experiment between two theories, the theories must *agree* in their description of the experimental situation and *disagree* in their description of the outcome of the experiment. The issue thus becomes whether or not incommensurable theories can compete with each other in such a way that there are rational or empirical grounds for choosing between them.[23] What is the nature of the contact between theories which have no common external subject matter and whose statements cannot be put into deductive relation with one another? This is another focal question which will be faced below.

Assuming for the moment that critical contact of some sort is possible between incommensurable theories, Feyerabend argues that the use of incommensurable competitors is vital to ensure the thoroughness of theory testing. After all, all observations contain a margin of error and all theories are supported or tested only by means of a finite set of experiences. Alternatives are needed to make agreement between a theory and experience significant. Alternatives are needed if divergences between a theory and experience are to count as evidence *against* the theory rather than serving as the source for mere normal science puzzles to be accommodated sooner or later within the theory. Clearly, the stronger an alternative to a given theory is, the more powerful a weapon it will be for testing that theory. Since incommensurable theories do not agree anywhere, it follows that they will make the strongest possible alternatives provided only that they can be made to compete with one another. Since testability is agreed on all sides to be desirable, the use of incommensurable alternatives must also be desirable. This is the heart of Feyerabend's argument for the desirability of employing incommensurable theories. It should be noted that this argument depends essentially on the claim that incommensurable theories can be made to criticize each other. I shall dispute that claim shortly.

V

It is time to gather a few threads together. I shall list the central departures from the old orthodoxy with which the rest of this paper will be concerned. Except where noted, these departures are common to both Kuhn and Feyerabend.

(1) The treatment of theories as unfinished and developing entities which must be evaluated while in flux, rather than as finished products to be evaluated in the artificial context of justification.

(2) The treatment of scientific discourse as involving a thoroughly unified language, that language being theoretical throughout. Considered from another angle, this departure from standard two-language theories may be conceived as the denial that there is any theory-neutral subject matter for highest-level theories.[24]

(3) The treatment of change of highest-level theories as altering the *whole* of scientific discourse, including the categories of that discourse.

(4) The treatment of scientific revolutions as drastically altering the experienced world. This comes about because observations are described in the category scheme of the language of science and because both Kuhn and Feyerabend treat the categories of observational discourse as the categories of experience. As Feyerabend puts it, "each theory will possess its own experience" ([20], p. 214), or as Kuhn puts it, "we may want to say that after a revolution scientists are responding to a different world" ([33], p. 111).[25] So much for the orthodox notion of the continuity of scientific progress!

(5) The explicit denial, peculiar to Feyerabend, that there is any theory-independent component of meaning for observation sentences. This denial brings with it the claim that, in the case of a disagreement between two highest level theories, no retreat to a less theoretically committed description of the relevant observations is possible.

(6) The adoption of scientific realism together with the insistence that realism requires a complete redescription of the world in accordance with whatever highest level theory one adopts. (Kuhn is not as explicit about this as Feyerabend.)

(7) The insistence on the availability (and, in Feyerabend's case, the desirability of using) incommensurable theories, and, correspondingly, on the completeness of the overthrow of the abandoned theory *and its experience* in scientific revolutions.

VI

These departures from orthodox empiricist positions raise a host of methodological problems. Before embarking on a discussion of some of these problems, however, it would be well to point out that the difficulties with which I will be concerned arise only in cases of revolutionary change or theory competition. Kuhn's 'normal science' with its relatively stable criteria of success for theories and its gradual fitting of theory to experience, can be readily fitted with the logical empiricist account of science. Similarly, though he seems entirely unaware of it, Feyerabend's account of theory competition is very much like the orthodox account except when the competing theories are at the highest level. Competing subordinate theories, after all, pertain to the same external subject matter, namely the internal subject matter belonging to the highest-level theory in which they are embedded. Accordingly, those theories can be connected to a common external subject matter by correspondence rules and compared with each other in essentially the manner suggested by two-language models. It is only when highest-level theories are supposed to compete, only when revolutionary changes are supposed to occur, only when incommensurability is in question, that we face a new situation. I now turn to some of the difficulties of that situation.

Many of the problems we will consider center on the difficulty of comparing incommensurable theories. The difficulty is fundamental, and it has not gone unremarked in the literature.[26] But it is important to assess these problems with great care since there is little hope of resolving the basic problems facing philosophy of science today without coming to grips with them.

Although Feyerabend no longer considers incommensurable theories to be in critical contact with each other, he has offered a number of procedures which he once thought would allow such theories to be compared. Limitations of space do not permit me to deal with all of these procedures on this occasion,[27] but it will be instructive to see how, in a couple of typical examples, they fail.

One of the procedures which Feyerabend suggests for comparing two incommensurable theories, T and T' is to "construct a model of T within T' and consider its fate. ... Within physics the construction, in T', of a model $\bar{T}$ of T, usually is not possible without violation of highly confirmed laws. This, then, is sufficient evidence for the rejection of $\bar{T}$ and, via the isomorphism, of T ([23], p. 233)."[28] Against this procedure, it need only be pointed out that the terms entering the model of T within T' will, on Feyerabend's account, be governed by the theoretical principles of T' and hence, according to Feyerabend, cannot be used to make the same claims as are made with the isomorphic terms of T. Therefore the fate of T and of the model may be entirely different. For example, it would not be difficult to construct an isomorphic model of the wave theory of sound (in which the waves are longitudinal) within a wave theory of light and test for the consequences (e.g. absence of plane polarization). When these tests refute the model *within the theory of light*, this evidence clearly does not give grounds for abandoning the original theory of sound. If the internal subject matter of a theory determines the meaning of its claims, and if no constraints on meaning are generated by the pragmatics of the theory, then no structural similarities will suffice to establish the requisite linkage between theories. By excluding *all* logical contact, incommensurability blocks comparison.

It should be added that the procedure of theory comparison just discussed can be of considerable use once incommensurability is abandoned. Thus, if it could be determined that the entities (or properties, behaviors, etc.) referred to or picked out by the model were in an appropriate sense 'the same' as those picked out by T,[29] then this procedure may, indeed, be used to compare T and T'. The critical point is this: the isomorphism would be used either to demonstrate *similarity* of meaning between certain terms of T' and certain terms of T or it would be used only after such similarity had been demonstrated. Feyerabend's procedure can, therefore, be used only if the incommensurability of T and T' is foresworn.

Another procedure by which Feyerabend hoped to be able to compare incommensurable theories calls for "the invention of a still more general theory" which will supply "a common background that defines test statements acceptable to *both* [of the incommensurable] theories." ([20], pp. 216–17). Again, there are a number of objections to this procedure. For present purposes, one of the most fundamental ones, already stated by Shapere ([52], p. 58) will suffice. According to Feyerabend's views on meaning, this procedure would entirely change the meanings of the terms in the theories being tested. After all, meanings are supposed to be a function of the highest-level theoretical principles. It follows that two theories which are both embedded in a third in such a way that the meanings of some of their terms are determined by the principles of the third theory (as is required if we are to have 'test statements acceptable to both theories') cannot be the same theories as the two incommensurable ones with which we started, for no logical contact is possible between incommensurable theories. This procedure works only if Feyerabend abandons his doctrines of meaning and incommensurability.

All of Feyerabend's abstract procedures for comparing incommensurable theories are as faulty as these two. But the situation is even worse than this suggests. When he comes to deal with particular cases, Feyerabend often tries to show that the theories in question are both incommensurable and competing. It turns out, however, that the very arguments which show that the theories compete establish that they are *not* incommensurable, at least not in Feyerabend's sense of being incapable of being brought into deductive relationships. None of Feyerabend's purported examples of incommensurable competitors are genuine.[30]

Because of the limitations of space, one example will have to suffice.[31] In the very paper in which Feyerabend definitively adopted the account of incommensurability under discussion, he treats Newtonian mechanics and relativity theory as incommensurable competitors, and offers the following as his principal reason for the nondefinability of Newtonian concepts within relativity theory: "any such definition must assume the absence of a limit for signal velocities, and cannot therefore be given within relativity theory."[32] Now, if the definitions of Newtonian terms within relativity theory are genuinely barred for this reason, it must be because there is no limit to signal velocities in Newtonian mechanics while there is such a limit in relativity theory. But this is to say precisely that the notion of a signal velocity is common to both theories and that the two theories can therefore be brought into deductive relationships. It follows that the theories are not incommensurable in Feyerabend's sense.

This situation needs to be spelled out further. I think that Feyerabend has mistaken the incommensurability (non-interdefinability) of a number of key *concepts* of the two theories for the incommensurability (lack of logical relation) of the theories themselves. The incommensurability of *concepts* (or of terms within a theory) is a matter of their interdefinability *within the appropriate object language*. Thus, the square root of two, being indefinable in the system of rational numbers, is an 'incommensurable number' within that system.³³ Similarly, the incommensurability of the concepts under discussion, if genuine, is a matter of the indefinability of the Newtonian concepts (or of terms obeying the Newtonian rules) within the technical language of relativity theory. The commensurability of *theories*, in contrast, is a matter of the similarities and differences that may properly be ascribed to those theories *in a metatheoretical language*. The commensurability of theories is therefore much more a matter of degree and much more dependent on the purposes of comparison than is the commensurability of concepts. For Feyerabend's purposes, the incommensurability of Newtonian mechanics/relativity theory must amount to the lack of *relevant* similarities between the (concepts of the) theories *as described in some appropriate meta-language*.

But what similarities are relevant? One answer to this question is obtained by recalling Feyerabend's arguments for theoretical pluralism. Theoretical pluralism is desirable because theories may be employed to criticize each other. Those similarities which further criticism are therefore desirable. And Feyerabend has provided us with an excellent abstract account of such criticism, though it is an account which he himself has misunderstood:

> We shall diagnose a [relation of strong criticism] either if a new theory entails that all concepts of the preceding theory have extension zero or if it introduces rules which cannot be interpreted as attributing specific properties to objects within already existing classes, but which change the system of classes itself ([18], p. 168).³⁴

This passage enables us to eliminate incommensurability very quickly. If two theories have no concepts which are related in meaning, neither can entail anything about the extension of the concepts of the other.³⁵ Furthermore, without any similarity of meanings, neither can be interpreted as offering any grounds for changing the system of classes of the other. It follows, therefore, that similarities in meaning are required for Feyerabend's program of theory comparison.

I should add that this passage demonstrates the incoherence of Feyerabend's account of incommensurability. The incommensurability of theories cannot be demonstrated by proving that certain entailment relations hold between them, for entailment relations are obtained only where concepts are commensurable.³⁶ A precondition of the use of the tools which Feyerabend employs here is that there be some logical contact between the theories in question, i.e. that there be some stability of meaning. This shows both the futility of the above account of incommensurability and the importance of logical contact in those cases in which one theory is used to criticize another. Logical contact is required if a new theory is to serve as an improvement on an old one rather than as a mere replacement for it. Without logical contact, it is indeed true, as Feyerabend put it in *Against Method*, that "anything goes". Once again, Feyerabend's arguments for the desirability of incommensurability between Newtonian mechanics and relativity theory have failed: it is undesirable for them not to be in logical contact with each other.

VII

Once it has been granted that some overlap in meaning can be established between the claims of competing theories, the situation is greatly improved. This admission will encourage an analysis according to which Newtonians and relativists, for example, have different beliefs about *the same* objects (e.g. the Earth, the Sun, Jupiter, and the stars). Not every fundamental difference in belief will be seen as resulting in complete incommensurability. Different concepts entering different conceptual networks will be treated as referring to the same objects or as having related senses. Such identity of extension or similarity of sense will be thought to occur in spite of false beliefs incorporated into the very concepts at stake, and will depend on the fact that, at one and the same time, concepts can be radically incorrect in their way of picking out something and nevertheless succeed in picking it out. Against the background of such commitments there is hope that it will be possible to give an account of the improvement in our conceptual schemes — an account which, I have been arguing, Feyerabend cannot hope to offer.

Such an account will seek to use alternative theories as a means of increasing testability, but its methods of comparing such alternatives will differ significantly from Feyerabend's. After all, the force of the preceding criticisms of incommensurability has concentrated on showing incommensurable theories to be indigestible wholes, insufficiently related to each other to be useful for purposes of criticism.

It is an elementary lesson in experimental technique that, in so far as possible, one should keep the background situation constant if one wishes to determine the interdependence of certain theoretical parameters or variables. I maintain that a similar claim pertains in methodology and that the employment of a common background is an essential part of theory comparison. Unless two competing theories are in logical contact, replacement of one by the other will not reveal which features, if any, of the new theory constitute improvements on the old one. Only if theories are compared against a common background (which may, of course, be fixed only temporarily) will it be possible to make a specific assessment of the virtues and faults of each theory as compared with the other.

This is the heart of the matter. If theories succeed one another as wholes, without any point-to-point contact, it will be impossible to assess the merits of any *particular* claims within a given theory. Only the theory as a whole will be acceptable or not. If, however, there are continuities in theoretical transitions, then the claims which are retained, or at least approximated, can undergo continued testing. After, all, a claim is not thoroughly tested unless it is taken together with differing auxiliary hypotheses (themselves testable in turn) in order to fully assess the impact its truth would have on experience or experiment. Since Feyerabend wishes to maximize the testability of fundamental theoretical principles, he ought to enable them to retain their force as long as possible in the face of both small and large theoretical changes so that they can be tested with different (and improved) auxiliary hypotheses.

This is the sort of procedure which was in fact followed in dealing with the Michelson–Morley experiments. If Newtonian ether theory were the indissoluble whole which Feyerabend makes of it, it would have been impossible for stationary-ether theorists, in dealing with these experiments, to proceed in a series of steps culminating in the claim that the round trip velocities of light along perpendicular axes of a body moving with respect to the ether are equal. Such a procedure would have been incoherent, for it would have resulted in experimental 'proof' of an analytical falsehood. Yet this is just what happened. A number of auxiliary hypotheses were proposed to save the ether theory. Some were shown to be *ad hoc*. Others, while not *ad hoc*, were shown to be false. While not all possible saving hypotheses were investigated, no satisfactory alternative to the conclusion of the equality of the velocities was found.

This pattern of development fits much better with Kuhn's account of the piecemeal progress of normal science than Feyerabend would lead us to believe. Without the attempts to check on the various *ad hoc* and false auxiliary hypotheses which were employed in the attempt to save the ether theory, the 'crucial' experiment would not have been decisive. Even given those attempts, the experiment is in no way sufficient by itself to determine the choice between the competing theories, but it enables Newtonians and relativists to agree on the directional independence of the round-trip velocity of light within a fixed reference frame. Such limited agreement would have been impossible were the theories at stake genuinely incommensurable. And if such agreement were blocked, the significance of the experiments as support for this limited claim — and hence as support for relativity theory over Newtonian mechanics — would have been lost. This example illustrates perfectly my contention that specific claims within a theory may be isolated, tested and carried over into radically new theoretical contexts. This feature is of critical importance, for it enables the test of an old theory to serve as a guide in the formation of new theories, to suggest what ought to be retained and what abandoned in constructing alternatives to accepted points of view.

Another aspect of both Kuhn's and Feyerabend's treatment of incommensurable theories which is methodologically unsound is their insistence that the testing of such theories requires their prior acceptance. Both men argue that one cannot *entertain and use* a highest-level theory without *accepting* it because of the thoroughgoing way in which the adoption of such theories revises experience (or, at least, our conception of what it is that is experienced). Indeed, one of the major props in Kuhn's arguments for the desirability of mono-theoretic periods of normal science is the instability of experience under theory change. In order to progress, Kuhn claims, science needs the stability guaranteed by *acceptance* of a single highest-level theory.

But Kuhn and Feyerabend are wrong about this. Radical alternatives *can* be entertained and used before they are accepted. If this were not so, no reasons could ever be given for the initial adoption of a revolutionary new theory. A greater number of case histories support my contention that consideration, and even testing, of a revolutionary theory can occur independently of the acceptance of that theory. Perhaps the most notorious example of the kind is provided by Einstein's contributions to quantum mechanics in spite of the distaste and disbelief he manifested with respect to that theory. Conceptually more interesting, however, is Planck's original development of the quantum hypothesis. Planck favored phenomenological over statistical thermodynamics. Nonetheless, when he was unable to account for the blackbody spectrum on the basis of the phenomenological theory, he *used* the statistical theory in conjunction with the well-developed Maxwellian theory of linear harmonic oscillators. It was the elaboration of this approach, as is well known, which led him to the quantum hypothesis.[37]

Planck, the reluctant revolutionary, brought about a revolution by exploring the consequences of a theory he did not accept. The lack of logical contact between Feyerabend's incommensurable theories prevents them from being exploited in this way. If theories are to be used to criticize each other, we must ensure that their consequences can be explored *before* they are adopted. Once again, incommensurable alternatives, if they are possible, are methodologically undesirable.

One more point will conclude this section. Feyerabend's belief that he can insulate theoretical meaning from the pragmatics of theories is an important source of his confidence that competing theories can be incommensurable. But it is simply untrue that one can provide an adequate account of the meanings of the terms of an empirical theory without taking into account the world-language regularities which are exhibited by the sentences of the theory in use. In order to drive this point home, it is sufficient to consider the differences between certain 'pure' and 'applied' geometries. In most 'pure' geometries, the notions of 'point' and 'line' are mathematically dual, i.e. formally symmetric. In order, therefore, to distinguish the notion of a point from that of a line in *applied* geometry, it is necessary to take account of the pragmatics of the theory – i.e. to consider the circumstances in which sentences about points as opposed to sentences about lines are appropriate. *Pragmatic constraints on meaning are necessary to tell one of these notions from the other*! This suffices to show, Feyerabend to the contrary notwithstanding, that "the fact that a sentence belongs to the observational domain" *can* "have a bearing on its meaning." ([17], p. 39)

The importance of the pragmatic aspects of meaning may be seen more clearly by considering how we would react to the utterances of a person who consistently states mathematical truths (and no mathematical falsehoods) about triangles but who consistently responds to squares, rather than triangles, with sentences about triangles. (There should be no evidence that he is joking or trying to deceive us.) Surely the coherence of his statements about triangles with ours does not by itself suffice to show that he understands the full meaning of the term 'triangle'. On the other hand, consider a person who offers a coherent account of triangles in which some triangles have angle sums of 180° while others have angle sums of more or less than 180° and who responds to triangles (rather than squares) with sentences about triangles. Such a person, surely, should be said to use the term 'triangle' so that its meaning is continuous with the meaning it has in applied Euclidean geometry.

These cases show that pragmatic elements *do* enter into the meanings of even theoretical terms. They help show, I think, that empirical theories require at least a minimal form of external subject matter. But I do not pretend to have shown *how* the acknowledged continuities in pragmatics between alternative theories can be used to establish meaning continuities between the theories. That is a question for another paper.

VIII

In arguing against logical empiricists, Feyerabend maintained that their philosophy of science excluded scientific realism *a priori*, and that it was therefore unacceptable.[38] I maintain that Feyerabend's doctrine of incommensurability does not allow him to adopt a fully realistic position. By parity of reasoning, if I can make my claim stick, he should reject those aspects of his own position which are antithetical to realism.

Now, Feyerabend holds that the realist's position is characterized by his employment of a distinction between appearance and reality.[39] The realist wishes to explain the way the world appears to us in terms of the impact on us of a reality which 'underlies' or causes appearances. Appearance and reality may be of radically different types, and for the distinction between them to function properly adequate means must be found for describing both the appearances which are to be explained and the reality which explains them.

But this requirement is frustrated by Feyerabend's adoption of the doctrine of incommensurability. So far as a person's acceptance of a highest-level theory which purports to describe reality entails that *all* the descriptions he proffers contain a commitment to those fundamental principles of the theory which purport to govern 'the real', he will be unable to offer a description of 'appearance' independent of those principles. Whatever theory he accepts, he will not be able to describe, let alone explain, the 'appearances' experienced by those who hold incommensurable theories. ('Each theory will possess its own experience!') But this means that it will be impossible for anyone to describe, much less account for, the phenomena (the appearances) experienced by the holders of incommensurable theories. This cuts the essential interdependence of appearance and reality 'characteristic of realism' and guarantees *for conceptual reasons* the incompletability of any realistic program in the face of incommensurability.

Feyerabend attempts to accommodate this antirealist tendency to some extent by adjusting, or rather weakening, his realism, e.g. by removing explanation from the center of scientific methodology (see my [3] p. 251). If we take seriously his claims that knowledge involves an "everincreasing ocean of alternatives" ([23], p. 224) and that these alternatives consist of "unplausible conjectures which possess no empirical support and which are inconsistent with facts and well confirmed theories" ([22], p. 308), then all these alternatives must, in the end, be considered on a par — they are all equally irrefutable, all equally unverifiable. Thus Feyerabend's realism, as he explicitly admits (see [3], p. 25), no longer accepts the usual methodological principle of the realist that progress in science is measured by the convergence of scientific knowledge towards an 'ideal view', i.e. towards something like a uniquely correct description of some aspect of level of reality or a complete account of the way things appear on the basis of *the* reality or realities underlying these appearances.

Perhaps Feyerabend thinks that all that is required by scientific realism is that theories be interpreted literally in the sense that they be interpreted as claiming that the world is constituted out of the fundamental entities they name, mention or describe. Such a 'realism' would be, in effect, an 'operational negative' of instrumentalism. But such a realism would be of little methodological interest. By failing to give an account of the grounds for choice between theories and by severing the intimate explanatory connection between reality and appearance, a connection which, as we have seen, Feyerabend cannot accommodate, such a realism would be but a pale shadow of the realism he set out to defend.

IX

To sum up, let me suggest that this study has shown that there are a number of desiderata which an adequate account of theory change in science ought to satisfy.

(1) Theories ought, at least occasionally, to be able to penetrate and alter the categories of observational discourse. That they in fact *do* do so seems clear from a number of historical examples generated by the 'historiographical revolution'. I think that the claim that they *should* be allowed to do so amounts to no more than the claim that theoretical realism cannot be ruled out *a priori*.

(2) Yet theories must not be allowed wholly to control the categories of observational discourse. If that were allowed, it would be impossible to locate continuities between theories by retreating to less theory-committed accounts of what is observed. In short, we would be forced to open the Pandora's box of fully-fledged incommensurability.

(3) As against Kuhn, theories differing as radically from each other as possible, given the above constraints, should be regularly employed, at the very least as foils against which to test the theories we accept. The weight of the scientific community's approval of (say) Einstein's theory of relativity should count for little unless that community has seriously considered Whitehead's and Milne's alternatives — and, yes, Newton's too (as Prof. Dicke

has been doing recently).[40] After all, the use of alternatives does not prevent the full elaboration and articulation of the theories we accept, and it is that full elaboration which Kuhn seeks to ensure with his strictures in favor of mono-theoretic dominant traditions.

(4) The measure of the usefulness of radical alternatives should be in terms of their comparability (rather than their incomparability) with their competitors.

(5) In furtherance of the end mentioned in (4) above, a new account of the ways in which (for example) wave and particle theories of matter are comparable is called for. Such an account may not rely on a categorially frozen observation language, but must require that a *pro-tempore*, relatively theory-neutral external subject matter be available to the competing theories. (Eight years ago, Sellars suggested that the conceptual framework of common sense supplied just such external subject matter.)

(6) Finally, a new articulation of the realist's regulative ideal of a perfectly perfect theory,[41] i.e. of reaching *the* truth is called for. Such an ideal must be recast in terms of the new account of continuity between theories. But that is a topic for another day.

NOTES

* We regret the delay in publication of this paper. See also subsequently published articles by the author, in particular 'Conceptual Change, Cross-Theoretical Explanation, and the Unity of Science' in *Synthese* 32 (1975), pp. 1–28, and 'More Than a Marriage of Convenience: On the Inextricability of History and Philosophy of Science' in *Phil. of Sci.* 44 (1977), pp. 1–42. – Ed.

[1] J. J. C. Smart [53]; Wilfrid Sellars [51]; Hilary Putnam [44]; and Paul Feyerabend [23]; all in Cohen and Wartofsky [7], pp. 157–261.

[2] Cf. for example, Sections 2 and 3 of [19]. Here, as in a number of other places, Feyerabend treats the logical empiricists as if they were committed to the incorrigibility and quasi-logical certainty of basic claims in the observation language. But this is simply false, as I show in my [3].

[3] It should be stressed that the distinction between L_T and L_O is thought to be the result of conceptual analysis; these languages are *not* historically distinct and need not be separable in practice. Cf., for example, Nagel in [41], p. 106 ff.

[4] A very useful exposition of these matters is offered by W. Sellars in [49].

[5] Cf. *ibid*, para. 6 or Hempel [26], pp. 33–34 and n. 26.

[6] For example, [15], Section 2. See also [22].

[7] The substantive points in this paragraph are covered by Carnap in [5], Section 24 and by Hempel in [27], p. 696, and [28], Section 10. In this section of [28] Hempel cites some pros and cons connected with the policy of employing partially interpreted terms in the semantic meta-language.

⁸ Note that a semantic meta-language in which the theoretical terms do not have any empirical content cannot be chosen, for then no evidence *could* be pertinent to the question whether the truth conditions of theoretical claims are met. On the other hand, given that some theoretical claims have some observational content, there is no reason to expect their truth conditions to be statable solely at the observational level. Indeed, Hempel has argued in many places (cf., e.g. [26], Section 7) that there cannot be wholly observational truth conditions for large classes of theoretical claims. (I would like to thank Prof. P. Machamer for suggesting this note.)

⁹ Cf. Nagel [41], pp. 91, 96 (n. 4), and 141 ff.; Hempel [26], n. 26, where Hempel says: "The conjunction of all the postulates of an axiomatized theory may be construed as a sentential function in which the primitives play the role of variables. But a sentential function cannot well be said to 'define' one particular meaning of (i.e., one particular set of values for) the variables it contains unless proof is forthcoming that there exists exactly one set of values for the variables which satisfies the given sentential function." Such a proof in general cannot be offered in a partial interpretation in Hempel's sense, since all that is closed is the observational content and structural interrelations of the theoretical concepts. Cf. also [28], n. 64, p. 220. (This note, numbered 68 in the original edition, formulates the issue in terms of the Ramsey sentence of the theory.) The critique developed here was suggested by Section II, paras. 11–18, of W. Sellars [51].

¹⁰ In order to achieve observational equivalence with the original theory, it is necessary to conjoin its axioms and add all the appropriate quantifiers to this conjunction. The resultant multiply quantified sentence is known as the Ramsey sentence of the theory. It allows the deduction, by means of the logical apparatus of the theory, of an appropriately quantified analogue of any theoretical sentence as well as generating precisely the same observational consequences as the original theory. For technical details, cf. Scheffler [47], Section 21.

¹¹ Cf. Sellars at [51], pp. 175–176: "If ... the predicate expressions of a deductive system are *essentially* variables, the statement functions of the system *could* have no statement counterparts with predicate *constants* for which they were true – surely an incoherent notion."

¹² It is by no means clear that logical empiricists must treat every instance in which a theory yields a falsified prediction as a falsification of the theory. There are *ceteris paribus* clauses built into every theory. And there is considerable slack in the fit between the observation language and the theoretical language as they are meshed together under the correspondence rules (cf. Hempel [26], p. 37.) However, the price that must be paid for 'loosening' the fit between theory and experience is high; it becomes unclear whether any experiment or set of experiments can ever falsify a theory. For some related topics, cf. Section 2 of Lakatos [35].

¹³ Cf., e.g., Sections X and XII of Kuhn [33] and Section 15 of Feyerabend's [20], esp. pp. 213–214.

¹⁴ Cf. n. 5 of Feyerabend's [23] for an explicit defense of his broad construal of theories. I am deliberately slurring over Kuhn's distinction between paradigms and theories.

¹⁵ Deep differences between Kuhn and Feyerabend can be found in practically any pair of their articles, but the most explicit treatment of their differences is Feyerabend's [16] together with Kuhn's [34].

¹⁶ Cf. Section 5 of [16].

[17] There are further complications to the argument. Cf. Section 10, pp. 300-301 of [21], pp. 40-41 of [14], and the discussion between Feyerabend and Hanson at pp. 235-236 and 251-252 of the same volume. On pp. 159-165 of [35], Lakatos argues, in effect, that the Michelson-Morley experiments provide another illustration of this claim. (He puts the matter in terms of his model of competing research programs.)

[18] Cf. also [34], p. 264.

[19] Cf. the last Section of [34].

[20] Cf. [17], p. 39, for one instance of the claim that "the fact that a statement belongs to the observational domain has no bearing on its meaning" and n. 27 of [23], *loc. cit.*, for Feyerabend's most succinct account of the way in which theories are supposed to determine meanings.

[21] At first, Feyerabend hesitated between two different accounts of incommensurability. The one he later rejected treated two theories as incommensurable when they presupposed inconsistent principles. The one he later accepted treated two theories as incommensurable if not a single term of either theory could be defined within the other. Cf. [17], p. 59, where *both* accounts appear within a single short passage, n. 19 of [20], which favors the first account, and [23], pp. 231-232, where Feyerabend definitively chooses the second account under the pressure of some criticisms offered by D. Shapere.

[22] This formulation comes from a letter of Feyerabend's dated 27 October 1969. It seems to me to offer the clearest account of what Feyerabend intended in 'Reply to Criticism.' I am grateful to Prof. Feyerabend for permission to quote from our correspondence.

[23] In fairness to Feyerabend, I should point out that as of [14] he no longer holds that incommensurable theories can be thus compared. Accordingly, he thinks the grounds of theory choice are more or less aesthetic. This is a major element in his 'epistemological anarchism' and lends support to his insistence on the need for propaganda techniques in cases of theory competition.

[24] Highest-level theories are those theories which are not subordinate to any other theory. The external subject matter of subordinate theories is supplied by the theories to which they are subordinate on this account.

[25] But see the last two sections of [34].

[26] A modest sample of critical writings on this topic includes the following: Achinstein [1]; Butts [4]; Giedymin [24]; Kordig [30], [31]; Leplin [37]; M. Martin [38], [39]; Scheffler [48]; and Shapere [52].

[27] I believe that I have shown all of them to fail on pp. 166-183 of my [3]. The procedures in question are stated on pp. 216-217 (with 214-215) of [20] and pp. 232-233 of [23].

[28] The passage omitted from this quotation seems utterly confused. It purports to offer an example from number theory of the procedure under discussion. The example concerns the proof of the irrationality of $\sqrt{2}$. "To start with, rational and irrational numbers form two distinct domains, R and I, so that it does not seem to be possible to relate a single entity, $\sqrt{2}$, to both of them. However, we may select a subdomain, R', of I which is isomorphic to R with respect to all the properties we think essential of the latter, and interpret the 'experiment' as being between I and R'." First of all, the proof normally occurs in a context in which no distinct (and certainly no closed) domain I is available. In contemporary practice, the distinct domains in question are established only

when I is formed *by constructive techniques* out of the resources of logic and of R. But this means that I and R are *not* incommensurable. Secondly, *both* the claim, 'There is no number in R and R' such that its square equals two' *and* the claim 'There is a number in I whose square is two' are true. If the 'experiment' is to answer the question 'Is there such a thing as a number whose square is two?' the answer is, 'Yes, there is – and it belongs to I, not to R or R'.' In any event, the model plays no essential role in *this* argument. On the other hand, if the 'experiment' occurs (as it did in Greek days) in the context of the background assumption that all numbers belong to R (or, perhaps, that all numbers are integers and that all numerical relations may be expressed by ratios), the answer seems paradoxical – 'There is a number which is *not* a number' – precisely because I and R' are not available. But Feyerabend's procedure is not available then either. The whole example is phony; it concerns not the comparison of incommensurable theories, but the comparison of commensurable theories about arithmetically incommensurable numbers.

[29] The sense of sameness required here depends on the theories in question. Thus *formal* similarities of one sort or another suffice in mathematics. (For many purposes, the series of rational integers is 'the same' series as the series of integers.) In general, pragmatic similarities will be required in comparing empirical theories. (One thing wrong with the model of the theory of sound within the theory of light is that the pragmatics of the model – which are the pragmatics of testing claims about light – differ radically from the pragmatics of the theory of which it is a model.)

[30] To the extent that Kuhn is willing to allow overlap in meaning between incommensurable theories, these strictures will not apply to his notion of incommensurability. I must confess that I do not understand the extent of overlap that Kuhn allows, even after a number of rereadings of [34].

[31] Another example is discussed in a similar vein by Dudley Shapere in the last section of [52]. (The example concerns the transition from impetus to inertial dynamics.) Peter Machamer, in [38], offers an extended case study in which he rebuts Feyerabend's treatment of Galileo in [21].

[32] [23], p. 231. Feyerabend has discussed this case in a similar vein a number of times. Cf. [17], pp. 80–81; [20], pp. 219–221; and esp. [18]. The fourth chapter of my [3] offers a more extended treatment of this case.

[33] For my purposes, the crucial point is that the *system* of rational numbers is not therefore incommensurable with the system of real numbers. See above, n. 28.

[34] The original has the phrase "*change of meaning*" in lieu of the bracketed phrase. In context, "change of meaning" amounts to incommensurability of the old and the new theory.

[35] But see Martin [39] and [40] for some arguments that the use of auxiliary hypotheses enables one to produce such entailments within limited contexts. The issue at this point will turn on the possibility of introducing such auxiliary hypotheses without changing the meanings of the terms in at least one of the theories. Feyerabend would deny that this can be done. See above, pp.16–17.

[36] An independent argument to the same conclusion is offered by Kordig in Section II of [30].

[37] Certain aspects of this case were called to my attention by Gary Gutting's [25]. A very helpful discussion of the case is offered by M. Klein in [29]. Cf. also Planck's [43].

[38] In Section VI of [22], Feyerabend makes an important distinction between a *philosophical* position which excludes the realistic interpretation of theories *in general*, and scientifically based arguments against the realistic interpretation of particular theories. (A theory which is notoriously open to such scientific arguments is that of the Bohr–Rutherford atom, for it can be shown that such atoms would be immensely unstable.) Scientific arguments can legitimately show that particular theories can only be legitimately used as predictive calculi, but, in Feyerabend's view, philosophical arguments are incompetent to show that theories in general cannot be legitimately given a realistic interpretation.

[39] Cf., for example, [18], p. 164.

[40] This comment was suggested by a similar one on p. 112 of R. Palter's [42].

[41] The term is John Stachel's, and refers to an all-encompassing and absolutely true theory. Such a theory is obviously unattainable. The issue of realism is whether the *conception* of such a theory serving as a regulative ideal provides a proper foundation for methodology. There are some obscure hints about this question in the appendix of my [3].

REFERENCES

[1] Achinstein, Peter, 'On the Meaning of Scientific Terms', *J. Phil.* **61** (1964) 497–509.

[2] Bunge, Mario, ed., *The Critical Approach to Science and Philosophy* (Free Press, New York, 1964).

[3] Burian, Richard, 'Scientific Realism, Commensurability, and Conceptual Change. A Critique of Paul Feyerabend's Philosophy of Science', unpublished Ph.D. Dissertation (University of Pittsburgh, Pittsburgh, 1971).

[4] Butts, Robert E., 'Feyerabend and the Pragmatic Theory of Observation', *Phil. Sci.* **33** (1966) 383–394.

[5] Carnap, Rudolf, 'Foundations of Logic and Mathematics', *International Encyclopedia of Unified Science* **I**, No. 3 (University of Chicago Press, Chicago, 1939).

[6] ———, 'The Methodological Character of Theoretical Concepts', in Feigl and Scriven [12].

[7] Cohen, R. S. and M. W. Wartofsky, eds., *Boston Studies in the Philosophy of Science*, Vol. II (Humanities Press, New York, 1965).

[8] Colodny, Robert G., ed., *Beyond the Edge of Certainty* (Prentice-Hall, Englewood Cliffs, N.J., 1965).

[9] ———, ed., *Mind and Cosmos* (University of Pittsburgh Press, Pittsburgh, 1966).

[10] ———, ed., *The Nature and Function of Scientific Theories* (University of Pittsburgh Press, Pittsburgh, 1970).

[11] Feigl, Herbert and Grover Maxwell, eds., *Minnesota Studies in the Philosophy of Science*, Vol. III (University of Minnesota Press, Minneapolis, 1962).

[12] Feigl, Herbert and Michael Scriven, eds., *Minnesota Studies in the Philosophy of Science*, Vol. I (University of Minnesota Press, Minneapolis, 1956).

[13] ——— and Grover Maxwell, eds., *Minnesota Studies in the Philosophy of Science*, Vol. II (University of Minnesota Press, Minneapolis, 1958).

[14]   Feyerabend, Paul, 'Against Method: Outline of an Anarchistic Theory of Knowledge', in Radner and Winokur [45].
[15]   _____, 'An Attempt at a Realistic Interpretation of Experience', *Proc. Arist. Soc.* **58** (1958) 143–170.
[16]   _____, 'Consolations for the Specialist', in Lakatos and Musgrave, eds. [36].
[17]   _____, 'Explanation, Reduction, and Empiricism', in Feigl and Maxwell, eds. [11].
[18]   _____, 'On the "Meaning" of Scientific Terms', *J. Phil.* **62** (1965) 266–274.
[19]   _____, 'Das Problem der Existenz theoretischer Ertitäten', in Topitsch, ed. [54].
[20]   _____, 'Problems of Empiricism', in Colodny, ed. [8].
[21]   _____, 'Problems of Empiricism, Part II', in Colodny, ed. [10].
[22]   _____, 'Realism and Instrumentalism: Comments on the Logic of Factual Support', in M. Bunge, ed. [2].
[23]   _____, 'Reply to Criticism', in Cohen and Wartofsky, eds. [7].
[24]   Giedymin, Jerzy, 'The Paradox of Meaning Variance', *BJPS* **21** (1970) 257–268.
[25]   Gutting, Gary, 'Feyerabend's Attack on Method', unpublished.
[26]   Hempel, Carl G., 'Fundamentals of Concept Formation in Empirical Science', *International Encyclopedia of Unified Science*, Vol. II, No. 7 (University of Chicago Press, Chicago, 1952).
[27]   _____, 'Implications of Carnap's Work for the Philosophy of Science', in Schilpp, ed. [46].
[28]   _____, 'The Theoretician's Dilemma: A Study in the Logic of Theory Construction', in Feigl, Scriven, and Maxwell, eds. [13].
[29]   Klein, M., 'Max Planck and the Beginnings of the Quantum Theory', *Arch. Hist. Exact Sci.* **1** (1962).
[30]   Kordig, Carl R., 'Feyerabend and Radical Meaning Variance', *Nous* **4** (1970) 399–404.
[31]   _____, 'The Theory-Ladenness of Observation', *Rev. Metaphys.* **24** (1971) 448–484.
[32]   Kuhn, Thomas, 'Historical Structure of Scientific Discovery', *Science* **136** (1962) 760–764.
[33]   _____, *The Structure of Scientific Revolutions*, 2nd ed. (University of Chicago Press, Chicago, 1970).
[34]   _____, 'Reflections on my Critics', in Lakatos and Musgrave, eds. [36].
[35]   Lakatos, Imre, 'Falsification and the Methodology of Scientific Research Programmes', in Lakatos and Musgrave, eds. [36].
[36]   _____ and Alan Musgrave, eds., *Criticism and the Growth of Knowledge* (Cambridge University Press, Cambridge, 1970).
[37]   Leplin, Jarrett, 'Meaning Variance and the Comparability of Theories', *BJPS* **20** (1969) 69–75.
[38]   Machamer, Peter, 'Feyerabend and Galileo: The Interaction of Theories and the Reinterpretation of Experience', *Stud. Hist. Phil. Sci.* **4** (1973) 1–46.
[39]   Martin, Michael, 'Ontological Variance and Scientific Objectivity', *BJPS* **23** (1972) 252–256.

[40] ───, 'Referential Variance and Scientific Objectivity', *BJPS* **22** (1971) 17–26.
[41] Nagel, Ernest, *The Structure of Science* (Harcourt, Brace and World, New York, 1961).
[42] Palter, Robert, 'Philosophic Principles and Scientific Theory', *Phil. Sci.* **23** (1956) 111–135.
[43] Planck, Max, *A Scientific Autobiography and Other Papers* (Philosophical Library, New York, 1949).
[44] Putnam, Hilary, 'How Not to Talk about Meaning', in Cohen and Wartofsky, eds. [7].
[45] Radner, M. and Winokur, S., eds., *Minnesota Studies in the Philosophy of Science*, Vol. IV (University of Minnesota Press, Minneapolis, 1971).
[46] Schilpp, P. A., ed., *The Philosophy of Rudolf Carnap* (Open Court, La Salle, Ill., 1963).
[47] Scheffler, Israel, *The Anatomy of Inquiry* (Knopf, New York, 1963).
[48] ───, *Science and Subjectivity* (Bobbs-Merrill, New York, 1967).
[49] Sellars, Wilfrid, 'The Language of Theories', in Sellars [50].
[50] ───, *Science, Perception and Reality* (Routledge and Kegan Paul, London, 1963).
[51] ───, 'Scientific Realism or Irenic Instrumentalism: Comments on J. J. C. Smart', in Cohen and Wartofsky, eds. [7].
[52] Shapere, Dudley, 'Meaning and Scientific Change', in Colodny, ed. [9].
[53] Smart, J. J. C., 'Conflicting Views about Scientific Explanation', in Cohen and Wartofsky, eds. [7].
[54] Topitsch, Ernst, ed., *Probleme der Wissenschaftstheorie: Festschrift für Viktor Kraft* (Springer, Vienna, 1960).
[55] Heisenberg, W. *Physics and Beyond* (Harper and Row, New York).

Professor R. M. Burian
Dept. of Philosophy
Virginia Polytechnic Institute and State University
Blacksburg
Va. 24061
U.S.A.

MICHAEL MARTIN

HOW TO BE A GOOD PHILOSOPHER OF SCIENCE: A PLEA FOR EMPIRICISM IN MATTERS METHODOLOGICAL

INCOMMENSURABILITY AS METHODOLOGY

It has been argued that philosophy of science in our time is basically of two types[1] — speculative and analytic. Speculative philosophy of science constructs theories of the world based on the findings of the specific sciences and is continuous with science on the one hand and metaphysics on the other. Analytic philosophy of science analyzes the concepts and modes of reasoning characteristic of science and eschews speculation. However, Paul Feyerabend's philosophy of science does not seem to fit easily into either category. For Feyerabend, philosophy of science has a rather different task than either speculation or analysis, namely to expound, criticize and defend large methodological programs for scientific progress and growth.

One of the most significant insights of Burian's paper[2] (and of his excellent doctoral thesis[3] which I have had the privilege to read) is closely connected with this point. Feyerabend's doctrine of incommensurability is not just a description of certain theories in the history of science, for example, Newtonian mechanics and the theory of relativity; it is not merely a criticism of the neopositivistic analysis of reduction, explanation and theories. Feyerabend's doctrine of incommensurability is part of a larger methodological program. For Feyerabend scientists have a choice: they can choose to make their theories incommensurable with one another or they can choose not to make them incommensurable. On methodological grounds they should make their theories incommensurable. Only with such a choice is dogmatism prevented and testability and criticism are increased to the maximum.

Seen in this light Feyerabend's doctrine of incommensurability is an extension and refinement of Popperian philosophy of science. Popper argues that the essence of science should be conjecture and refutation;[4] the conjecture of bold theories that are easily criticized and refutation of these theories. Feyerabend's doctrine of incommensurability is, at least in this view, a way of maximizing testability and criticism; it is Popperianism carried to its logical extreme.

Feyerabend claims that a major difficulty in testing scientific theories is to bring the problems with these theories into relief. There are always problems with scientific theories. The difficulty is to *see* them *as problems*. This is, in general, impossible unless alternative theories which bring to light the problems of the old theory are considered. Since the hardest problems of all to detect are the fundamental ones, the only theories which will bring them to light are theories which depart in the most fundamental ways from the old theories. And the theories which depart in the most fundamental ways are incommensurable. Hence, incommensurable theories will bring into relief the most basic problems of the old theories and provide the most severe and penetrating criticisms of them.

Burian correctly argues that if theories are really incommensurable in the way Feyerabend seems to be claiming, they cannot bring problems of the old theories into relief since they make no logical contact with them. Maximum testability is therefore not enhanced by Feyerabend's proposal.

THEORETICAL INCREMENTALISM

What does Burian substitute for Feyerabend's Extreme Theoretical Incomparabilism with its maxim 'Conjecture new theories as incommensurable as possible with the old theories?' This is not so clear as it might be. Burian has a quotation from Heisenberg at the beginning of his paper which one would naturally suppose represented his own views: "... only those revolutions in science will prove fruitful and beneficial whose instigators try to change as little as possible ...." This suggests a position diametrically opposed to Extreme Theoretical Incomparabilism which, for convenience, I will refer to as Theoretical Incrementalism. On this view scientific change should depart from old theories in small increments, the smaller the better. The position can be stated in terms of the maxim: "Conjecture theories that are continuous as possible with old theories'.

Students of Popper will notice an interesting similarity between Theoretical Incrementalism and Popper's doctrine of piecemeal social planning. Popperian philosophy of science which advocates bold conjecture is sometimes contrasted with Popper's social philosophy in which bold conjecture is minimized and small incremental and piecemeal change is stressed.[5] Ironically, Theoretical Incrementalism and not Extreme Theoretical Incomparabilism seems to adopt the Popperian methodology for social change as the methodological model for theoretical change in science.

Another analogy may bring home the point. Kuhn makes the distinction between normal science and revolutionary science. Normal science proceeds by articulating the prevailing paradigm which in actual practice may come down to small piecemeal and incremental change. However, in revolutionary science new incommensurable paradigms appear that are radically different from the old paradigms. To interpret Kuhn in Feyerabendian terms, in revolutionary science new incommensurable theories are conjectured. But Extreme Theoretical Incomparabilism takes as its model for *all* scientific change what, according to Kuhn, is characteristic only of revolutionary science. Theoretical Incrementalism, on the other hand, seems to take as *its* model for *all* scientific change what Kuhn thinks is characteristic of normal science.

RADICAL CHANGE AND COMPARABILITY

Yet Burian, despite his quotation from Heisenberg, does not seem to advocate Theoretical Incrementalism. For in the conclusion of his paper he says "as against Kuhn, theories differing as radically from each other as possible, given these restraints, should be regularly employed ..." (p. 24). The restraints Burian seems to have in mind are those which prevent complete incommensurability. But Burian's statement seems to give rather different advice from Heisenberg's methodological advice "to change as little as possible." The position suggested by Burian's closing statement might be called Moderate Theoretical Incomparabilism and its maxim might be stated as follows: 'Conjecture theories that are as incommensurable with the old theories as possible without being completely incommensurable'.

But even this does not seem to be Burian's final position. At the end of his paper he says "the usefulness of radical alternatives should be in terms of their comparability (rather than incomparability)." Here the position might be called Radical Theoretical Comparabilism and the maxim might be stated as: 'Conjecture theories that are as radically different from old theories and as comparable as possible'. Although this is not completely clear, Burian's reason for advocating Radical Theoretical Comparabilism seems to be that radically different, though highly comparable, new theories provide the severest test of old theories. Whether this is true we will consider in a moment. However, one thing should be clear. It is necessary to distinguish two aspects of theory change: (1) how radically a new theory departs from the old; (2) (2) how comparable the new theory and the old theory are. Some radical departures may make for easy comparability but others may make comparison difficult.

ALTERNATIVE METHODOLOGICAL POSITIONS

Now there are many logically possible alternatives to the four methodological positions suggested by Burian's paper and by Feyerabend's work. For example, one might advocate theories with moderate ease of comparability and moderate differences from old theories in all branches of science at all stages of science. On the other hand one might advocate such moderation only in certain sciences and at certain stages while advocating Theoretical Incrementalism at other stages and in other sciences. Or one might advocate Theoretical Incrementalism in certain sciences at certain stages, Moderate Theoretical Incomparabilism at other stages in other sciences, and Radical Theoretical Comparabilism at still other stages and in other sciences. Alternatively, one might advocate following Theoretical Incrementalism as well as Radical Theoretical Comparabilism in all branches of science at all stages of development, i.e., one might advocate that old theories be developed in small steps and radically different theories that are easily comparable be introduced. Indeed this may be Burian's position, for at one point in his paper he seems to advocate the simultaneous use of *both* methodologies. In any case, this list does not begin to exhaust the alternatives; there is an indefinite number of possibilities.

Feyerabend advocates the proliferation of scientific theories for scientific growth.[6] One way of looking at the present suggestion is that it advocates the proliferation of methodological points of view for the purpose of comparison. Once we see that the four methodological positions considered above are not the only alternatives, the question of which one science should follow arises. The answer given to this question will depend on the criteria of scientific progress one has adopted. Feyerabend seems to have adopted the criterion of maximum testability and Burian, at least for the purposes of his internal criticism of Feyerabend, adopts this as well. However, two questions must be raised about this criterion: (1) which of the alternative methodological positions sketched above would bring about maximum testability?; and (2) is maximum testability the only criterion, or even the most important criterion, of scientific progress?

INCOMMENSURABILITY AS A FUNCTION OF RADICAL MEANING CHANGE

Before we consider these questions more needs to be said about incommensurability itself. For unless we are clear on what sort of departures new theories can make from old theories, alternative methodological positions of the sort we have in mind will not be understood. According to Feyerabend — at least at one point in his career — incommensurability is a function of the changed meanings of the terms in the theories. Furthermore, he has argued that theoretical change — at least of a radical kind — brings about change of meaning; hence, theoretical change of a radical kind brings about incommensurability.

Whether a Theoretical Incrementalist need accept Feyerabend's views on this point is unclear. However, it is important to notice that Feyerabend's position on meaning can give some support to Theoretical Incrementalism. For the Theoretical Incrementalist can argue that since radical change of theories brings about change of meaning and change of meaning brings about incommensurability and incommensurability hinders testability, radical change of theory should be discouraged.

Feyerabend's position on meaning seems also to lend support to our interpretation of Burian's position at the end of his paper as Radical Theoretical Comparabilism. For, if radical departures in meaning bring about incommensurability and this in turn adversely affects testability, Burian surely must advocate that *these* radical changes in theories be discouraged. Whether he does suppose that radical change in meaning brings about incommensurability is unclear. But at least some of his remarks suggest that he does.

However, one might well challenge the idea that radical change of meaning necessarily involves incommensurability. Indeed, there are three types of radical change of meaning that one can think of, yet in each case comparison is still possible: (1) change in the sense or connotation of terms, for example, all terms in theory T_1 may have a different sense from the same terms in theory T_2; (2) change in sense and reference, for example, all terms in T_1 may have a different reference as well as a different sense from the same terms in T_2; and (3) change in the basic ontology, for example, the variables in T_1 range over different types of entities from the variables in T_2, despite the fact that the same terms are used in the two theories.

In Case (1) comparison of T_1 and T_2 is still possible, since the reference of the terms can still be the same. The theories T_1 and T_2 may for all their changes of sense still contradict or entail one another, since contradiction and entailment may turn on extensional relations and not on considerations of sense.[7] In case (2) comparison is also possible. Change of reference is under certain circumstances compatible with logical comparison.[8] Even in case (3) comparison is possible. Radical ontological change may make comparison via some referential relation impossible. However, ontic bridge laws which contain mixed variables – one variable which ranges over one sort of entity – may be found. Relative to these ontic bridge laws two theories differing completely in their ontologies may be shown to be in logical conflict.[9]

Even though comparison may be possible given radical changes of sense, reference, and ontology, the comparison may be difficult to make. Hence anything analogous to a crucial experiment or test may be difficult to bring about. Nevertheless, such comparisons may provide the severest sort of test for the old theory. Short of complete incommensurability fundamental changes in sense, reference, and ontology combined with strenuous efforts of comparison may throw into relief the most fundamental problems of an old theory. I say 'may', since whether these problems will in fact come to light or not given radical shifts in meaning is in part a *psychological* question. It is a question of what sort of new cognitive framework will best enable scientists to recognize problems with the old cognitive framework. This is an empirical question that philosophers of science have no particular competence to answer, although they may offer certain working hypotheses.[10]

In any case, the refutation of Feyerabend's very extreme position leaves open the possibility that a somewhat less extreme position, for example, Moderate Theoretical Incomparabilism, may still be correct. Since the radical shifts in meaning that we have reviewed do not prevent comparison on logical grounds, the psychological effect of such shifts on problem recognition remains to be determined. Burian seems to assume (at the end of his paper) that new theories that are both radically different from old theories and highly comparable with the old theories are the most useful in testing the old theory. But this is by no means clear. On psychological grounds it may turn out that the new theory, which is very difficult but not impossible to compare with the old, is potentially the most illuminating of the old theory's problem.

MAXIMUM TESTABILITY AS THE CRITERION OF SCIENTIFIC PROGRESS

So even if one does not challenge the criterion of maximum testability and criticism as the sole criterion of scientific progress, neither Theoretical Incrementalism nor Radical Theoretical Comparabilism, nor a combination of both, is established as the most plausible methodological guide for science. However, this criterion can, and should, be challenged. Surely more is wanted in science than maximum testability. Even if radical change of meaning decreases testability, such change may nevertheless be desirable relative to other more important criteria. What might these other criteria be?

Simplicity might be one and maximum testability is often compatible with maximum simplicity. The simplest hypothesis, *pace* Popper, is not always the most testable.[11] Furthermore, one may want simplicity in science over maximum testability, other things being equal. Moreover, theoretical science is also interested in depth, i.e., in theories that go beyond the appearance of things to their innermost structure. Burian argues in his paper that Feyerabend's requirement of incommensurability conflicts with the goal of scientific realism, i.e., the search for deep theories that explain appearances. Now it might also be the case that the criterion of maximum testability conflicts with this goal. It is not implausible to suppose that deep theories are the very hardest to test and criticize. Yet it might be argued that these theories are the most valuable; that the confirmation of deep theories, although difficult, gives us the greatest scientific understanding.

It seems to me that a more plausible criterion for scientific growth than maximum testability — at least in the advanced sciences — is the confirmation of simple yet deep theories. Such theories must be testable, of course, but such theories need not be the *most* testable. If we wanted the most testable theories, low level ones might be more suitable.

THE EVALUATION OF DIFFERENT METHODOLOGIES

The question remains as to which of the many methodological maxims we considered earlier is most conducive to this goal. I would like to make a brief plea here for experimentation and empiricism in deciding the issue. Feyerabend once wrote a paper called 'How to be a Good Empiricist: A Plea for Tolerance in Matters Epistemological' in which he advocated the proliferation of theories as a way of maximizing scientific growth. Ironically his arguments for this position are rather *a priori*. To be sure, he did appeal to cases from the history of science. But it seems to me that these cases were used for illustrative purposes given a thesis independently arrived at. A really good empiricist should give empirical grounds for the alleged benefits of following such a program.

Which of the many methodological programs outlined above will maximize scientific growth where this is defined in terms of the creation of well-confirmed deep simple theories is something that cannot be decided on *a priori* grounds. Empirical evidence is necessary. Unfortunately such evidence is not to my knowledge available at the present time. But it seems to me that such evidence could be found.

First of all, historical evidence could be used. However, historical evidence is often indecisive unless some sort of comparative study is made. Thus if it were shown that well-confirmed deep, simple theories were achieved in some domain of science when Theoretical Incrementalism was used, this would not show that such results could not have been achieved or surpassed if a less extreme methodological position had been adopted. What is needed, it seems to me, is more comparative historical research and less citing of isolated cases. One might compare the growth of some science during periods where one methodological position seemed to dominate with other periods where other methodological positions prevailed. As far as I know no studies of this kind have been done.

Cross-cultural comparison might also be made. Does Russian social science operate on a principle similar to Theoretical Incrementalism whereas American social science operates on a different principle? If so, has Russian social science achieved more well-confirmed, deeper and simpler theories than American social science? If it has, this may lend some support to Theoretical Incrementalism if no plausible alternative explanations were available.

Another possibility is that of setting up something like a controlled experiment. Let several groups of scientists work in a particular rather well-defined branch of science. Encourage one group to develop theories continuous with the old theories, another group to develop theories that depart radically from the old in such a way that makes comparison easy, and still another group to develop alternative theories that depart radically from the old in such a way that makes comparison difficult. After a length of time compare the results of the groups' activities; see what they have achieved in terms of well-confirmed simple deep theories. To be sure, such a study has large methodological problems but something like it is possible in principle and might well give us much more reliable knowledge than we have now.

CONCLUSION

I have contrasted several methodologies: Feyerabend's and a number of positions that are suggested, if not held, by Burian. I have argued that many alternatives to these positions exist. I have argued that radical shifts of meaning do not entail incommensurability but may make comparison harder. Nevertheless, such comparison may make visible the problem with theories that could not be made visible in other ways. I have suggested that whether or not this is so is in part a question in the psychology of problem recognition and that more empirical investigation is needed. But even if one of these alternatives does provide for maximum testability it may, on other grounds, still not be preferred. I have suggested an alternative criterion to maximum testability: the confirmation of deep, simple theories — and I have argued that it is unclear which of the various methodological alternatives will achieve this goal. However, I have urged that empirical evidence must decide the issue.

To return to the beginning. Feyerabend is surely correct about one thing: philosophers of science need not merely analyze scientific concepts or be speculative. Philosophy of science has as one of its tasks the formulation, justification, and criticism of alternative methodologies. Feyerabend was wrong to suppose that such a task can ultimately be achieved by armchair methods and anecdotal historical evidence. Once alternatives are opened up and positions clarified by philosophers of science this task must be turned over to scientists of methodology — scientists who investigate the effects on science of following different methodologies. Such a clearly defined group of scientists does not exist at present. When it does, philosophers of science may well become good empiricists and a plea for empiricism in matters methodological will not be an empty dream.

REFERENCES

[1] But see May Brodbeck, 'The Nature and Function of the Philosophy of Science', *Readings in the Philosophy of Science*, ed. H. Feigl and M. Brodbeck (Appleton-Century-Crofts, New York, 1953), pp. 3–7. Brodbeck distinguishes four meanings of 'philosophy of science': the socio-psychological study of science; the moral evaluation of the scientist's role; the 'philosophy of nature'; the logical analysis of science. Brodbeck rejects all but the last. However, it is difficult to fit Feyerabend into any of Brodbeck's categories. Indeed, it is difficult to fit Bacon, Mill, and Popper into her categories. For Feyerabend as well as Bacon, Mill, and Popper are not merely logically analyzing science; they are recommending programs of scientific growth and procedure. Such recommendations could hardly be called metaphysical speculations, ethical evaluations or sociological investigations of science.

[2] Richard M. Burian, 'Scientific Realism and Incommensurability: Some Criticisms of Kuhn and Feyerabend', this volume, pp. 1–31.

[3] Richard M. Burian, 'Scientific Realism, Commensurability and Conceptual Change: A Critique of Paul Feyerabend's Philosophy of Science', unpublished Doctoral Dissertation, University of Pittsburgh, 1971.

[4] Karl Popper, *The Logic of Scientific Discovery* (Basic Books, New York, 1959).
[5] For an illuminating extended discussion of this contrast, see Barry Hallen, 'Boldness and Caution in the Methodology and Social Philosophy of Karl Popper', unpublished Doctoral Dissertation, Boston University, 1970.
[6] For a critique of the method of proliferation, see Michael Martin, 'Theoretical Pluralism', *Philosophia* 2 (1972) 341–349; reprinted in Michael Martin, *Social Science and Philosophical Analysis* (University Press of America, Washington, D.C., 1978), pp. 101–109.
[7] See Israel Scheffler, *Science and Subjectivity* (The Bobbs-Merrill Company, Inc., Indianapolis, 1967), pp. 45–66.
[8] See Michael Martin, 'Referential Variance and Scientific Objectivity', *British Journal for the Philosophy of Science* 22 (1971) 17–26.
[9] See Michael Martin, 'Ontological Variance and Scientific Objectivity', *British Journal for the Philosophy of Science* 23 (1972) 252–256.
[10] Cf. Michael Martin, 'Anomaly-Recognition and Research in Science Education', *Journal for Research in Science Teaching* 7 (1970) 187–190.
[11] See Marguerite Foster and Michael Martin, eds., *Probability, Confirmation and Simplicity* (Odyssey Press, New York, 1966), pp. 233–249; Michael Martin, 'The Falsifiability of Curve-Hypotheses', *Philosophical Studies*, pp. 56–60 (1965).

Professor M. Martin
Dept. of Philosophy
Boston University
Boston, Ma. 02215
U.S.A.

ROGER J. FABER

FEEDBACK, SELECTION, AND FUNCTION: A REDUCTIONISTIC ACCOUNT OF GOAL-ORIENTATION*

TABLE OF CONTENTS

INTRODUCTION	44
I. FEEDBACK: A SURVEY OF THE PROBLEM	50
II. BECKNER'S EXPLICATION OF FEEDBACK: SUMMARY AND CRITIQUE	
Summary	56
Critique	57
III. TOWARDS AN IMPROVED DEFINITION OF FEEDBACK	
Asymmetric Causal Links	63
Reducible Variables and Fundamental Boundary Conditions	68
Physical Distinctness	70
Necessary Relations	76
Negative Feedback	77
A Revised Explication of Feedback	80
Informal Summary	82
IV. FUNCTION AND PURPOSE	
The Problem	84
Selection: a Necessary Condition?	86
Selection: a Sufficient Condition?	89
Goals in Cybernetic Systems	94
The Context of Functional Activity	101
V. THE ROLES OF 'SELECTION' AND 'FEEDBACK' IN FUNCTIONAL EXPLANATIONS	
What Does a Functional Explanation Explain?	103
Is a Functional Explanation Reductionistic?	111
SUMMARY AND CONCLUSION	116
APPENDIX	
A.1. Electronic Voltage Regulator	117
A.2. Allosteric Enzymes	119
A.3. Oscillating Chemical Reactions	122
A.4. Temperature Control in Mammals	124

R. Cohen and M. Wartofsky (eds.), *Methodology, Metaphysics and the History of Science*, 43–135.
© 1984 *by D. Reidel Publishing Company.*

    A.5. Is Natural Selection a Feedback Process?   125
    A.6. Limitations on the Size of a Population   129
    A.7. Learning by 'Variation and Selection'   131

NOTES   132

REFERENCES   134

ABSTRACT. The theme of this essay is the mechanism of goal-orientation in non-conscious organisms and in machines. The first portion presents an explication of feedback which, because it is done in the language of analytical physics, is claimed to be a reductionistic explication. Applications are made in the appendix to examples from physics, chemistry, biology, and psychology.

    The second portion examines the rival claims of two theories of functionality and purposiveness in biology: a selectionist theory and a cybernetic theory. It is argued that natural selection is neither a necessary nor a sufficient condition of functionality, and that feedback, in the context of survival, is a sufficient condition.

    Finally, functional explanation is examined in the light of the earlier sections and it is argued that functional explanations are not used to explain the origin of functioning entities, but are used to explain the survival of goal-oriented systems and to elucidate the interconnections among their parts.

INTRODUCTION

The success of Darwinism as an explanatory method and the concomitant demise of the argument from design have made the existence of teleological systems in nature seem puzzling. If a mechanistic explanatory scheme (one in which any element of a system merely responds to local and immediately prior causal influences) is correct, how are we to make sense of the apparent goal-directedness of living things? When we speak of them in purposive language, when we ascribe functions to their parts, when we say that they or their parts perform certain actions in order to achieve certain results, are we illicitly imposing human categories of consciousness and will upon undeserving collections of chemical reactions? Or is there something about the nature, organization, and context of living things — something compatible with mechanism — which justifies our use of this vocabulary?

    Recent studies of teleological language and explanation have accomplished a considerable advance in the clarity with which we may view questions about means, ends, functions, and goals; both in a broader perspective that includes consciously purposing agents and in a more narrow focus upon living organisms. Several questions recur throughout the discussion. How shall we explicate the concept of function? What do, or ought, we to mean when we claim that a functional member is necessary for some property or ability of a larger system? What kind of end do functions serve? How, and what, do functional explanations explain? In particular, how, if ever, do they explain why a functioning member of a system is there? What is the relation between functional explanations and ordinary causal explanations? Are teleological concepts and explanations reducible to physiological or physical concepts and explanations? And if they are reducible, in what sense, and by what sort of general program could that reduction be effected?

The last two questions have received rather less attention recently than the others. Yet they deserve close attention, because the debate over the possibility of reducing upper-level science like biology and psychology to the physical sciences could profit from an enrichment of our stock of examples of inter-level reduction. If the reduction is ever to go through, or if we are even to believe it possible in principle, we must have a programmatic but plausible story to tell about how concepts and explanations characteristic of the upper floors of the scientific edifice can be connected with those that operate lower down. Clearly, we need some additional examples to set alongside the paradigmatic reduction of thermodynamics to statistical mechanics, and the currently debated reduction of Mendelian to molecular genetics.

The question of reducibility may be put in several ways (Beckner [1968] and Schaffner [1967]):

(1) We may ask whether biological organisms are ontologically reducible to physical entities; that is, whether the list of basic things compiled by physicists for use in explaining non-living matter will do for biologists as well. Now it seems to me that, if a rose or a honey bee were analyzed into its constituent protons, neutrons, and electrons, there would be no important parts left over; nevertheless, I shall not be directly concerned with that issue in this essay.

(2) Even if the bee is composed exhaustively of the physicist's particles, it does not follow that they have in the insect just those properties that they display in the cyclotron, or that they still obey the physicist's laws. So we may ask whether the laws of biology are derivable from the laws of physics, possibly with the aid of bridge postulates and suitable approximations and idealizations.

(3) In their theories and explanations biology and other upper-level sciences characteristically employ concepts that are not found in physics. A

prerequisite, then, for a successful theory reduction is a reduction of such concepts. Four examples of upper-level concepts with which this essay will be concerned are: (a) *Selection*. More accurately described as blind variation and systematic elimination, this concept has its most obvious application in the neo-Darwinian theory of natural selection; but there are other examples of its use, such as the theory of anti-body production (Edelman [1974]) and certain theories of human problem solving (Campbell [1974a]). (b) *Feedback*. The peculiar kind of organization called negative feedback is a pervasive feature of systems at all levels, from the molecular to the ecological; in addition, feedback is built into a wide variety of human artifacts such as thermostats and the guiding mechanisms of space vehicles. (c) *Function*. Another characteristic shared by living systems and the products of engineering is the seeming appropriateness of their being described as healthy or sick, or as working properly or not; that is, as functioning or malfunctioning.[1] (d) *Purpose*. Normally we ascribe purposes only to conscious agents or, derivatively, to their tools. But the physicalist theory of mind may, if it proves successful, require us to say that having a purpose is identical to being organized in a certain way or having a certain kind of structure. And even a dualist theory may require us to say that having a purpose is identical to the conjunction of consciousness with a certain kind of organizational pattern. Under either theory we are led to ask what kind of physical structure is involved in a system being purposive. A corollary of this way of explaining purposiveness — as involving a certain pattern of organization — is that the concept can be adapted to fit any material instantiation of the pattern; so purposiveness is not necessarily restricted to the human nervous system or even to living matter.

Questions about teleology can be approached along two somewhat different lines. Some studies aim at the broadest possible understanding of the way we use functional terms and explanations, drawing examples from ordinary usage and the sciences of life and society, including biology merely as a special case. Recent contributions to this enterprise, which received an early articulation by Carl G. Hempel (1959), have been made by Berent Enc (1979), Peter Achinstein (1977), Christopher Boorse (1976), Andrew Woodfield (1976), and Robert Cummins (1975). Other studies, of which Ernest Nagel's (1961) and Morton Beckner's (1959) are early examples, focus more narrowly and intensively upon the specifically biological uses of teleological language and functional explanation. More recent contributions to this effort have been made, for example, by Nagel (1977), David Hull (1974),

Michael Ruse (1973), and Francisco J. Ayala (1970). Among those who emphasize the biological context of functionality, approaches differ according to which type of biological process supplies the guiding example. Those who take homeostasis or regulation as a paradigm analyze functions cybernetically by reference to goals, following an early suggestion by Rosenblueth, Wiener, and Bigelow (1943); whereas those who, like William Wimsatt (1972) and Larry Wright ([1973] and [1976]), find a paradigm in natural selection relate functions to the criterion according to which a selection is made, or to the causal sequence which leads to a functional part being present in its containing system.

The cybernetic theory has drawn criticism (Manier [1971], Ruse [1973], and Woodfield [1976]), much of which is summarized and sharpened by Wimsatt (1971), who also proposes an alternative theory (Wimsatt [1972]).[2] Wimsatt offers two arguments to support his contention that attempts to analyze the concepts of function and purpose in terms of the structure of a self-regulating system "cannot succeed." In the first place, he has produced an example of a self-regulating system, one of whose parts has a function which can be inferred only from information about the intentions of the designer of the device (Wimsatt [1972], p. 20, ff.). Secondly, he charges those who would base function on feedback with having failed to present an acceptable analysis of feedback and of goal. As a remedy for this situation Wimsatt prescribes an analysis of function based on selection procedures and their criteria. He admits that selection processes are probably only a sufficient criterion for the use of functional language, and presents considerations (weighty ones, in my estimation) which make it seem likely that they cannot be necessary conditions.

But Wimsatt's first argument against the cybernetic theory establishes only that some functions of the parts of a system may be determined by factors outside the system — by, for instance, its operating context or mode of origin. Briefly, his example describes a homing mechanism which tracks down targets in its vicinity. There is a part of the device which limits the amount of time it spends tracking a single target and causes it to transfer its attentions to others. Unfortunately for the device, some of its targets have the property that, if the device homes in on them, it will be destroyed. Thus the limiter produces at least two effects: it guarantees that the device will remain active by tracking many different targets, and it guarantees that sooner or later the device will end its career by encountering one of the lethal targets. As Wimsatt points out, whether the function of the part is to provide an impulse for variety or a suicide impulse cannot be determined merely by observing the device — we must ask the designer what he intended to accomplish by inserting that part into the mechanism.

This argument leaves open the possibility that some other functions may nevertheless be determined by the internal structure of the system itself. In fact Wimsatt's terminology suggests this possibility. I should imagine that the part he discusses does other things besides limiting the tracking time; for example, it probably dissipates energy and makes clicking noises. Yet, independently of any speculation about the intentions of the designer, Wimsatt refers to it as a limiter; thereby, it seems, implicitly ascribing to it as a function one of its several causal consequences. The justification for calling it a limiter and not a clicker may lie in the fact that the limiting action has further consequences in the feedback loop whereas the clicking does not. I shall present arguments to support this free-standing assertion in the second half of this essay.

Wimsatt's second argument I choose to take not as a deterrent but as a challenge. In Sections I–III I shall attempt to produce an acceptable analysis of feedback, in the general tradition of Rosenblueth, Wiener, Bigelow, Beckner, and Manier.

Unlike the analysts of teleological concepts who aim to construct an explication which covers all or most of the ways we use them both in ordinary language and in scientific speech, I shall proceed from the particular only a short way outward. Concentrating on the cluster of terms which biologists apply to the processes of individual survival and reproduction, I shall analyse the biological use of function-ascriptions according to what Peter Achinstein calls the good-consequence doctrine. This theory includes what he terms the goal doctrine as a special case when we interpret the term *goal* mechanistically, apart from conscious intentions. Focussing still more narrowly on survival, I shall begin by proffering a reductionistic explication of what I take to be the central example of goal-directed mechanism in biology, namely homeostasis or self-regulation, arguing that in this instance a good consequence may appropriately be called a goal. That position is in line with most of the general analyses mentioned above. Achinstein, for example, recommends an eclectic attitude toward good consequences, goals, and etiology; and Enc understands functions as directed toward stability or the maintaining of some property of a containing system. But when we concentrate on biological systems we can be quite specific as to what the good consequence produced by the functioning part is, just what is being maintained, and how. A cybernetic analysis illuminates in the first instance the internal workings of an organism with respect to the functioning of its parts toward a self-contained organismic goal, survival. But, although I shall not carry it further here, the analysis need not stop at that point: William T. Powers' (1973) cybernetic study of behavior — that is, of the particular goals of an organism with respect to external objects — sets up a program for extending the concept of feedback to such activities as the pursuit of prey and the grasping of objects by hand.

The concept of function may not be easily pinned down. Perhaps a combination of approaches will prove to be necessary. In any case, I shall not offer a candidate for the role of necessary condition. Wimsatt claims that selection processes — especially natural selection — provide a sufficient condition for ascribing functions. Now this claim must be taken seriously because, despite all that can be said in favor of the goal doctrine, analyses which link the concept of function to a major theme of modern biology have much to recommend them. On this score, I shall have something more favorable to say about Wimsatt's and Wright's selectionist accounts than recent authors have conceded. Nevertheless, in sections IV and V I shall argue that, although some selection processes may indeed provide purposive contexts, natural selection, except in a thinly metaphorical sense, does not; thus I shall argue that natural selection is not a sufficient ground for what I shall recommend as the primary use of functional language in biology. I shall also argue that feedback, taken in a larger cybernetic context, is a sufficient condition and that it provides some promising leads toward the solution of problems that Wimsatt's approach leaves unanswered, especially questions about goal attainment, the force of *ceteris paribus* in functional analysis, and counterfactual conditions. Moreover, an analysis based on feedback provides a comfortable framework within which to discuss other kinds of functional explanations than those considered by Wimsatt — explanations of other matters than the presence or origins of functional behavior.

Finally, let me alert the reader to my line of approach toward understanding goal orientation. Consistent with my purpose of exploring the possibility of reducing biological concepts to physical, I shall present an analysis of self-regulation not by direct reference to the goal-oriented system itself, but explicitly in terms of what we say about it. This strategy has been employed also, though perhaps less self-consciously, by other analysts of negative feedback, such as Nagel and Beckner. There are two reasons for adopting such a method. In the first place, it will prove impossible to formulate a criterion of self-regulation that specifies merely a finite sequence of a system's overt behavior; we must refer also to what it would do in counterfactual circumstances, speaking of its abilities, capacities, or propensities. And we cannot speak of such things without at least implicitly recognizing nomic necessity, laws of nature. So any criterion by which we are to recognize homeostasis must be applied against a background of physical theories. Secondly, and more decisively (for the laws of physics might be left unacknowledged in the background of our analysis), my primary concern with reduction requires me to show how characteristically biological descriptions and explanations of an organism can be deduced from, discovered in, or otherwise obtained from a lower-level description of it. For that purpose we require an account of functionality in biology which is formulated explicitly in terms of the merely mechanical, non-purposive language we apply to the objects of our study.

I. FEEDBACK: A SURVEY OF THE PROBLEM

What is a feedback system? The form of the answer depends upon the precise import of the question. Let us consider four alternative ways of putting the question more precisely. We might ask: what pattern of observed behavior characterizes a feedback system? This is the way Rosenblueth, Wiener, and Bigelow (1943) originally posed the question about goal-directedness. Their point was that, regardless of the materials of which a system may be constructed, if we see it behaving like a person (or other living thing) seeking a goal, then we must ascribe a goal to the system. They explained goal-directedness in terms of feedback, but their criterion of goal directedness was behavior. Taylor (1950), among his other criticisms of this approach, pointed out that in a finite sequence of behavior many systems may successfully mimic a genuinely goal-directed one without in fact being so. A sequence of chance occurrences may produce the right sort of behavior, so long as the sequence is finite. On this point, I think, Taylor is right. And his caveat applies equally to the explication[3] of feedback. The explication must be concerned with the causal connections in the system, with its state of organization; not merely with what it happens to do while under observation, but also with what it would have done had circumstances been different and with what it would do in the future with respect to a wide range of possible circumstances, only some of which could be realized in fact. So a second, and better, way of putting the question is this: What causal connections and structural features, and what patterns of actual and potential behavior arising from these features, characterize a feedback system?

But how can one assert a counterfactual conditional statement about an object? Only on the basis of a rule by which other possible worlds are imagined; that is, only on the basis of a theoretical model of the system – a model of the sort that it is the business of physics and biology to construct.[4] In response to the textbook command, 'analyze the following object', the physicist, relying heavily on intuition, proceeds to construct an abstract model by selecting relevant details, ignoring others, inventing variables, identifying mechanisms, writing down laws and boundary conditions, imagining 'input' functions, etc. The object fades from view in his mind as it is replaced by a logical network of symbols and relations among them. And biologists, too, though less theoretically inclined than physicists, have their theoretical models of biological systems. Whether these ingenious constructions really describe their objects is a question to which no finite set of observations can provide an infallible answer. Nevertheless, we take these models as true and on them we base our assertions of the possible behavior of the systems we study. If a model is available, it is possible to translate back and forth from an object-centered language (a theory), in which one makes assertions about the actual or possible behavior of the system, to a theory-centered language (a meta-theory), in which one makes assertions about the derivability of certain statements from the theoretical model. To say that the object would do A under circumstances C is equivalent, *via* this translation, to saying that a statement describing A is deducible from the conjunction of the theoretical model and descriptions of C. Consequently, when we calculate what a system would have done under different circumstances, or would do if certain circumstances should arise, we are drawing deductions from, and thus are demonstrating properties of, the theoretical model. So a third and meta-theoretic way of formulating our basic question is this: What are the characteristic properties of the theoretical model of a feedback system?

Although the latter two formulations of the question are equivalent (because they are intertranslatable), there is this advantage to be gained by adopting the meta-theoretic form: it is the natural one to use when questions of theory reduction are mooted, for these questions, too, concern the derivability of statements at one level of theory from others at another level. As part of the larger program of reducing biology to physics we must face the peculiarly biological concepts of function and purpose. Even the concept of feedback, by no means the exclusive property of biologists, does not have a natural place in the conceptual armory of physics; hence it, too, is in need of a reductionistic treatment. And the problem of reducing feedback acquires added interest if, as I shall argue later, feedback plays an important role in functional and goal-directed phenomena.

Because I am concerned in this paper with working out a stage in the reduction of biological theories to physics, I shall assume that the reducing theory lies ready to hand; that (at least in principle) there is available for the systems of which we speak a complete physical analysis based on a theoretical model of the system. Although the theoretical model of a system may be formed at any of several levels, the third, or meta-theoretic, form of the question may be given the appropriate reductionistic thrust for this context by specifying that the model be a physical one. The fourth, and reductionistic, form of the question, then, is this: Given a theoretical model of an object, constructed according to the ordinary canons of physical analysis, what features characterize the model as referring to a feedback system? This question asks for properties, not of the object itself, but of the description that a physicist would ordinarily apply to it. The question assumes that the stage has been set for the derivation of the behavior of the object, by means of the appropriate physical laws, from a set of initial and boundary conditions. The answer to this question must be cast in the language of analytical physics — in terms of variables, functions, boundary conditions, interactions, etc. If the reductionist thesis is true (in some form or other) then an answer to this formulation of the question would imply an answer to the object language form as well, since all possible information about the system would be derivable from the basic physical analysis (augmented perhaps by bridge hypotheses). If the reductionist thesis is not true, or requires modification, one way to find this out is to try to effect a reduction. If the search for an answer to the fourth formulation of the question is successful, then the concept of feedback will have been 'reduced' to physical principles; further, if the concept of function is linked to the feedback pattern, then functional language will have been similarly reduced. Of course, the search may not be successful, or it may be only partially so.[5] But it seems to me that the question of the reducibility of higher order theories to physics is more convincingly answered, not by abstract proofs of possibility or impossibility, but by determined attempts at reduction in fairly concrete cases.

Most authors who have offered explications of feedback have employed a combination of the approaches I have tried to distinguish above. The explication which comes closest to answering the reductionistic form of the question, and the most systematically worked out explication I know of, is that of Beckner in *The Biological Way of Thought* (1959). Beckner's analysis has been criticized and extended by Manier (1971) and by Wimsatt ([1971] and [1972]), but, as I shall demonstrate below, even with their suggested emendations the explication does not work.

In following section I shall give a brief summary of Beckner's definition of feedback and, by applying it to a variety of concrete examples, demonstrate both its virtues and its faults. By combining my criticisms with those of Manier and Wimsatt, I shall construct a revised, reductionistic version of Beckner's explication of feedback, which I believe captures what we mean by feedback and does so in the language of physics.

Before proposing and testing formal criteria, let us first try to clarify, in an intuitive way, exactly what feedback is. The feedback pattern is easily represented by a block diagram, a kind of flow chart of causal connections, as shown in Figure 1.

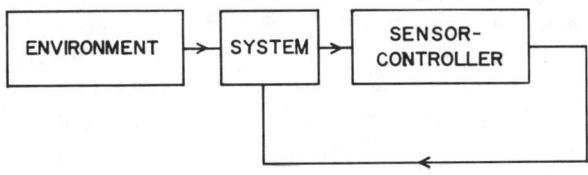

Fig. 1.

Here the idea is that the system is subject to environmental influences which produce changes in its internal state. The state of the system, or some aspect of it, is monitored by a sensing, controlling device which causes further changes in the system. These changes are such as to maintain the system in, or cause it to approach, some goal state despite the perturbing effects of the environment. Examples of feedback systems are the 'homing' mechanism which guides a missile and the homeostatic system which maintains body temperature in mammals. An adequate explication of feedback must be a kind of logical sieve which passes the sort of system just described, but strains out such apparently goal-oriented systems as a pendulum, a stream carving a channel down a hill, or a thermodynamic system approaching equilibrium.

Because the term feedback is *not* precisely defined, it is necessary to specify carefully the rather narrow sense in which I intend to employ it. In particular, I want to construct a logical sieve which will strain out some kinds of system to which the term frequently is applied, but which lack a certain degree of organizational complexity. As an example of what I mean, consider a safety valve; to be precise, let us take the regulator weight on a pressure cooker. How does this device work? Suppose that at some initial time there is an equilibrium; that is, the weight is floating on the rising stream of escaping vapor. There is an equilibrium pressure inside the cooker which is jointly determined by the rate of input of heat from the stove and by the rate at which steam escapes through the orifice in the lid. This rate of escape, in turn, is partly determined by the size of the orifice, and that depends on the height at which the weight is floating. Now suppose there is a sudden increase in the rate of heat input — how does the system respond? The pressure rises, the outflow of steam increases, the regulator weight is lifted higher, thereby increasing the size of the orifice, until a new equilibrium is achieved. The new pressure will be greater than the old, but not as great as it would have been if the weight had been held in a fixed position. Because the weight responded to the increased pressure by allowing the orifice to grow, the original disturbance has been partially compensated for.

What is objectionable about this system is that it is too simple — so simple, in fact, that systems exhibiting this kind of organization abound in inorganic nature. The geysers of Yellowstone fit this model (with the ground water playing the role of the regulator weight); so does any saucepan fitted with a lid; so does a volcano which periodically blows out a plug of cooled lava. The system does not seem to be one which seeks out a goal state or *actively* compensates for external perturbations. Rather, it seems to be totally passive; to be, in fact, merely a system whose equilibrium state is determined by a comparatively complex boundary condition. Whether 'feedback' can be explicated so as to include a thermostat and exclude a safety valve remains to be seen, but that is what I shall try to do.

When it is working properly, a negative feedback system maintains itself in a certain state; it achieves a kind of stability. A typical inanimate object merely endures in a passive way, as a system at or near equilibrium or as a steady state in a flow process. Living systems (and certain simpler objects, among them feedback systems) also show stability over time, but not by mere passive endurance. They actively maintain themselves by making corrections to counteract the structure-destroying influences in their environment. A boulder may endure relatively unchanged for millennia, whereas a gnat may live for only a day; yet the gnat, poised in a state very far from equilibrium and subject to various forces which, if unchecked, would destroy it quickly, represents a feat of balance not to be compared with that of the rock. We may, then, distinguish active stability from passive stability. An actively stable system *survives*, whereas a passively stable system merely *endures*.

The same contrast may be drawn between these two kinds of stability in simpler systems. There are two ways that the temperature of a small object may be maintained at a fairly constant level in the face of a considerable range of environmental temperatures: one is to attach the object to a massive system of so large a heat capacity that the fluctuations of the environment are averaged out; another way is to construct a thermostat to regulate the temperature of the object. The first way produces a system which is always in a state of quasi-equilibrium, too sluggish to change very much; the second produces a system in a state which can be maintained only because it is far from equilibrium — whose operation depends on the flow of energy through it because dissipative processes are of the essence of its inner workings. It is possible to build a massive heat reservoir which keeps a more constant temperature than a given thermostat. The difference between them is not a matter of the degree of stability; the difference is one of delicacy, of flexibility. The contrast is activity *vs.* stolidity and, when sufficiently complex, interlocking networks of regulation and control appear on the scene, life *vs.* non-life.

Many authors have emphasized the fact that a feedback system consists of several parts; specifically that it can be analyzed into a main system which has some variable property, h, whose value is controlled or regulated, and a disconnectable part, coupled to the main system by a relatively small amount of energy, which acts as a sensor to detect deviations of h from some preferred state and initiates a chain of events that leads to a partial restoration of h to a less deviant value. The sensor-controller mechanism must be physically separable, i.e. disconnectable (Beckner [1959]).

The variable h has a range of possible values, depending on conditions both internal and external. Without the sensor-controller mechanism in operation, h will range relatively widely; with it, h is restricted to a small portion of its possible range. Thus a feedback system may be thought of as a thermodynamic system which remains in a state of low entropy; i.e., it retains a relatively improbable set of values of one of its state variables. And it does so because of its structure or pattern of organization — a structure which involves the flow of energy through the system and which employs irreversible, dissipative processes.

Glansdorff and Prigogine (1971) have employed the term 'dissipative structure' to designate a class of systems which maintain a definite structural pattern in a dynamically stable state far from equilibrium, in a region where linear non-equilibrium thermodynamics cannot be applied. The application of non-linear thermodynamics to systems involving dissipative processes far from equilibrium has led to the identification of many stable structures which display temporal order (e.g., oscillations), spatial order (e.g., patterns of convection cells) or both (e.g., waves of chemical activity propagating through a medium) (Winfree [1972]). Abner Shimony (1975) has suggested that feedback systems may be thought of as special kinds of dissipative structures; certainly one of Glansdorff and Prigogine's examples, the Zhabotinsky reaction, turns out to be a feedback system, as we shall see in the Appendix.

II. BECKNER'S EXPLICATION OF FEEDBACK: SUMMARY AND CRITIQUE

Summary

In this section I present not just a summary of Beckner's (1959) analysis, but a translation of it into somewhat different language. My excuse for doing so is that there is no uniformity in the terminology of the various authors whose work I wish to draw from; my reason for doing so is that I prefer to employ throughout this paper the terminology best suited to the explication I shall eventually argue for.

Beckner lists three elements in what he calls "the pre-analytic content of the idea" of feedback. They are these:

(1) The system is in, or tends toward, a particular state such that fortuitous variations outside the state or deviations from the path to it are compensated for by suitable changes within the system itself.

(2) The system's behavior, *qua* teleological, involves the expenditure of energy derived from a local source.

(3) A specialized structure ('hookup') is involved, disruption of which stops the teleological behavior but does not completely change the nature of the system.

Beckner's formal analysis consists of the following definitions:

(1) System, system variables. A system is represented by, or may be said to be, a set of variables which together determine its state.

(2) Environmental variable. A variable g is an environmental variable with respect to a system S if and only if g is not one of the system variables and there is a system variable which is a single-valued function of g (and possibly of other variables as well).

(3) Feedback.[6] A system S is a feedback system with respect to a variable h_1, if and only if all the following conditions are met:

(i) The system variables are functionally independent.

(ii) The variable h_1 is a single-valued function of the system variables: $h_1 = \varphi(f_1, f_2, \ldots, f_n)$.

(iii) There is an environmental variable g_1 which is a single-valued function of a set C of variables which includes h_1: $g_1 = \psi(h_1, h_2, \ldots)$.

(iv) One of the system variables, call it f_1, is a single-valued function of a set K of variables which includes $g_1 : f_1 = \chi(g_1, g_2, \ldots)$.

(v) "If h is regarded as a function of time, then h is either restricted within a specified range of variation, or else h shows a regular pattern of change."

(vi) The system "is so constituted that it is nomically possible" to produce changes in S or the environment so that condition (v) is abrogated but (ii) remains in effect.

Critique

(1) The main virtue of Beckner's approach is that it attempts to make a directively organized system recognizable as such even when it is not being successful, as when external conditions make the goal state unattainable. In other words, the characteristics of a feedback system are, for the most part, expressed as structural and organizational features, not behavioral ones.

(2) Points 1 and 3 of Beckner's 'pre-analytic' treatment seem to me to be all that one can say about the phenomenon at that level. The second point is also emphasized by Manier (1971), who argues that a feature of feedback systems is their 'autonomy'. By this he means that energy stored in the system is 'triggered' by a process which itself consumes relatively little energy. Nevertheless, I believe this requirement is too restrictive; indeed, it does not reappear, even implicitly, in Beckner's formal criteria. His own description of a magnetically activated current regulator provides, I think, a counter-example. Perhaps a clearer case would be a greenhouse whose temperature is regulated by a shutter mechanism which obtains its energy by intercepting a little of the sunlight it controls.

(3) The first four points of Beckner's six-point definition of feedback clearly are in language appropriate to the fourth, or reductionistic, form of the question posed in Part I. It is not clear to me, however, that his last two points are also intended in this way. In point (v) Beckner refers to the output behavior of the system, treating h as a function of time. It is not quite clear whether the reference is to the observed behavior of the system during some finite sample of its history or to the set of all possible outputs (including what it would have done under conditions contrary to fact or not yet realized). To say "h is *restricted* within a *specified* range ..." (emphasis mine) at least suggests the latter interpretation; the words "shows a regular pattern of change" suggest the former. If the reference is to a finite sample of behavior this condition is too weak, for during a finite time most physical quantities

will be observed to traverse only a subset of their possible values. If the reference is to all possible outputs of the system, the condition is too vague, for it does not make explicit how the allowed range is specified. Conditions (v) and (vi) appear to be attempts to formalize points 1 and 3 of the 'preanalytic content'. I shall attempt to make both of these points more explicit in Part III where I discuss 'physical distinctness' and what it is that makes feedback negative.

(4) The relations among the several variables introduced in criteria (i)–(iv) are conveniently represented by another block diagram, as shown in Figure 2.

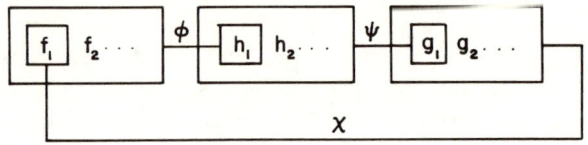

Fig. 2.

It is clear from the diagram that the variables f_1, g_1, and h_1 participate in the overall pattern on an equal footing, as do the sets of variables S, C, and K. Whether h_1, with respect to which S is said to be a feedback system, can be given a distinctive place in the analysis remains to be seen. If it can be done at all, some additional factors will have to be brought in, for it is clear that criteria (v) and (vi) will not do so.

(5) Beckner does not forbid, and in two examples actually employs, analytic relations for the function $g_1 = \varphi(h_1, \ldots)$. That this practice can have disastrous consequences may be seen from the following example. Consider a single-loop electrical circuit consisting of a battery and three resistors. (See Figure 3.) Here E is the electromotive force of the battery and V_1, V_2, and V_3 are the potential drops in the resistors. The physical law for this system requires that $E = V_1 + V_2 + V_3$. Let us now apply Beckner's six criteria to this system:

(i) $S = \{V_1, V_2, V_3\}$. The system variables are not functionally related.
(ii) $h_1 =_{\mathrm{df}} E$; so $h_1 = V_1 + V_2 + V_3$.
(iii) $g_1 =_{\mathrm{df}} E - V_3 = h_1 - V_3$; so the function $g = \varphi(h, \ldots)$ is analytic.
(iv) $f_1 =_{\mathrm{df}} V_1$; so $f_1 = g_1 - V_2$.
(v) While the battery is fresh, E remains close to its initial value, E_0.
(vi) Overheating the battery can cause E to drop well below this value without altering the functional relation (ii).

Thus, all six criteria are satisfied. Clearly, it would be in the spirit of

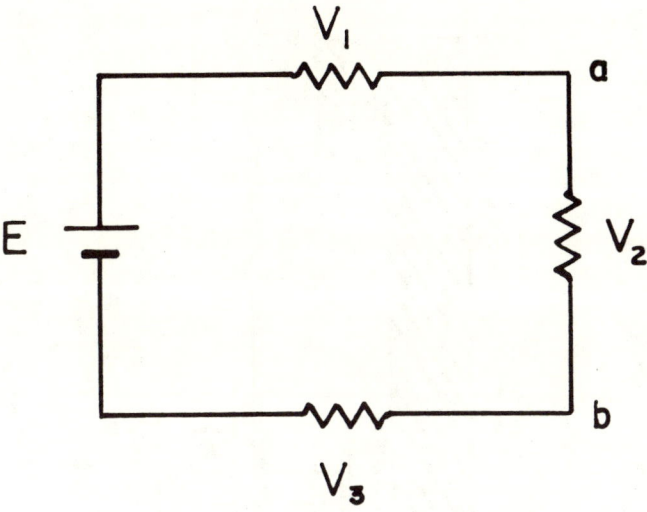

Fig. 3.

Beckner's approach to correct this fault by requiring all three relations, $h_1 = \varphi(f_1, \ldots)$, $g_1 = \psi(h_1, \ldots)$, and $f_1 = \chi(g_1, \ldots)$, to be true not by definition, nor as a consequence merely of the definitions and the laws of physics. They must be true only because of the particular circumstances that constitute the system. Let us define a *necessary* relation as one that follows from the conjunction of the laws of physics and a set of definitions; and let us say that a relation is *contingent* if it is not necessary. So the three relations must be contingent.

(6) Beckner permits the use of an invertible function for the relation $g_1 = \varphi(h_1, \ldots)$. That this is too lax a criterion may be seen from the following example. Consider a double piston–cylinder system. (See Figure 4.) The cylinders are joined by a conductor of heat so that their temperatures remain equal, and the pistons are connected by a rigid rod. The pistons and rod are of negligible weight, but the lower piston has a weight W resting on it. N, P, V, and T are the mole numbers, pressure, volume, and temperature of the gas in the lower cylinder, and n, p, v, and t are the corresponding quantities in the upper cylinder. A and a are the areas of the lower and upper pistons, respectively. The appropriate physical law is the ideal gas equation: $PV = NRT$ and $pv = nRt$, where R is the gas constant. The six criteria can be applied thus:

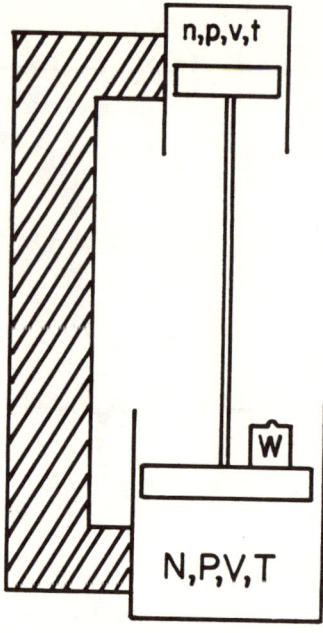

Fig. 4.

(i) $S = \{N, T, V\}$. The system variables are independent.
(ii) $h_1 =_{df} P$; so $h_1 = NRT/V$.
(iii) $g_1 =_{df} p_{g_1}$; so $g_1 = (PA - W)/a$.
(iv) $f_1 =_{df} T = t$; so $f_1 = pv/nR = g_1 v/nR$.
(v) P is constant or oscillates with decreasing amplitude.
(vi) By changing W, P may be put outside its previous range, yet relation (ii) remains true.

Again all six criteria are satisfied, yet the system does not actively seek a goal state. The problem here seems to be that the causal connection between g_1 and h_1 is reversible: causal influences can and do pass in both directions around the loop. Intuitively it is clear that the causal connections should be asymmetric; i.e., the flow chart should contain an arrow indicating that causal influences flow clockwise around the feedback loop. Wimsatt's main criticism of Beckner is also directed at this problem (1971). He suggests that the required irreversibility may be built into the formal definition by adapting Simon and Rescher's (1966) method for establishing the order of causal influences. I shall return to this problem in the next section.

(7) However, Beckner's specification of single-valued functions in criteria (ii), (iii), and (iv) also results in the exclusion of systems which clearly do display the kind of organization one wishes to include. Any system in which hysteresis is present would be ruled out, as when g_1 is a double-valued function of h_1. Also, any system would be ruled out in which, for example, h_1 is a function not of f_1, but of the integral of f_1 over time; i.e. f_1 controls, not h_1 directly, but the rate of change of h_1. Both of these difficulties are exemplified by an ordinary thermostat system, shown schematically in Figure 5(a).

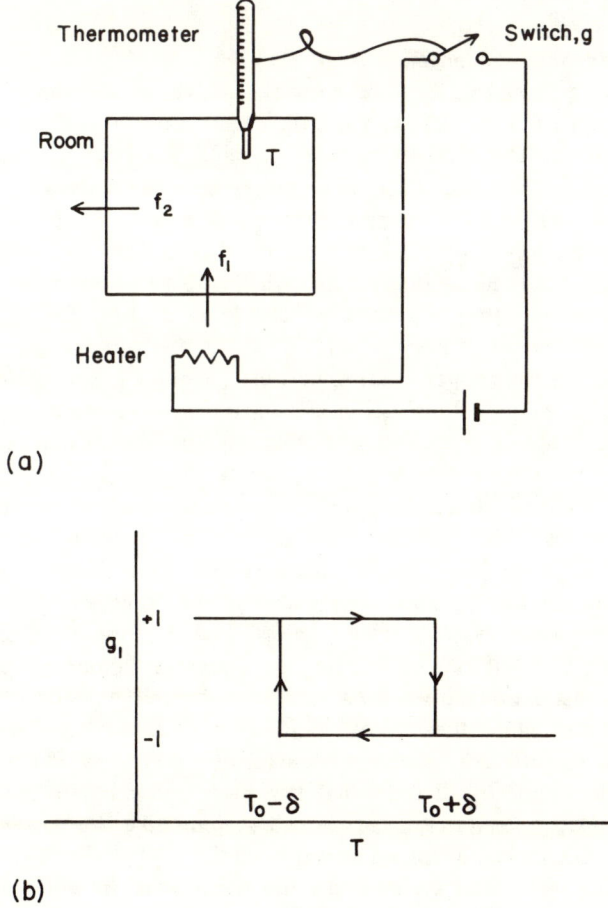

Fig. 5.

In Figure 5(a), T is the temperature of the room, f_1 is the rate of energy input from the heater, f_2 is the rate of loss of energy to the surroundings, C is the heat capacity of the room, and g_1 represents the state of the switch: $g_1 = +1$ for ON, and $g_1 = -1$ for OFF.

Once more, the six criteria:
(i) $S = \{f_1, f_2\}$. The system variables are independent.
(ii) $h_1 =_{df} T = T_0 + \frac{1}{C} \int_0^t (f_1 - f_2)\, dt \neq \varphi(f_1, f_2)$.
(iii) The dependence of g_1 on h_1 can be represented most clearly by a hysteresis curve as shown in Figure 5(b).
(iv) $f_1 = ag_1 = a$ or 0.
(v) T remains between $T_0 - \delta$ and $T_0 + \delta$.
(vi) If f_2 is too great, T drops below $T_0 - \delta$, yet (ii) remains true.

So criteria (ii) and (iii) are violated, even though this is an active goal-seeking system. The relation between h_1 and the state variables is not a simple mapping from the space of f_1 and f_2 onto the space of T, as required by criterion (ii); and the relation between g_1 and h_1 is not single valued. From the violation of criterion (ii) the moral is easily drawn: the definition of feedback should allow one of the variables to be a time integral or other non-elementary relation of another, at least in some cases. It is not so obvious what change is required in criterion (iii). As the previous example also shows, the question of the proper relation between g_1 and h_1 is a subtle one, and I shall postpone a detailed discussion of it until the next section.

One thing that does seem clear at this stage, though, is that an acceptable definition of feedback should not artificially restrict the variety of mathematical entities which can be used to represent causal connections. Some connections may be represented by algebraic or transcendental functions; others, as we have just seen, by integrals over time. There is, apparently, no limit to the variety and complexity of the 'engineering' that may be found in feedback systems. Nevertheless, I shall propose, purely for later convenience and without attempting to suggest a taxonomy of feedback systems, a distinction between two common types. The behavior of a feedback system in which the rate of change of h_1 is determined by the system variables is so different from the behavior of a system in which the system variables determine h_1 directly that it is often necessary to treat these two kinds of system separately. Let us call systems like the current regulator circuit, in which φ stands for a mapping from $f_1 \ldots f_n$, to h_1, Type I feedback systems; and let us call systems like the thermostat, in which φ represents time integration of the f_i, Type II systems. Another example of Type II feedback is an automatic steering mechanism where the rudder angle of a ship, a system variable, determines the rate of change of the vessel's direction of motion.

III. TOWARDS AN IMPROVED DEFINITION OF FEEDBACK

Asymmetric Causal Links

Having pulled apart Beckner's definition of feedback, let us now see what can be done with the pieces. All the problems we raised in the preceding section have fairly obvious solutions, save one. The major question remaining is: How can we guarantee that causal influences pass in just one direction around the loop? As the example of the coupled pistons and cylinders shows, and as Wimsatt (1971) has also pointed out, the causal relations are not symmetric. In some sense which has yet to be made clear, g_1 is controlled by h_1; as a 'sensing' variable g_1 is determined by h_1 without acting back upon it via the same physical interaction. Still at the intuitive level, we can express this asymmetry by means of the revised flow diagram of Figure 6.

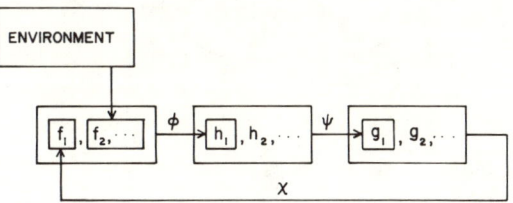

Fig. 6.

Here the arrows indicate the asymmetric causal influences. Although the environment may certainly affect other variables, at least one system variable (other than f_1) must be subject to external influences, for it is just the fortuitous changes in some of the system variables which the feedback loop compensates for, by producing changes in f_1. Moreover, the symbol f_1 in this disagram may be taken to stand for more than one (but not all) of the system variables.

So the question is: How can we give an explication of the causal asymmetry in the feedback loop *using the language of analytical physics*? I shall consider two possible solutions to this problem, both of which may be adequate, but which are quite different in spirit. The first illustrates a fact about the causal relations which must not be overlooked in any eventual theory, but the second, which may be more general, is the one I shall adopt.

(1) *Dissipative processes*. Before proposing a possible formal definition of the proper relation between causally connected variables, let us consider two examples, to serve as guides to the intuition. First, consider the governor of a steam engine (Figure 7). As a result of the rotation of the shaft the weights

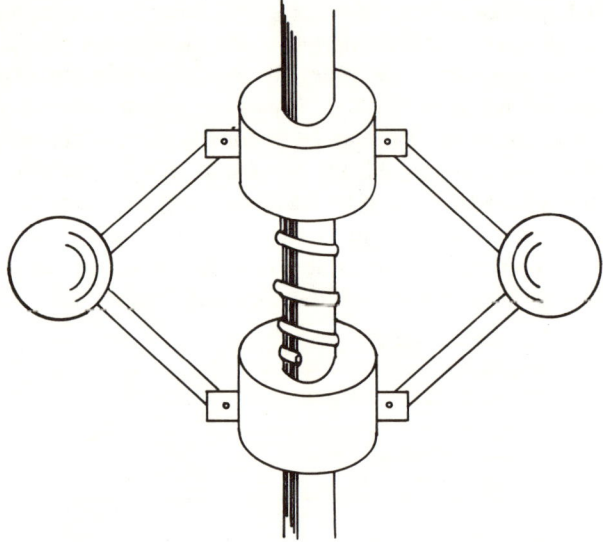

Fig. 7.

move outward, lifting the collar to which they are attached. This motion is then transmitted to the throttle of the engine. Now, a dynamical treatment of the governor in the absence of friction would show that the collar oscillates up and down (and the weights oscillate out and in) with phase and amplitude determined by the initial conditions. It is only because this system is so strongly damped that it is possible to establish a mapping (in this case single-valued) from the variable v (the rotation speed of the shaft) to the variable x (the position of the collar). Moreover, the mapping is not contemporaneous: if the engine should suddenly shift to a new rotation speed, a characteristic time must elapse (a relaxation time) before a new, equilibrium value of x is established.

FEEDBACK, SELECTION, AND FUNCTION

Second, consider the switch which controls a thermostat. Here, too, only friction (strong damping) keeps the switch from 'oscillating', that is, from bouncing back and forth between its 'off' and 'on' positions. Something like a relaxation time must elapse also in this device before the switch settles into a new state in response to a change in temperature. Here, too, it is possible to establish a mapping onto the variable g, but in this case, because of hysteresis, the mapping is from a set of variables consisting of T *and of* g itself. That is, the values of T and g at a time t uniquely determine the value of g at a time Δt (equal to the relaxation time) later. These two examples suggest this possible formal restriction on the relation ψ: There is a relation ψ between g and a set C of variables which includes h such that the value of g at time t is an equilibrium value achieved by means of a dissipative process and is determined by the values of h and the other members of C at time $t - \Delta t$, where Δt is the characteristic relaxation time of the dissipative process. The set C may include g itself.

(2) *Causal independence.* We wish to construct a formal criterion to guarantee that any exogenous perturbation of the system will set off a sequence of responses that travels around the feedback loop in just one direction. We have suggested one way to state such a criterion, *viz.* by requiring that there be a time lapse (a relaxation time) between the occurrence of a new value of h and the occurrence of a new, equilibrium value of g, the equilibrium being achieved by some dissipative process.

Now let us consider an alternative possibility. If we require that the set S consists of two or more variables, $f_1, f_2, \ldots f_n$, and that only a proper subset, $f_1, f_2, \ldots f_{m<n}$, of these n variables is related to g_1 via the relation χ, and that the class of external influences with respect to which the system is a feedback system consists of influences producing changes in the coset of variables, $f_{m+1} \ldots f_n$, then, with the addition of one more requirement, to be discussed directly, we can guarantee that environmentally produced changes in the coset will directly cause a change in h_1, but not a change in g_1; thus that the sequence of changes will be: environment $\rightarrow f_{m+1}$, etc., $\rightarrow h_1 \rightarrow g_1 \rightarrow f_1$, etc. and not the reverse. The additional requirement is that the variables $f_1 \ldots f_n$ must be unaffected by changes in the coset. Let me give a few examples of what I mean. In a simple battery-resistor circuit, i is functionally related to E and R via the circuit equation (which merely expresses the law of conservation of energy for this system): $i = E/R$. Mathematically the equation may be inverted so as to make any one of the three variables the dependent variable, and the other two the independent variables;

any pair of them are functionally (i.e. mathematically) independent, in the sense that knowledge of the value of one member of the pair is insufficient to determine the value of the other. But there is another kind of 'independence'– 'dependence' distinction to be made here. A change ΔR in R produced, say, by turning the shaft of the variable resistor, or by inundating it with X rays, or by changing its temperature, is followed by Δi, but not by a ΔE. Again, a change ΔE in E produced by switching in additional cells, or by chemical alteration of the electrolyte, or by changing the temperature of the battery, etc., is followed by Δi, but not by ΔR. Finally, a Δi produced, say, by a varying magnetic field normal to the loop formed by the circuit, results in no change to E or R.[7] There is, then, a physical asymmetry in the relations among i, E and R which is not caught by the mathematical form of the equation, for the equation is totally symmetric. The asymmetry lies in the possible causal influences, the connections joining the system to other systems in its environment. In other and vaguer words, the asymmetry resides in the boundary conditions. Because of this asymmetry, if the circuit were to be chosen to participate in a Type I negative feedback arrangement, a suitable choice of roles for the variables would be $f_1 = R, f_2 = E, h_1 = i$; but an identification of i as one of the system variables and R as h_1 would be inappropriate.

These characteristics of the circuit are reflected in its theoretical model. In the spirit of our reductionistic approach let us assume that the system is characterized by a set B of boundary conditions. That is, from B and a set P of fundamental physical theories and by means of the ordinary procedures of physical analysis it is possible to *obtain*[8] the relation $i = E/R$. It is possible, without contradicting B, to introduce additional boundary conditions to represent interactions between the system and an environment. For example, one such condition, B_R, might represent a mechanical linkage between an external device and the variable resistor. From B, P and B_R the ordinary procedures of physical analysis lead to the triple conclusion that R changes by a certain amount, that i changes by a certain amount, but that E remains unchanged. Note that the third conclusion follows from the *absence* of a boundary condition linking E to external influences. Alternatively, the condition B_R may be nullified and another environmental boundary condition, B_E, may be imposed, representing an external influence upon the battery. Again, the ordinary physical analysis leads to the result that E and i change but that R remains constant because of the absence of a boundary condition on R. Finally, it happens that the two boundary conditions are not inconsistent. Both may be imposed simultaneously with the result that from

B, P, B_R, and B_E it is possible to obtain the result that R and E both change and that i changes by an amount that combines the changes it undergoes when B_R and B_E are applied separately.

By way of contrast consider an ideal gas confined to a cylinder with a moveable piston. From P and B for this system one can obtain the relation $p = kT/V$ where p is the pressure of the gas, T its temperature and V is the volume of the container. If an external mechanical connection is fitted to the piston and the cylinder is immersed in a constant temperature bath, then the corresponding boundary condition, B_V, leads to the result that V may change while T remains constant, producing a change in p. Alternatively, if the piston is locked in place and a flame is applied to the cylinder (boundary condition B_T), the temperature changes while the volume remains constant, again producing a change in p. But these two boundary conditions are not simultaneously realizable — B_V and B_T are inconsistent. Thus, in this system (relative to the physical theories P and the boundary condition B) the variables T and V are not causally independent.

With these examples to guide our intuition, let us attempt a definition of causal independence of variables. I have alluded several times to the fact that strict logical entailment is often not the tool by which the practicing physicist extracts an analysis of a particular system from boundary conditions and physical theories. Rather, his 'deductions' are a judicious mixture of entailment, idealization and approximation. In what follows, then, I shall find it convenient to let the symbol $\xrightarrow{P}$ represent this process; so the symbolic statement $R \xrightarrow{P} Q$ is to be read as 'Q is obtainable from R by the ordinary methods of analytical physics'. Given fundamental physical theories P and a system S which is characterized by boundary conditions B and variables f_1, f_2, and h_1 and a relation $h_1 = \varphi(f_1, f_2)$ which are obtainable from P and B by the ordinary procedures of physical analysis:

Then the variables f_1 and f_2 are *causally independent* with respect to S and an environment E if and only if additional boundary conditions B_1 and B_2, which are not subsets of B, representing interactions between S and E, can be formulated such that

(i) $P \;\&\; B \;\&\; B_1 \xrightarrow{P} \Delta f_1 = a, \;\&\; \Delta f_2 \simeq 0$

(ii) $P \;\&\; B \;\&\; B_2 \xrightarrow{P} \Delta f_2 = b, \;\&\; \Delta f_1 \simeq 0$

(iii) $P \;\&\; B \;\&\; B_1 \;\&\; B_2$ is consistent and $\xrightarrow{P} \Delta f_1 \simeq a \;\&\; \Delta f_2 \simeq b$.

Here a and b are non-zero and can assume a range of values. Note that I have not required the causal influence represented by B_1 to have an absolutely null effect on f_2 — merely that the effect be small compared with that of B_2. A limited amount of 'cross-talk' between the causal channels is permissible.

The concept of causal independence provides, I think, the key to the puzzle with which this section is mainly concerned; namely, how to specify that the causal influences proceed around the feedback loop in one direction only. A relation $u = \varphi(x_1, x_2, \ldots)$ is *causally asymmetric* with respect to the variables x_1 and x_2 if and only if the variables are causally independent. This definition of causal asymmetry does not specify or limit in any way the physical mechanisms by which the asymmetry is achieved. No doubt dissipative processes are involved — usually, if not invariably. But I do not see how a definition in terms of mechanisms or types of mechanisms can be made general enough, for the simple reason that the 'engineering' options available, both to human designers and to evolutionary variation, are effectively unlimited.

But what about relations of the form $u = \varphi(x)$, which involve only one independent variable? Surely there are some functions of this form which are causally asymmetric. Take the earlier example of the governor of a steam engine, where u represents the amount of extension of the weighted arms and x represents the speed of rotation of the shaft. Changing the rotation speed alters the extension of the arms, but changing the extension of the arms produces a negligible (but not quite zero) effect on the speed of rotation. How can the position of the arms be adjusted independently of a change in rotation? One way of doing so is to adjust the stiffness of the spring against which the centrifugal force of rotation acts. So we see, on more careful examination of the device, that u *does* depend on more than one variable. The extension of the arms is determined both by the rotation speed and by the setting of the spring, so the relation may be written as $u = \varphi(x, s)$, and it is easy to show that the definition of causal asymmetry is satisfied in this case. I suggest that every apparent example of a causally asymmetric function of just one variable involves in fact at least one other which under the given boundary conditions is ignorable because it normally remains constant.

Reducible Variables and Fundamental Boundary Conditions

We are attempting to produce an explication of feedback which is in the spirit of the fourth, or reductionistic, form of the question posed in Part I: not an explication in terms of experimental tests, behavioral characteristics or laboratory operations, but an explication in terms of some theoretical model of the system and of the standard analysis that a physicist would apply to the model. Accordingly, both the boundary conditions which specify what the system is and the variables which are used to describe its behavior must be formulated in this reductionistic mode. Not just any boundary condition that a physicist would normally write, nor any variable he might wish to define will do for this special purpose.

It seems almost a truism to state that a reductionistic description of a system will take a form which is relative to the *level* to which the reduction is made. Depending on the system being reduced, it is not always either necessary or practical to reduce right down to the level of neutrons, protons, electrons, etc. (Indeed, at this scale of things it is not clear just what is fundamental.) Frequently a reduction to a macroscopic level is sufficient. But whatever the level, there will be sets of concepts, explanatory principles, and basic entities appropriate to it. For instance, a satisfactory reductionistic analysis of a mechanical feedback system such as a self-steering apparatus may be made in terms of the laws of mechanics and of such basic entities as levers and wind vanes. To produce a reductionistic analysis of an allosteric enzyme system it is necessary to become only as fundamental as the laws of reaction kinetics and entities such as chemical species.

Before placing some formal restrictions on acceptable boundary conditions and variables relative to the level to which the reduction is being made, let us consider some examples. Suppose the reducing theory is Newtonian mechanics and gravitation and the basic entities are Newtonian mass-points. Then acceptable boundary conditions would specify the possible initial values of the positions and velocities of these particles. Thus B for this system consists of sets of values of initial positions and velocities. But the reduction of another system might be to the level of the laws and entities of practical electronics. The entities would be circuit elements such as resistors and capacitors and the laws would include a.c. and d.c. circuit theories. Acceptable variables would be resistances or capacitances of specific circuit elements, currents at specific points, etc., and acceptable boundary conditions would specify what the circuit elements are, how they are connected together and the possible initial values of their variables.

However, it is common in ordinary physical analysis to state a boundary condition or define a variable in much less specific terms. For example, in analyzing a simple battery-resistor circuit one may wish to allow for the possibility of adding other circuit elements in parallel with the resistor; therefore, instead of introducing as a variable the resistance R of the resistor, one may introduce the impedance Z between the points a and b where the resistor is connected into the circuit. This variable is a kind of generalization of R, because Z reduces to R when only the resistor is present. Thus it would be a very useful variable to define. But it is not a basic variable, for it is not defined in terms of the properties of specific basic entities. The variable Z stands for 'the impedance of whatever is connected between a and b', so it is not a property of a basic entity nor is it a definite function of such variables. Thus it is not suitable as a variable in a reductionistic analysis.

Similarly, an unacceptable boundary condition might be a statement like: 'A battery is connected to some (unspecified) load', or 'the resistance between points a and b is less than 1000 ohms'. Acceptable statements would be: 'This battery is connected to that resistor', and 'this variable resistor is connected between points a and b and its range is 0 to 1000 ohms'.

So the variables must be properties of the particular basic entities which compose the system (relative to the level of reduction), and the boundary conditions must be statements which provide complete physical specifications of these components, how they are arranged, how they interact and by what mechanisms. Let us call these sorts of variables and boundary conditions *reducible variables* and *fundamental boundary conditions*.

Of course the set B of fundamental boundary conditions may be divided into subsets in many ways. But there is one possible method of partitioning B which is particularly appropriate for the analysis we wish to make, as will become evident in the next section. The method is this: B is divided in such a way that certain subsets specify only some of the interactions of these components; but for any component that is specified in the subset a complete physical description must also be included. For example, if a subset B_φ specifies that a certain wire-wound resistor is present, it would also (by this method of partitioning B) specify the length and diameter of the wire, the alloy of which it is made, how it is wound, how it is insulated, etc. I shall designate as a *component-wise decomposition* of B a partition into subsets each of which includes the complete physical specification of the intrinsic properties of each component it mentions. And the subsets produced in this way I shall call *component-wise subsets* of B.

Physical Distinctness

An interesting and, I think, instructive test of the ability of any definition of feedback to 'strain out' spurious cases is provided by one of the simplest possible electrical circuits — a resistor connected across the terminals of a battery. Suppose that the resistor is made of thin metal wire, so that its conductance decreases as its temperature rises. The temperature of the system stabilizes when the rate of joule heating of the resistor by the current passing through it is balanced by the rate of heat loss to the environment. A drop in environmental temperature produces an increased rate of flow of heat out of the system, resulting in a drop in the temperature of the resistor. This drop, in turn, causes the conductance to rise, thus producing a compensating increase in the current and in the rate of joule heating. The overall result is that the temperature of the resistor doesn't drop as much as it would have done if the conductance had remained constant. A flow diagram (Figure 8) emphasizes the similarity of this system to a proportional thermostat.

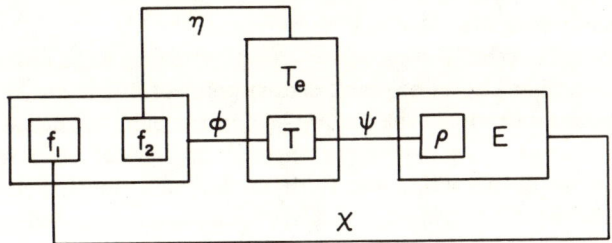

Fig. 8.

In the figure, T is the temperature of the resistor, T_e is the environmental temperature, ρ is the conductance of the resistor, E is the voltage of the battery, f_1 is the rate of joule heating, and f_2 is the rate of heat loss to the environment. The relations represented by the Greek letters are: $T = \varphi(f_1, f_2) = \int (f_1 - f_2)\,dt$; $\rho = \psi(T)$ = a monotonically decreasing function of T; $f_1 = \chi(\rho, E) = E^2 \rho$; and $f_2 = \eta(T, T_e) = k(T - T_e)$. Obviously, the formal analysis of this simple circuit, at least as developed so far, is essentially identical to what would obtain for a proportional thermostat. Equally obviously, some additional formal criterion must be devised to exclude this circuit or the idea of explicating feedback in terms of the formal physical analysis must be abandoned.

An intuitive distinction is not hard to find. In the thermostat the variables g and h refer to properties of physically distinct entities (the room and the switch), whereas in the battery-resistor circuit they refer to properties of the same entity (the resistor). Moreover, it is possible to 'disconnect' the causal relation represented by ψ in the first case but not in the second. The problem facing us is whether we can express these distinguishing points in the language of functions and variables.

The distinction is nicely summed up, and a solution to our problem hinted at, by Manier (1971) in his discussion of what he calls his "first" boundary condition" for the applicability of the feedback model:

That the function of x in S is P implies that 'P' designates a particular target value from a range of physically possible system outputs; and that some components of S carry or translate information concerning the difference between the output of S at a given time and the value P; that these information carrying (I) components are physically distinct from the operative (O) components which directly determine the output; and that this distinction is reflected in the fact that pathologies of the system making it incapable of achieving P do not imply a disruption of the lawlike relationships between the O components and the output of the system.

I shall argue below that Manier's requirement that the system respond to the difference between the output of S and P is certainly too restrictive. Nevertheless he has expressed the essential point of difference between the proportional thermostat and the battery–resistor circuit, and he has suggested that there is a distinctive *logical* property of the laws of the system which (if it works) would be within the spirit of our reductionistic program. He says, in effect, that the property of 'physical distinctness' is reflected by the fact that there are states of affairs ("pathologies of the system") such that goal achievement is impossible, yet $h = \varphi(f_1, f_2, \ldots)$ is still true. This is reminiscent of, but stronger than, Beckner's condition (vi); because Beckner requires that there be states of affairs such that, as a mere matter of fact, P is not achieved.

Promising as Manier's suggestion seems, it raises a formidable problem; namely, what kinds of states of affairs are allowable? For it is easy to see that not just any physically possible state of affairs *is* allowable. To see this, consider that the resistor of our example may become a superconductor at temperatures below a certain value, T_0. Then any state of affairs such that T is less than T_0 implies that the activity of 'compensation' will not occur, yet $dT/dt = f_1 - f_2$ remains true.

Before proposing a solution, let us elaborate the problem. The difficulty can also arise in cases of Type I feedback, and it turns out that the same simple battery–resistor circuit can provide an illustration of this case, too. By considering its behavior when the external conditions vary so slowly that the temperature is always nearly at equilibrium ($f_1 \simeq f_2$) the circuit may be treated as an example of Type I feedback, with a definite single-valued function relating T and i. Because of the essential ambiguity of Type I systems, either T or i may be chosen to be the 'h_1' variable. For the sake of variety let us look upon the circuit as a current regulator with the symbol h_1 assigned to i. (See Figure 9.)

There is a feedback of sorts in this system (not the sort we want), for a rise in current increases the temperature of the resistor, thereby increasing

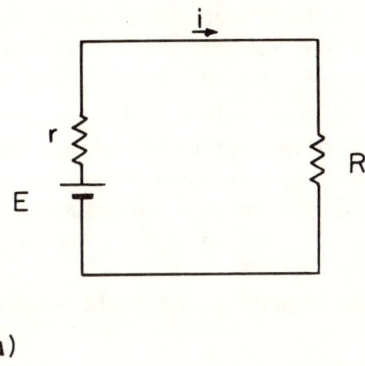

(a)

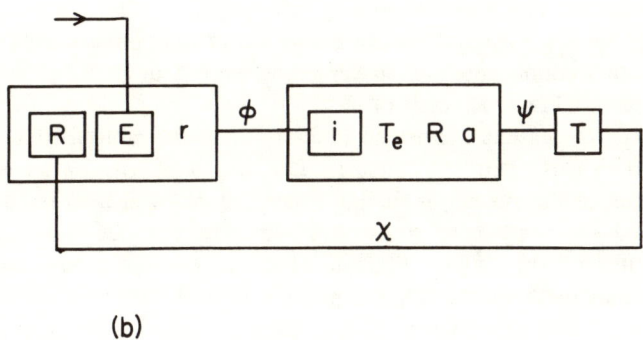

(b)

Fig. 9.

its resistance (if metallic conduction is the physical process), resulting in a decrease in current. External perturbations of the voltage of the battery will result in smaller changes in current when the resistance responds in this way than would be the case if the resistance remained constant.

In the figure, E is the voltage, R the resistance, r is the internal resistance of the battery, i the current, T the temperature of the resistor, T_e the temperature of its environment, a represents the amount of thermal insulation around the resistor, and C is a proportionality factor characteristic of the metal of which the resistor is made. φ is the circuit law, ψ expresses the equality of the rate of electrical heating and the rate of cooling via radiation and conduction to the environment, and χ expresses the proportionality between resistance and temperature for metallic conductors.

As before, there is a possible state of affairs, namely the system's being so cold that the resistor becomes a superconductor, in which the relation χ is no longer true, so that the compensating action of the resistor can no longer occur; yet the relation φ remains valid. If this state of affairs counts as a "pathology of the system," then Manier's criterion is satisfied by a device which does not have the physically distinct parts the criterion was designed to guarantee. Yet Manier is surely on the right track. The question is how to formulate the criterion.

Let us follow the approach outlined in Part I. We can choose to employ either object-centered or theory-centered language. The criterion may be more simply stated in the former mode. Take the battery–resistor circuit as an example. If component A (a certain resistor) and component B (a certain battery) and components C, C' (the wires) are present and connected, then in the nature of things the relation φ will obtain — but so will ψ and χ. Moreover, there is no simpler set of *components and connections* (although there are other conditions, such as very low temperatures) such that ψ and χ will cease to obtain but φ remains true.

However, for reasons I outlined in Part I it is desirable to translate this analysis into theory-centered language. Let us attempt it. We have a theoretical model of the system, including a set P of physical laws and a set B of fundamental boundary conditions. A component-wise decomposition of B will allow us to form a subset B_φ from which the relation φ may be obtained by the ordinary methods of analytical physics. In the battery–resistor example it turns out that any component-wise subset of B from which φ can be obtained is also one from which ψ and χ can be obtained; but in a proper feedback system that would not be the case. In the former case the relations ψ and χ are not independent of φ, but in the case of true feedback they are. Let us state this more formally by means of a definition: A relation ψ is *physically independent* of a relation φ in a system S if and only if there exists a component-wise subset B_φ of B such that $B_\varphi \xrightarrow{P} \varphi$ and not $B_\varphi \xrightarrow{P} \psi$. In theory-centered language, then, a criterion of physical distinctness or disconnectability may be stated thus: The relations ψ and χ must be physically independent of the relation φ, and χ must be physically independent of ψ.

Consider how this criterion works out for a genuine feedback system, e.g., a thermostat. In the theoretical model of a thermostat a component-wise subset B_φ of B specifies that there is a room with a certain heat capacity, that heat flows out of the room to a cooler environment and that heat pours in from a furnace. From just this partial characterization of the system an analytical physicist can produce the relation φ connecting the temperature of the room to its heat capacity and the two rates of heat flow. But the physicist would not be able to conclude that the rate of heating by the furnace is correlated with the state of a certain switch (relation χ), nor the state of the switch to the temperature (relation ψ), for the subset of boundary conditions does not specify that the switch is present. So ψ and χ are physically independent of φ. Moreover, there is another component-wise subset B_ψ which describes the switch and the wires connecting it to the thermometer but makes no reference to the wires connecting the switch to the furnace. From B_ψ one can obtain ψ but not χ. So χ is physically independent of ψ.

The requirement of physical independence serves to rule out a large class of devices, ranging from the simple pendulum to others which may tempt us with a greater initial plausibility. Let us therefore take one more example, this one drawn from another class of physical phenomena. Consider a gas confined to a container with an orifice through which gas diffuses out, and some means by which new gas is added to the container. This latter process may occur by the boiling of a liquid which is also present in the container (note that I hereby fulfil the promise made in Part I to discuss the pressure cooker) or it may be pumped in through an inlet pipe, etc. Assume that the container is made of some elastic material (every material is elastic to some degree) so that the dimensions of the container, including the size of the orifice, depend on the pressure of the gas and on the elastic constant of the material. Let φ relate the pressure p to the rate f_1 at which gas is added to the vessel and the rate f_2 at which it diffuses out. Let ψ relate the size A of the orifice to p and the elastic constant K. Finally, let χ relate f_2 to A and p. I think it is clear that this system is not a feedback device, since it lacks a separate sensor-controller sub-system.

One natural way to construct a block diagram for this system is shown in Figure 10(a). This diagram does not conform to the standard feedback pattern, thus arousing our suspicions. However, owing to the wide latitude one normally has in setting up a theoretical analysis, we can rearrange the variables and their connecting relations without altering what is asserted about the system, to produce the diagram comprising Figure 10(b). Here we find the requisite number of distinct variables and relations among them. But do the physical connections represented by these relations exhibit the required independence from one another? As in the preceding example, we

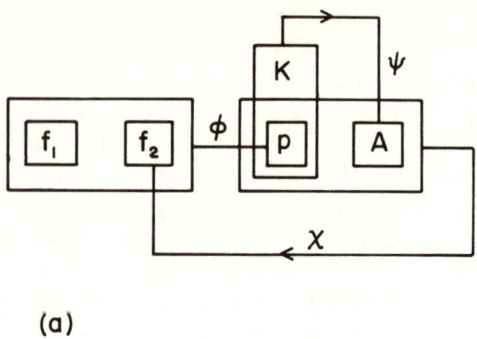

(a)

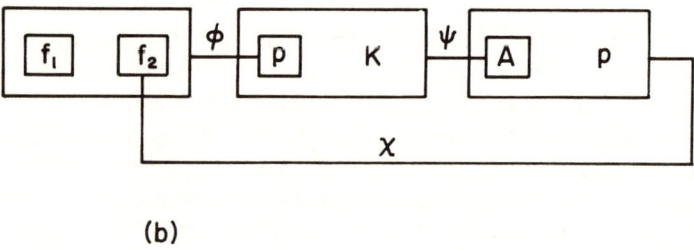

(b)

Fig. 10.

find that they do not. In particular, the response of the orifice to a variation in the pressure and the action by which it counteracts the change are not two physically distinct processes, but one and the same, namely an increase in size. Consequently, the relation χ is not physically independent of the relation ψ.

Necessary Relations

In item (5) of my critique of Beckner's analysis, I noted that we must not build our block diagram out of relations which are true merely by definition or by reason of the general laws of physics. The three relations φ, ψ, and χ must express physical, causal connections, not merely logical ones. If we place a sufficiently subtle interpretation upon the requirement of physical independence formulated above, we may read out of it this further prohibition. For if χ, for example, is a proposition which holds true simply by definition, or if it follows from the general laws of physics together with definitional stipulations, then it also follows from the laws of physics conjoined to any of the boundary conditions of the system, including B_ψ. A similar argument applies to the relation ψ and the boundary conditions B_φ. And the relation φ is guaranteed to be contingent by virtue of the causal independence of the variables f_1 and f_2. Nevertheless, I shall express this prohibition explicitly, for clarity, in the formal explication below.

Negative Feedback

Finally, we must add a requirement that the feedback be negative rather than positive. Goal directedness is consonant with negative feedback, which has the effect of controlling or stabilizing a certain variable, but not with positive alone, which would destabilize the variable. Even self-running oscillators, some of which depend on positive feedback to drive their oscillations, rely on some mechanism which terminates the effect of the positive feedback and reverses the direction of change of the oscillating variable.

Before attempting a general formulation of a condition that the feedback be negative, let us survey a few special cases. For a Type I feedback system the three relations φ, ψ, and χ are all elementary functions. In systems of this type an external perturbation may cause a change, δh_1, in h_1, which in turn produces a change in g_1: $\delta g_1 = (\partial \psi / \partial h_1) \delta h_1$. This then produces a change in f_1: $\delta f_1 = (\partial \chi / \partial g_1) \delta g_1 = (\partial \chi / \partial g_1)(\partial \psi / \partial h_1) \delta h_1$. Finally, a new change, $\delta h_1'$, occurs such that $\delta h_1' = (\partial \varphi / \partial f_1) \delta f_1 = (\partial \varphi / \partial f_1)(\partial \chi / \partial g_1)(\partial \psi / \partial h_1) \delta h_1$. For negative feedback $\delta h_1'$ must be opposite in sign to δh_1; so for Type I feedback systems we require that the product of the three partial derivatives be negative.

But feedback is not limited to cases where the three relations are elementary functions. We have designated as Type II systems those for which the relations ψ and χ are elementary functions but φ involves a time integral over some of the f_i. The relations may, of course, be even more complicated than this. A discussion of systems other than Type I will be facilitated by the introduction of some technical terms. I shall designate as *trajectories* those statements which describe the dynamical behavior of a system, that is relations expressing the variables as functions of time. A relation of this sort, for example $h = h_0 e^{-at}$, where h_0 and a are given definite constant values, will describe a system, if it describes it at all, only under a particular set of circumstances and only for a portion of its history. I shall call relations like φ, ψ, etc., which describe permanent causal connections among the variables of a system, *enduring relations*. These typically relate some of the variables of the system to others, and the time variable does not appear in the relation. For some systems (Type I) the enduring relations are elementary functions, for example, $g_1 = kh_1$; here no reference is made, even implicitly, to the fact that the variables may be expressed as trajectories. I shall refer to this sort of relation as a *non-dynamical enduring relation*. In other systems the enduring relations may involve differentiation and/or integration with respect to time, or other operations which explicitly treat the variables as functions of time. I shall call these *dynamical enduring relations*.

For Type I systems it is easy to express a condition that the feedback be negative, as we have just seen, and to state just what effect the feedback loop has upon the system; that is, how and in what respects the system behaves differently because of the causal connections specified by the sets of boundary conditions B_ψ and B_χ. But for Type II systems both tasks are more difficult. Before I suggest a general criterion that the feedback be negative, one applicable to Type II systems and others, let us examine two instructive examples of how the ordinary physical analysis of such a system would proceed.

In Type II feedback systems the relation φ has the form $h_1 = h_1(0) + \int_0^t (f_1 - f_2)\, dt'$. This relation may be rewritten in terms of the time derivative of h_1: $dh_1/dt = f_1 - f_2 = \varphi'(f_1, f_2)$, and in this form we do have an elementary function mapping values of f_1 and f_2 onto values of the derivative, dh_1/dt. It is easy to see that for this system the condition for negative feedback becomes: $(\partial\varphi'/\partial f_1)(\partial\chi/\partial g_1)(\partial\psi/\partial h_1) < 0$.

Unfortunately, it is not a straightforward task to generalize this sort of analysis. Consider the following relation $\varphi(h_1; f_1, f_2, \ldots)$: $dh_1/dt = af_1 + bf_2 + c(df_1/dt)$, where a, b and c are constant coefficients. The integral form of this relation is: $h_1 = a\int_0^t f_1\, dt' + b\int_0^t f_2\, dt' + cf_1$. From this we can see that the first and third terms may represent opposing influences of f_1 upon h_1. In particular, if a is negative and c is positive, an increase in f_1 will tend to produce an increase in h_1 (according to the analysis of the preceding paragraph) via the first term, but it will also tend to produce a decrease in h_1 via the third term. If f_1 is connected by a feedback chain to h_1, the loop will have both positive and negative tendencies. Which will dominate? An analysis will show. Let us combine this relation with the following elementary functions for χ and ψ: $f_1 \propto g_1$ and $g_1 \propto h_1$, so $f_1 = kh_1$, where k is also a

FEEDBACK, SELECTION, AND FUNCTION 79

constant. Substituting kh_1 for f_1 in the relation φ, assigning f_2 a fixed but arbitrary value, and integrating, we obtain:

$$h_1 = \left(h_0 + \frac{bf_2}{ak}\right) \exp\left(\frac{akt}{1-ck}\right) - \frac{bf_2}{ak}.$$

If the term $ak/(1-ck)$ in the exponent is negative (this is analogous to the earlier requirement that the product of the three partial derivatives be negative) and if f_2 varies slowly compared with the relaxation time $\tau = \left|\frac{1-ck}{ak}\right|$, then h_1 will be controlled within limits such that the derivative dh_1/dt will remain approximately zero. However, if f_1 is constant (this is equivalent to abrogating the boundary condition B_χ or B_ψ), then h_1 increases or decreases without limit.

A system obeying these equations might also satisfy all the criteria so far laid down for a feedback system. Clearly, if the term $ak/(1-ck)$ is negative for such a system, its behavior would also fit our intuitive notions of what a negative feedback system is. But how can we describe in a generalizable way just what makes it negative?

As we can see from these examples, the analysis of Type II systems and, indeed, of any system whose enduring relations are dynamical, must be an explicitly dynamical one. The boundary conditions of the problem specify which variables are input variables and which are response variables. In our notation, the f_i are the input variables and h_1 is the response variable. In a dynamical analysis an overall relation is written relating h_1, considered explicitly as a function of time, to the f_i. Initial conditions must also be specified. Families of input functions for $f_2(t)$, together with initial and boundary conditions are 'fed in' to the analysis, and families of functions $h_1(t)$ emerge.

Let us define the natural, or undriven behavior of the system as the family of trajectories $h_1(t)$ generated by the dynamical analysis for fixed (constant) input values of the f_i and a range of initial conditions. If the feedback is negative, a range of initial conditions leads to asymptotic behavior for h_1 – either to a constant, finite value or to a value which varies within finite bounds which are themselves independent of the initial conditions. But if the boundary conditions B_ψ and B_χ are abrogated, a dynamical analysis based only on B_φ, but with the same range of initial conditions, leads to a new family of trajectories for $h_1(t)$. In our example these trajectories were unbounded.

We are now ready to state the criterion of the feedback's being negative.

From the laws of physics and the complete set of boundary conditions for the system, a physical analysis leads to dynamical solutions for $h_1(t)$ which are asymptotic trajectories with a bounded, finite value; but from the laws and only the boundary condition B_φ the corresponding dynamical analysis leads to solutions for $h_1(t)$ which are either unbounded or exhibit asymptotic, bounded variations within wider limits than displayed by the system when operating under the full set of constraints.

A Revised Explication of Feedback

I shall now attempt to pull together the various strands of the preceding discussion by formulating a revised definition of a feedback system. I shall assume that there is given a set P of physical theories (such as quantum electrodynamics or electrical circuit theory), a set B of fundamental boundary conditions which characterizes a system S both with respect to its internal constitution and with respect to its interaction with an environment E, and a set D of basic definitions of reducible variables $f_1, f_2, \ldots, h_1, h_2, \ldots$ and $g_1, g_2, \ldots$, which have been introduced by the ordinary procedures of physical analysis.

In the definition of feedback which I am about to propose, several terms are used in fairly technical senses. Some of them have been introduced and defined above, but for the sake of convenience I list the terms and their definitions here.

(1) *Relation.* This term is used as defined in standard mathematical usage. A relation establishes a correspondence between mathematical objects by means of a specified rule. The objects so related may be the numerical values of physical quantities, functions of such quantities, functions of time, etc. A relation between a mathematical object u and a set of objects $x_1, x_2, \ldots$ may be an elementary function which maps values of the x_i onto contemporaneous values of u; or it may be a mapping from values of the x_1 at time t to values of u at a later time $t + \Delta t$, where u represents an equilibrium or steady-state value of a physical property achieved by means of a dissipative process and Δt is the characteristic relaxation time of the process; or the rule may involve differentiation and/or integration with respect to time, etc.

(2) *Necessary and contingent relations.* A relation $\varphi(u; x_1, x_2, \ldots)$ is necessary if and only if it is entailed by D and P. It is contingent if it is not necessary.

(3) *Fundamental boundary conditions.* A set B of statements which describe a system at a given level of analysis qualify as fundamental boundary conditions only if B specifies which parts constitute the system, how they interact with one another and with an environment E, and what are the ranges of their variable properties. B is separable into a set B_{int} of internal boundary conditions which specify the interactions among the parts of the system, and a set B_{ext} of external boundary conditions which specify the possible interactions with the environment. The system, narrowly conceived, is specified by P and B_{int} alone. Note that the concepts of fundamental entity and fundamental boundary condition, and hence the definition of feedback that follows, are defined relative to the level at which a reductive analysis is made. If the reduction is made to the level of atomic physics, the fundamental entities will be atoms; if it is made to the level of practical electronics, they will be electronic components and hardware, etc.

(4) *System.* A system is a piece of the world consisting of interacting parts. The system may also interact with the rest of the world, which is considered to be the environment of the system. We characterize a system, in its environmental setting, by a theoretical model which consists of a set of physical theories, P, and a set of fundamental boundary conditions, B.

(5) *Reducible variables.* Reducible variables are those which can be defined in terms of the properties of individual fundamental entities.

(6) *Causally independent variables.* Two variables x_1 and x_2 are causally independent with respect to a system S, its boundary conditions B, and an environment E if and only if there are two sets of boundary conditions, B_1 and B_2, not contained in B_{int} and representing interactions between S and E, such that:
  (i) $P \mathbin{\&} B_{\text{int}} \mathbin{\&} B_1 \xrightarrow{P} \Delta x_1 = a, \mathbin{\&} \Delta x_2 \simeq 0,$
  (ii) $P \mathbin{\&} B_{\text{int}} \mathbin{\&} B_2 \xrightarrow{P} \Delta x_1 \simeq 0, \mathbin{\&} \Delta x_2 = b,$
  (iii) $P \mathbin{\&} B_{\text{int}} \mathbin{\&} B_1 \mathbin{\&} B_2$ is consistent and $\xrightarrow{P} \Delta x_1 \simeq a \mathbin{\&} \Delta x_2 \simeq b,$ for a range of values of a and b.

(7) *Causally asymmetric relation.* A relation $\varphi(u; x_1, x_2, \ldots)$ is causally asymmetric with respect to the variables x_1 and x_2 if and only if x_1 and x_2 are causally independent.

(8) *Physically independent relations.* A relation $\psi(v; y_1, y_2, \ldots)$ is physically independent of a relation $\varphi(u; x_1, x_2, \ldots)$ with respect to a system S

and its theoretical model if and only if there is a component-wise subset B_φ of B such that $B_\varphi \xrightarrow{P} \varphi$ but not $B_\varphi \xrightarrow{P} \psi$.

(9) *Feedback system.* A system S is a feedback system with respect to reducible variables $f_1, f_2, \ldots, h_1, h_2, \ldots,$ and $g_1, g_2, \ldots$ and an environment E if and only if the following criteria are satisfied:

(i) The variables $f_1, f_2, \ldots, h_1, h_2, \ldots,$ and $g_1, g_2, \ldots$ are not defined as simple functions of the time variable, nor are any of the variables f_1, f_2, h_1, and g_1 related to one another by a necessary relation.

(ii) Each set of variables $\{f_i\}$, $\{h_i\}$, and $\{g_i\}$ is non-empty and functionally independent. The set $\{f_i\}$ contains at least two members.

(iii) There are three (contingent) relations φ, ψ, and χ such that $P \& B \xrightarrow{P} \varphi(h_1; f_1, f_2, \ldots) \& \psi(g_1; h_1, h_2, \ldots) \& \chi(f_1; g_1, g_2, \ldots)$.

(iv) The relation φ is causally asymmetric with respect to the variables f_1 and f_2.

(v) The relations ψ and χ are physically independent of φ, and χ is physically independent of ψ.

(vi) E acts upon the entire system only, or primarily, through the variables $f_2, f_3, \ldots$; that is, the boundary conditions are such that these variables may be written as arbitrary 'input' functions of the time and there are no additional constraints on the variables.

(vii) For Type I systems the product $(\partial\varphi/\partial f_1)(\partial\psi/\partial h_1)(\partial\chi/\partial g_1)$ is negative. For Type II and other systems, one or more of whose relations φ, ψ and χ are dynamical: a dynamical analysis of the system, based on the full set of boundary conditions and considering one or more of the f_i other than f_1 as input variables, shows that the undriven trajectories $h_1(t)$ exhibit asymptotic values which either are constant and finite or vary within finite bounds; and a corresponding analysis based on B_φ but omitting B_ψ and B_χ yields undriven trajectories which either are unbounded or exhibit asymptotic behavior which varies within wider bounds.

Informal Summary

This formal explication of negative feedback is designed for just one purpose, namely to serve as an example of the reduction of an upper-level concept to a lower-level description of a certain kind of object. As such, the explication is not intended for, nor is it likely to prove useful in, settling questions about whether this or that piece of machinery operates by means of negative feedback. Yet the formal analysis fails if it does not express the insights we have gained in thinking about and distinguishing concrete examples. And those intuitions, to the extent that they are valid, should help us to sort out fresh candidates, to determine whether their modes of operation are like or unlike our paradigms of negative feedback in this important respect. Both to facilitate criticism of the formal explication, therefore, and to guide our attempts to apply it, let us set down in unavoidably imprecise but perhaps more manageable language what we have formulated above in technical, theory-referring terms.

A complete feedback system must consist of at least two physically distinct working parts, a controlled subsystem S and a sensing–controlling regulator R, in touch with an environment. The subsystem S possesses some variable property h_1 which is causally dependent on at least two other properties of S, f_1 and f_2; and these are causally independent of each other. (A simple paradigm of such a system is a single-loop electrical circuit consisting of a battery and a variable resistor connected by wires. By twisting the dial of the resistor we can alter the current passing through the wires without affecting the electromotive force of the battery. Similarly, a change in the battery also affects the current, but does not alter the angle of the resistor's dial. Thus the current depends causally on both the resistance and the voltage, but those two properties of S do not materially affect each other.) The subsystem S is thus a suitable subject for feedback regulation of its property h_1: it can receive perturbations from the environment's action upon f_2 and it can receive corrections through the regulator's action upon f_1.

The asymmetry of this causal dependence of h_1 on the f's guarantees that causal influences will pass in just one direction around the feedback loop, and that the sensing and controlling roles of the sensor-controller are kept distinct: R passively senses changes in h_1 and actively corrects for them by acting upon f_1. By this requirement we rule out systems which achieve or tend toward a quiescent end state as a mere equilibrium.

We further require a minimum degree of articulation in the feedback loop — the physical distinctness of S and R — by specifying that R must be disconnectable from S both at the point of sensing and at the point of controlling. This distinctness may coincide with spatial boundaries, and usually does; but S and R need not occupy separate regions of space, as we discover when we apply the formal criteria to a chemical oscillator (see Appendix). Whether the distinctness of the sensor-controller from the controlled subsystem manifests itself spatially or not, the causal link by which R produces changes in f_1 can be disconnected without disrupting either the means by which it detects changes in h_1 or the causal dependence within S of h_1 on the other properties f_1 and f_2. Similarly, the sensing link can be disconnected without abridging the basic relation in the controlled subsystem.

We also require that the variable properties of S and R to which we assign places in the feedback process be distinct properties of the component parts of the system; that is, none may be a merely mathematical construction out of the others.

Finally, in requiring that the feedback be negative, we stipulate that the system actively regulate or protect h_1. That requirement is easily stated for one simple sort of mechanism, what we have called Type I feedback, but in general the requirement cannot be so simply put. In fact, we have expressed this criterion in quasi-behavioral terms, not in terms of what the system actually does in any finite sequence of behavior, however long — we have seen that truly behavioral criteria fail to capture the intuition that a feedback system must operate as such in virtue of its inner propensities — but rather, we specify the negativeness by referring to what the structure of the system and the propensities of its parts (fixed by the boundary conditions and the laws of nature) determine that it would do in a potentially infinite variety of circumstances.

IV. FUNCTION AND PURPOSE

The Problem

We need an adequate theory of function. What is it about a natural system that makes the concept of function appropriate? And when it is appropriate, how do we choose among the various actions of a thing a subset to designate as its functions? Without referring to the intentions of designers or users, can we tell that turning the furnace on and off is a function of a thermostat, but making clicking noises is not?

According to a cybernetic theory of function, as developed by Rosenblueth *et al.* ([1943] and [1950]), Beckner (1959), and Manier (1971), the term is applicable because the thermostat participates in a feedback loop, and its switching action has further consequences within the loop, whereas the clicking does not. According to a selection theory of function, as developed by Wimsatt (1972), by Ruse ([1973], Chapter 9) and by Wright (1973), the term is applicable because the system has been singled out from among others by some process, and the switching action helps to satisfy the criterion of selection whereas the clicking does not. Wimsatt claims only that selection processes provide a logically sufficient condition for the appropriateness of functional explanation and language. He admits that they may not offer a logically necessary condition, although he would like to be convinced otherwise and speculates that it may be "a law of nature that all teleological systems and processes result from some species of 'blind variation and selective retention'." Clearly, the cybernetic and selection theories of function are based on distinct paradigms of human purposive behavior. The cybernetic theory takes as its model the helmsman, who makes continual adjustments to the wheel of his ship in order to keep her on course; the selection theory is patterned after the problem solver, who tests one proposed solution after another until he finds one that 'fits'. The strategy in both theories is to find in non-human systems exemplars which fit the basic model so closely that the use of the terms 'function' and even, possibly, 'purpose' seems natural.

We also need an adequate theory of 'purpose'. This task is more complicated. On the one hand, when we ascribe a purpose to anything, it seems that we imply consciousness, either within the thing itself ("His purpose is to make the punishment fit the crime") or within its designer ("The purpose of the spiked anklets is to guard against the bites of sharks" . . . that is, the White Knight's purpose is to prevent shark bites and the anklets serve his purpose). In the latter case 'function' may be substituted for 'purpose', but an implication of consciousness remains. On the other hand, many discussions of function in sub-human biological systems introduce 'purpose' or 'purposiveness' in contexts where consciousness, at least in the usual sense, is ruled out. Thus, as applied to feedback systems, the terms 'goal-directed' and 'purposive' are often used interchangeably. Moreover, Wimsatt has made the novel suggestion that purpose be identified with the criterion of some selection process when that process exhibits a certain kind of complexity.

If purposiveness is to be explicated without reference to consciousness it must be possible to identify structures or patterns of organization which, when present in a system, make it a purposive one. If a structure or pattern of this kind can be identified in a system, then its parts may be said to have functions insofar as they contribute to (promote) the attainment of the purpose. According to a selection theory of function some systems are so organized that they possess a criterion of selection, and their purpose is the satisfying of the criterion. According to a cybernetic theory some systems are so organized that they possess a goal, and their purpose is the attaining of the goal. It is essential to the success of either theory that the criterion or the goal be an objective feature of the system itself and that it can be specified without reference to objects or agents outside the system.

Robert Cummins (1975) calls for an objective, principled way of distinguishing a purposive system. He suggests that the criterion must have something to do with the way the capacities of a system's parts are 'programmed,' and something to do with the greater degree of complexity or 'sophistication' of the organism's program relative to the complexity of its parts (p. 764). Let us see whether selectionist or cybernetic accounts of goal-orientation in biology can supply such a principled criterion.

One prerequisite of an adequate cybernetic theory is, as Wimsatt (1972) points out, an adequate explication of feedback. In the first part of this paper I have attempted to answer his objection that no such explication exists. Pending further criticism, I think I have shown that the feedback pattern is an objective property of a system, specifiable in terms of the laws and boundary conditions that constitute its theoretical model.

In his second objection to a cybernetic theory, Wimsatt claims that not all functions, even of the parts of a feedback system, are internal to the system. Now both cybernetic theorists and selection theorists agree that functions can be ascribed to objects only in reference to their participation in some larger 'purposive' system. Whether a certain smooth, round stone is a paperweight, a doorstopper, or a cannonball depends on whether someone is using it to hold down paper, to prop a door ajar, or to annoy his foes. Similarly, it is true that a certain spring-loaded coil of wire may be called a sensing coil just because it participates in a voltage regulator circuit. In short, the function of a part derives from its membership in the whole. In the introduction to this paper I discussed Wimsatt's example of a suicidal robot and concluded that the example establishes at most that some uses of 'function' may not be accounted for by a cybernetic theory. His argument leaves it an open question whether the more inclusive purposive system (in Wimsatt's example this would include the intentions of the designer of the robot) may not also be organized according to the feedback pattern. I shall return to this question below in a discussion of the context of functional activity. Whatever the answer may be to that question, I think it is clear that the cybernetic theory is still a viable one.

Selection: a Necessary Condition?

Next I wish to argue that there are uses of 'function' that a selection theory cannot account for, but a cybernetic theory can — that there are some wholes which by virtue of their internal constitution are plausibly characterized as purposive systems, yet in these cases no selection process is present, or, if one is present, a selection theory of function is nevertheless not applicable. This claim needs support, so let me bolster it with three examples, two hypothetical and one actual, which raise some doubts about the adequacy of Wimsatt's program to account for all cases of functional language.

First, suppose a sculptor, in the tradition of the 'found object' school, decides to make an assemblage of electronics parts and hardware. Fascinated with this new material, he constructs thousands of different creations. Quite unknowingly he happens, on his nine hundredth effort, to put together what a more scientifically inclined person would identify as a voltage regulator circuit. But would he be mistaken? The object has been produced by some kind of 'blind' variation (that is, by a variation unrelated to the principles of electronic engineering) but it has not been selected by anyone with the aim of using it to regulate a voltage. Is it *merely* Opus 900 in the artist's Electronic Period? I think we must be willing to say both that it is a voltage regulator and that the function of the little spring-loaded coil is to adjust the setting of the variable resistor.

Still even here a larger context exists. There is a science of electronic engineering and there are devices whose voltages need to be regulated. Although the artist's serendipitous creation was not intended to be put to that use, it might have been, and it could be put to that use, for there are other systems into which it could be incorporated as a regulator. Without a context of this sort the function of the device would be merely potential. There may be degrees even of potentiality, as there are degrees of complexity of the context. I shall return in a later section to this question of the role of context in ascriptions of function.

My second example is the tiglon, an infertile hybrid offspring of the chance mating of a tiger and a lion. Unlike the mule, also a hybrid, the tiglon is not a beast of burden, nor does it have any other agricultural function. It has no evolutionary function either, for it is a genetic dead-end street. It was produced by chance, it can leave no progeny, and according to reports it lacks all sense of social responsibility. In short, it has no function at all. Now what about its pancreas? Well, let us see how far we can apply a theory of function based on natural selection, the criterion of which is differential survival in succeeding generations. Let us consider three possible forms such a theory might take when applied to the function of a particular organ in an individual organism.[9]

Theory one (T_1). A function of organ O in individual A is to do f if doing f contributes to the differential survival of the progeny of A.

Clearly, T_1 fails because it does not support ascriptions of function to the tiglon or, indeed, to any kind of organism which is incapable of producing progeny. A selection theory of function may, however, be put in a form which refers to the ancestors of an individual organism, thus:

Theory two (T_2). A function of organ O in individual A is to do f if doing f contributes to the differential survival of the progeny of A, or if the doing of f by the same kind of organ, O', in the ancestors of A contributed to the differential survival of their progeny, and hence to the existence of A today.

This form of the theory avoids the difficulty of infertile individuals or kinds of organisms, but raises a new difficulty, namely, how is O to be identified as the same kind of organ as O'? The problem arises because organs are identified by their functions. According to T_2, the ancestors of the tiglon had an organ, O', called a pancreas, whose function it was to regulate the blood sugar. The tiglon also has an organ which does affect the blood sugar (among other things), but is it a pancreas? It can be called an organ of the same kind as O' only if it is already known to perform the same function. So T_2 will not establish the function of the tiglon's pancreas. Let us, then, try a less restricted form of the theory; that is, let us remove the requirement that O' be the same kind of organ as O.

Theory three (T_3). A function of organ O in individual A is to do f if doing f contributes to the differential survival of the progeny of A, or if the doing of f by an organ O', which is an evolutionary precursor of O, in the ancestors of A contributed to the differential survival of their progeny and hence to the existence of A today.

This, I think, is what Ruse (1973) also intends when he says that f is a function of O if f is an adaptation because for Ruse adaptation is an *historical* concept; not merely a fact about the present state of the organism. The trouble with T_3 is that, although it allows us to assign a function to the tiglon's pancreas, it runs into difficulties with vestigial organs. Suppose, to invent an example, that a certain lung-breathing amphibian has vestigial gills through which no blood flows, although capillary flow did occur in the gills of its ancestors. The gills of this organism still possess, however, membranes which direct the flow of water over the gill surface. Since directing the flow of water was a function of the membranes in the ancestral gills, T_3 requires us to say that it is their function still.

A cybernetic theory of function avoids all these difficulties. For it is an objective fact about the tiglon, discernible without reference to its being the survivor of a selection process, that it is able actively to maintain itself in the face of the structure-destroying forces in its environment. As an individual organism it survives: not by merely enduring as a rock endures, but by repairing damage, renewing parts, adjusting its actions to respond to events, etc. Construing 'adaptation' as a non-historical concept, the organism is adapted to its environmental niche. And one of the important ways this adaptation is achieved is by means of feedback regulation and control at many levels. It is a present fact about the tiglon that it possesses a feedback system which regulates the level of its blood sugar and that its pancreas participates in that system. So a cybernetic theory would support the statement that a function of the tiglon's pancreas is to regulate the blood sugar level, independently of any considerations about whether the tiglon participates in a more inclusive purposive system. Nor would a cybernetic theory assign functions to the gill membranes when they no longer participate in the mechanism that regulates the level of oxygen in the blood.

Finally, consider the Zhabotinsky reaction (see Appendix). Here we see how a pattern of organization can arise spontaneously and without the mediation of selection processes from a mixture of chemicals which does not possess that structure (although the oscillating reaction does have higher entropy than the separate ingredients possessed). A selection theory of function must remain silent about the Zhabotinsky reaction but a cybernetic theory would require us to say, for example, that it is a function of the bromide ions to 'turn off' the reaction which converts $PhFe^{++}$ to $PhFe^{+++}$ as it is the function of MB to restore the PhFe back to the ferrous state. Wimsatt has added to his selection theory of function a requirement of "phenotypic complexity" which is designed to rule out autocatalytic molecules and, I should imagine, the Zhabotinsky reaction as well, because in such systems the "genotypic and phenotypic levels are identical or are separated at most by a very small number of levels" (Wimsatt [1974]). However, I find it quite natural to use functional language in this context. Indeed, if one did not speak in this way, one would have to adopt rather clumsy locutions in discussing pre-biotic evolution. Moreover, as I shall argue below, there are functional explanations not concerned with origins into which these statements would fit quite comfortably.

Selection: a Sufficient Condition?

The preceding examples tend to show that a selection theory of function does not provide a necessary condition for legitimate uses of functional language. Now let me try to raise the suspicion that a theory based on natural selection may not be satisfactory even as a sufficient condition. As Wimsatt points out, a selection theory must assign functions with reference to a selection criterion.

In the case of wild animals and plants, not subject to the ministrations of animal husbandmen and plant breeders, the criterion must be the one associated with natural selection. Here, I think, the currents of modern neo-Darwinism may ultimately undermine Wimsatt's position. *Is* there a criterion? There seems to be none. And if there is no criterion, the attractiveness of the selection theory is greatly diminished. Let me support this claim.

Unlike Ruse, Wimsatt does not treat natural selection as a feedback process. To do so would be to concede the argument to the cybernetic theory of function, by making natural selection a special case of cybernetic system. Yet if there is a plausibility to his analysis, it must be because natural selection can be treated as a more general kind of goal-directed process. Certainly it must be conceded that a selection process *can* be goal-directed, as the following examples of human problem solving illustrate. (1) According to certain psychological theories (Campbell [1974]) we solve problems by a mechanism in which the subconscious brain proposes in a fairly random manner many possible solutions, and an editing process selects the best for implementation. It is easy and perhaps instructive to supplement this example of purposive human behavior with two others which also involve selection but which are less closely tied to current psychological theorizing. (2) There were many pebbles of assorted sizes and shapes scattered at David's feet as he prepared to do battle with Goliath, but he selected one which suited his purpose, namely one which could be used as a projectile for his sling. (3) A gold prospector swirls a mixture of sand and water in his pan in order to separate the valuable gold from other less valuable minerals.

Clearly, all three of these examples involve purposive behavior. If we find in them some common logical or structural feature, then we can proceed to hunt for identical features in non-human behavior. I think it is essential to these actions being purposive that there is in every case both a criterion of selection and a mechanism of separation, and that the two are distinct. The processes are purposive because, in each example, the separation mechanism picks out an object which meets the criterion. To see that this is so, consider the following contrasting situation. Suppose that on another occasion, perhaps while composing a bit of poetry, David absentmindedly picks up another pebble. Here again a pebble has been separated from its fellows, and the mechanism of separation is the same as before, namely being grasped and lifted by a human hand; but there is in this instance no additional criterion served by the separation, so the process as a whole is not purposive. Similarly, gold prospectors are not alone in their ability to separate gold dust from quartz dust. The ordinary scouring action of a fast running stream quite frequently produces a separation of the particles of sand in its bed, grading them according to their density. The separation mechanism is the same in the stream as in the prospector's pan, but only the prospector's process is purposive, because it includes a selection criterion (the value of the gold) which is lacking in the example of the stream. Even if we ignore the prospector's consciousness, the distinction between separation mechanism and selection criterion is suggested by his behavior, for he will vary his strategy, that is his selection mechanism, while retaining his criterion. For instance, he may pick the larger lumps from the pan with his fingers. It is perhaps significant that variable strategy is a logical feature which is found also in some feedback systems; an example is the onset of shivering when mere vaso-constriction no longer suffices to maintain body temperature. I shall return to this point in my discussion of goal states below.

Let us now apply this analysis to natural selection and ask: is it a purposive process or not? If natural selection qualifies as a purposive process, it must be because it can be analyzed as a problem solving one. There may be other kinds of purposive systems, as I think Wright's (1973) analysis shows, but if natural selection is purposive at all, it must be as a problem solving process – not, for example, as a creative process, like certain human artistic activities. Since Wimsatt does not claim that natural selection is a feedback process, let us suppose that the notion of a problem solving system is more general, and that feedback is only a special instance of this type. Thus, we cannot require that all the characteristics of feedback be present. Nevertheless, there are, I claim, the following necessary, but not, I think, sufficient conditions that must be met by a selection process even to qualify as a problem solving one. To qualify as purposive it must have, first, both a separation mechanism and an independently identifiable selection criterion, which, according to Wimsatt's proposal, may be called its purpose. The separation mechanism in natural selection is not hard to identify: those genotypes which survive in the next generation are separated from those which do not. But what criterion does this separation mechanism serve? Well, if one looks at the amazing results of natural selection, one is strongly tempted to suppose that a sort of engineering criterion is being met. It seems that there is at work here a marvelous adaptation of means to ends, a continual improvement in efficiency and, in general, of 'fitness'. But seeming is not enough. The existence of the criterion must be established independently of the separation mechanism. A certain stream may separate gold from quartz more efficiently than a prospector does, but the prospector is trying to do so and the stream is not.

Second, there must be a causal connection between the separation mechanism and some object or event which represents the criterion. It is essential to a problem solving action's being purposive that it be elicited by the problem. It is not enough that the twitching of a person's facial muscles may happen to dislodge a fly from his cheek — the twitching is said to have occurred *in order to* remove the fly only if it is elicited by the fly's presence. A mere facial tic happening by chance to occur at the right moment does not count as a purposive or goal-directed activity.[10] To put it more generally, an action which is not caused by the problem or by some (possibly non-conscious) 'awareness' of the problem, is not goal-directed, no matter how well it may happen to solve it.

Finally, the response must be contingent with respect to the problem it solves. That is, it must be the case that, given other initial and boundary conditions, the problem would still have been present but the response would not have occurred. If the problem inevitably produces its own solution then it solves itself; the actions which lead to a solution, being inescapable consequences of the problem itself, cannot be said to have occurred in order to solve the problem.

Note that in the examples of purposive selection mechanisms, considered above, the selection process is a contingent solution to some problem or a means to some end, and it was set in motion because of that problem. The selection of a pebble of the right size and shape satisfies the need for a missile, occurs because of that need, yet might not have occurred if other conditions besides the presence of the problem had not also been present. But in natural selection, far from solving some other problem or achieving some goal, the selection process *is* the problem the species faces. Proto-giraffes with necks too short to reach the lowest branches of the trees starve, or at least fail to be sufficiently fertile. That is the problem. This separating out of the shorter-necked animals was not put into effect because of some other problem or in order to achieve some end (e.g., longer, more graceful necks). There is no prior event or state which can be identified as the criterion or problem situation and which elicited the selection process. Therefore the selection process is not a goal-directed one.[11]

The contrast with a feedback system is obvious. In a feedback system an environmental influence produces a change in f_2; this, if uncompensated, will lead to a change in the protected variable, h_1. A separate part of the system (represented by the variable g_1) senses this problem; that is, a change in g_1 is caused by the initial perturbation of h_1. This sensing of the problem, in turn, causes a change in f_1, as a response to the original problem (i.e., as a direct, but contingent consequence of the original disturbance) and in such a direction as to partially or wholly counteract the disturbance. Now I do not claim that only feedback systems show this kind of purposive behavior, merely that they are among the systems that do. And I think it is reasonably obvious that natural selection does not.

It is frequently said that the only 'goal' of natural selection is survival: 'what survives survives'. There is no evidence to support the statement that the separation mechanism, the survival, serves some additional criterion such as efficiency or any other standard of engineering quality. It is true that natural selection often does a good job of imitating a purposive selection in which engineering standards are being met, just a stream may imitate a prospector. But one can easily find examples which show the imitation for what it is:

(1) The peacock's extravagant investment in fancy tail feathers illustrates this point. Clearly the evolutionary strategy of its ancestors and their less successful rivals consisted in outbidding each other to provide whatever the market — the ancestral peahen — was having. It was the esthetic equivalent of an arms race. The winners achieved esthetic overkill by diverting domestic resources from the business of ordinary living, thus rendering the present-day organism less able to elude predators and in general less efficient as a survival machine than its ancestors. Because a wealth of highly efficient mechanisms has emerged out of the process of natural selection, we may be tempted to suppose that it aims at efficiently surviving organisms as a goal. But if natural selection can be said to aim at anything, even in a very loose sense of the term, its goal is simply to produce organisms which out-reproduce their rivals of the moment by whatever means come to hand. Efficiency, when it appears, comes as the unintended result of a process which 'aims' at something else.

(2) Examples drawn from nature are necessarily complicated and subject to uncertainty, so let me offer a hypothetical example which, by virtue of its artificial simplicity, illustrates this basic characteristic of natural selection. Suppose a population of seagulls has as its only source of food a feeding machine which shoots out food pellets. The pellets sail a short distance through the air and land on the water, where they float until picked up by the gulls. The gulls spend most of their time swimming about in the water and eating. They fly but rarely, and then only out of exuberance. Under these circumstances an equilibrium population of one hundred birds is established, the number being determined essentially by the unvarying food supply. But one day a mutation occurs which results in an adult bird with a new behavior

pattern. This mutant does not wait in the water for the pellets to drop, but, because of a unique ability to hover, it catches them on the wing. This bird has a competitive advantage over the others, grows to be well fed and prosperous, and produces several offspring similarly endowed. Gradually the mutant strain becomes the dominant one in the flock, simply because the hoverers monopolize the food supply. Hovering, however, requires more energy than does swimming, so each mutant must eat more than its predecessors did. Consequently, the food supply which once supported a population of one hundred swimming feeders now can support only fifty hoverers. So this particular community of gulls makes less efficient use of its food supply after it has undergone variation and selection than it did before.

A selection theory of function will work only if natural selection has a criterion which may be identified as the ultimate purpose to be served by the various particular functions in all living systems. Now, it would be very difficult to prove the non-existence of such a criterion, and I have not done so; but the foregoing arguments and examples seem to rule out the more obvious and natural candidates for that role, such as fitness and efficiency. Of course, natural selection may be meeting some other criterion; for instance, it may be serving the inscrutable purposes of the Deity, or it may be striving toward some as yet unrevealed ultimate goal. But I take it that the proponents of a selection theory have in mind a naturalistic criterion, and there seems to be no good reason to suppose that a naturalistic criterion exists.

Goals in Cybernetic Systems

But does a cybernetic theory of function fare any better than a selection theory when confronted by the requirement that a purposive system must have an objective, self-contained purpose which is served by the various functions of its parts? I have argued that there is no reason to suppose that an objective *criterion* can be found in natural selection; the question remains whether an objective *goal* can be found in feedback systems. Let us apply our analysis of feedback to this question.

(1) *'Promote' and 'ceteris paribus'*. When we ascribe a function to a part of a system we imply that the part promotes the attainment of some purpose or the meeting of some goal. But as Wimsatt (1972) indicates, there is ambiguity here: what exactly does 'promote' mean? In other words, we would like to say that the system as it is presently arranged causes the goal state [12] to occur, or the aimed-at effect to be produced, with greater probability, or greater

exactness, than it would under some alternative arrangement, other things being equal. But how can we specify the alternative situation? Wimsatt proposes and criticizes several possible answers to this question, finally concluding that there may be no general way to pick out a unique set of circumstances for comparison.

If natural selection, as selection, is to provide support for the use of 'function', then, as we have seen, a function of an organ is some actual or potential consequence of the organ's presence in the organism which contributes to a differential survival rate in the next generation; that is, an adaptation. Note that merely contributing to survival will not do: the contribution of a new mutation must be such that an increase in survival rate relative to the existing dominant phenotype(s) occurs so that the mutation is favored by selection. But suppose some wildly improbable mutation in an earthworm should produce an organ quite different from the standard rudimentary heart, which nevertheless causes the internal fluids to circulate — a system of cilia, for example. Suppose that these cilia perform less efficiently than the usual heart, so that the offspring of the mutant worms have a decreased survival rate relative to the standard model of worm. The cilia are maladaptive. Nevertheless it is easy to see that promoting the circulation of internal fluids is a function of the cilia — not because this circulating action produces an increased survival rate (i.e., not because of selection for this trait), but simply because it leads to the continued existence of the individual organisms which possess the trait; without it the mutant worms would die. A selection theory encounters in examples like this the problem of how to specify what *ceteris paribus* means, as Wimsatt points out. No such problem faces a cybernetic theory, however, for according to this model the circulating action is a function just because it is part of the system of interlocking feedback loops that constitute the worm's physiology. If the part in question were disconnected, leaving the rest of the mechanism intact, then the regulation of the controlled variable would become less effective or cease altogether.

Wimsatt's question can be given a definite, in fact a quantitative, answer for any case of Type I feedback; in fact, both the nature of the goal or purpose and the alternative situation can be deduced, once the system is known to exhibit this pattern of organization. If the relations φ, ψ, and χ are all well-behaved functions, the standard mathematical technique of partial differentiation provides the key to the puzzle.

Let h_1 depend upon both f_1 and f_2 and let f_1 and f_2 be causally independent. Suppose that f_2 is affected by an external perturbation, but that f_1

remains the same. What would be the effect on h_1? The answer is given by the partial derivative of φ with respect to the variable f_2:

$$\delta h_1 = \left(\frac{\partial \varphi}{\partial f_2}\right) \delta f_2 = \delta h_1 \text{ (no feedback)}$$

If the feedback loop did not produce compensatory changes in f_1, then changes in h_1 would be related to changes in f_2 according to this equation. But the actual state of affairs is such that both f_2 and f_1 change: f_2 because of external influences and f_1 because of the feedback loop. Thus the actual situation is represented by:

$$\delta h_1 = \left(\frac{\partial \varphi}{\partial f_1}\right) \delta f_1 + \left(\frac{\partial \varphi}{\partial f_2}\right) \delta f_2 = \delta h_1 \text{ (feedback)}$$

But f_1 changes in response to a change in g_1 which in turn responds to a change in h_1, so we have:

$$\delta f_1 = \left(\frac{\partial \chi}{\partial g_1}\right) \delta g_1 = \left(\frac{\partial \chi}{\partial g_1}\right)\left(\frac{\partial \psi}{\partial h_1}\right) \delta h_1$$

Substituting this equation in the previous one and solving for δh_1 we obtain:

$$\delta h_1 = \frac{\partial \varphi}{\partial f_2} \cdot \delta f_2 \bigg/ \left[1 - \left(\frac{\partial \varphi}{\partial f_1}\right)\left(\frac{\partial \psi}{\partial h_1}\right)\left(\frac{\partial \chi}{\partial g_1}\right)\right]$$

or

$$\delta h_1 \text{ (feedback)} = \delta h_1 \text{ (no feedback)} \bigg/ \left[1 - \left(\frac{\partial \varphi}{\partial f_1}\right)\left(\frac{\partial \psi}{\partial h_1}\right)\left(\frac{\partial \chi}{\partial g_1}\right)\right].$$

Now, if the product of the three partial derivatives is negative (which may be taken as distinguishing negative from positive feedback) the actual change in h_1, with the feedback loop operating, will be some definite fraction of the change that would occur in response only to an uncompensated change in f_2. Thus we may say that the internal 'purpose' of this system is stability in the face of external perturbations. So 'ceteris paribus' can be translated thus: the class of boundary conditions, B, which characterizes the system remains in force, with the exception of B_χ or B_ψ. And 'promotes' can be made quantitative: the effect of external perturbations is reduced by a factor $1/[1 - (\partial \varphi/\partial f_1)(\partial \psi/\partial h_1)(\partial \chi/\partial g_1)]$. Similar analyses can be made for Type II systems and systems with discrete states, but the results depend on the details of the 'engineering'.

(2) *Goal states*. In the discussion of teleological systems it is almost universally assumed that there are certain 'preferred' or special states of the system which may be identified as the 'goal' of whatever activity characterizes the system as teleological. Manier (1971), just to pick a typical example, speaks of a system responding to the "difference between the output of S at a given time and the target value P." Opinion may differ as to whether the actual attainment of the 'goal state' should be part of a definition of teleological systems (or whether a 'tendency' to achieve the goal states is sufficient) but that there are such states seems to be taken for granted.

The fact is that a large class of negative feedback systems have no identifiable state or set of states which are special in any way. For Type I feedback systems our analysis of 'promote' in the preceding section makes it clear that what is reduced by the operation of the feedback mechanism is the amount of change in h_1 produced by a given change in f_2, not the difference between h_1 and some unique target value. Feedback systems of Type II cannot be treated in such a general manner, so I shall discuss two special cases which, I believe, illustrate typical results. Consider first the proportional thermostat. (Because their feedback patterns are isomorphic, this discussion applies equally to the allosteric enzyme system.) If the thermostat is designed to maintain the room at a temperature well above the environmental temperature, T_e, no provision needs to be made for cooling. The heater output will be some monotonically decreasing function W of the temperature of the room, varying between a maximum value for sufficiently low temperatures and zero for sufficiently high (see Figure 11(a)). With this arrangement the room will stabilize at a temperature $T > T_e$ over a wide range of external temperatures. From a given response curve $W(T)$, and from a given function relating the rate of cooling of the room to the outside temperature (e.g., Newton's Law of Cooling) it is possible to derive a function relating the equilibrium temperature of the room to the outside temperature when feedback is at work: $T_{\text{feedback}} = \text{fn}(T_e)$. Similarly, if the feedback mechanism did not 'compensate' for changes in T_e, for example, if the heater were maintained at a constant setting, one could derive a different function relating $T_{\text{no feedback}}$ to T_e. Typical relations are illustrated in Figure 11(b). For a given external temperature, T_e, the system seeks a steady-state temperature, T, which is related to T_e as shown by the curved line.

Even the automatic pilot mechanism which guides a torpedo to its target or maintains a ship on its course is less 'goal oriented' than one might at first suppose. If the ship, slapped by a wave or blundered into by an absent-minded fish, is deflected from its old course to a new one, the feedback

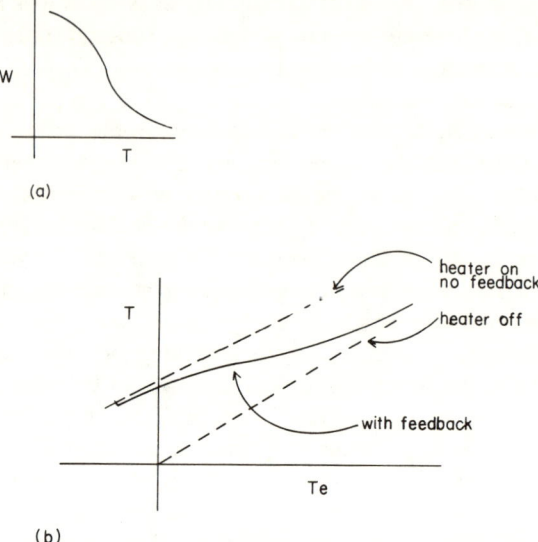

Fig. 11.

mechanism will bring it back to its old heading. With respect to this class of perturbations — stochastic ones — the system may indeed be said to possess a goal state. But there is another large class of perturbing forces — steady ones, such as may be produced by contrary winds or currents — to which the mechanism responds just as the thermostat does to external temperatures. The ship yields to the force, yielding more to a greater force; as with the thermostat, the feedback mechanism renders the ship less sensitive to variations in wind or current than it would be with a fixed rudder angle.

Yet the absence (in some cases) of a target value for the variable h_1 does not mean that goal-directedness is lacking in feedback systems. Our analysis makes it clear that active survival of h_1 is promoted by the structural organization of feedback; so we may say that the goal or purpose of a feedback system is to render the system relatively insensitive to the effects of its environment upon the variable h_1. I shall return to this point when I discuss the *context* of functional systems.

(3) *Complexity*. So far our discussion of cybernetic systems has concentrated on single feedback loops. Biological systems are, of course, far more

complicated than this, yet their complexity reinforces the point we have just made that the stability of the system (or the organism) in the face of environmental perturbations is the central purpose of cybernetic systems. I remarked earlier that a characteristic feature of human purposive action is the ability to vary the strategy while retaining a single goal. The temperature regulating mechanism in mammals also displays this feature, for when vasoconstriction is insufficient to maintain body temperature shivering sets in. There are in this case two feedback loops which aim at the same goal — stability of the temperature variable. Obviously, this same structure can be built into mechanical feedback systems, too. A simple diagram of a double feedback loop is shown in Figure 12.

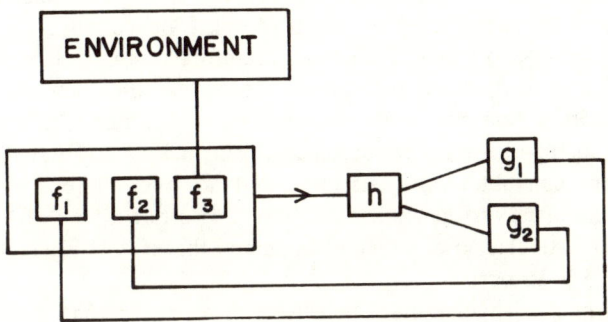

Fig. 12.

In criterion (vi) we required that the environment of a feedback system affects the values of the variables $f_2, f_3, \ldots$ but not the sets of variables $\{g_1\}$ or $\{h_i\}$. If this condition is met, then the system *regulates* the variable h_1 by causing compensatory changes in the variable f_1. However, if the environment affects, say, both f_2 and h_2, no regulation is possible. Thus a necessary condition of a system's functioning as a self-regulating system is that some of its variables not be affected by environmental 'input'. The boundary conditions of S determine whether this condition is met, but, although S operates so as to regulate h_1, it does nothing to regulate or maintain its own boundary conditions. The system is passive with respect to alterations of its boundary conditions, though, given those conditions, it is active with respect to alterations of $f_2, f_3, \ldots$. So a single feedback system is not a self-maintaining system.

Yet it may be the case that some sub-class of B is true because of the operation of another feedback system. For example, the Zhabotinsky reaction system will not 'function' if the solution is frozen solid or if it is vaporized, so an experimenter may arrange that the temperature be maintained between the freezing and boiling points by the action of a thermostat. I take it to be the rule in complex systems, especially living systems, that there are interlocking networks of feedback systems maintaining each other's boundary conditions. So a combination of feedback loops may be a self-maintaining system, at least within a limited range of environmental conditions.

Although an increase in complexity beyond that of a single feedback loop is compatible with our analysis of goal-directedness in cybernetic systems, a decrease in complexity will not do. Beckner's sixth criterion and Manier's first 'boundary condition' for the applicability of the feedback model are both designed to guarantee that the feedback loop contains a mechanism which is physically distinct from, and therefore disconnectable from, the system which is being regulated. I have attempted to accomplish the same end by requiring that the class of boundary conditions which define the system and distinguish its parts be separable in such a way that from a subclass, B_φ, one can obtain the relation φ but not the relations ψ or χ. This approach lends itself to generalization and thus offers a way of 'measuring' the degree of articulation of a feedback system, that is, the degree of physical independence of its parts.

If we form a triangular diagram whose corners represent the three subclasses of B and their associated relations, we can let arrows along the sides represent the obtainability of a relation from a subclass of B (see Figure 13). Thus the diagram of Figure 13(a) represents the lack of disconnectability that we have attempted to exclude by criterion (v). Diagrams of this type are easily formed for some of the other systems we have discussed. For instance, the thermostat may be represented by a diagram with no arrows at all, the allosteric enzyme system by Figure 13(b) and the Zhabotinsky oscillating reaction by Figure 13(c). The battery—resistor circuit which we have discussed as an example of a too-simple system is represented by a diagram with all six arrows present as in Figure 13(d). So we can say that the thermostat is more articulated than the enzyme system, which, in turn, is more articulated than the Zhabotinsky system, and that the battery—resistor system is not articulated at all. Criterion (v), then, may be thought of as a requirement of a minimum degree of articulation in a system, if feedback is to serve as a model on which ascriptions of function may be based.

FEEDBACK, SELECTION, AND FUNCTION   101

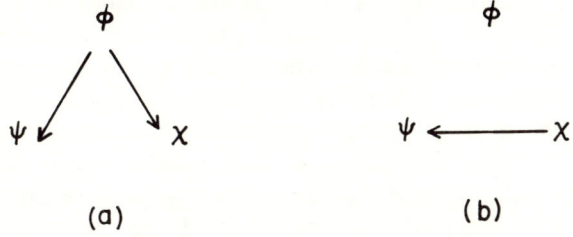

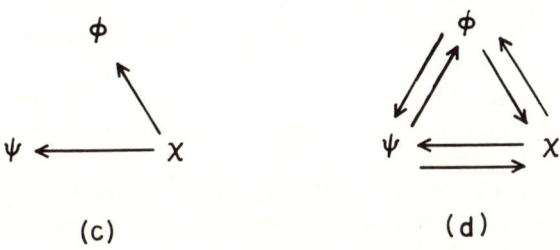

Fig. 13.

The Context of Functional Activity

As I conceded in the Introduction, Wimsatt (1972) correctly interprets his example of a suicidal robot as showing that the context of a feedback system determines at least some of the functions of its parts. I shall argue in this section that the context helps determine functionality in several ways; nevertheless, it is important to note to how great a degree the functions within a feedback device are determined by the system itself. First, a feedback system, to a certain extent, defines its own potential use, in that only some of its state variables could be the regulated one. If anyone should wish to use the system of Figure 6 as a feedback system the choice of the controlled or regulated variable is limited to f_1, h_1, or g_1 at most. Second, if (for reasons to be discussed below) the variable h_1 is the controlled variable, then the internal structure of the system determines that the function of the part whose state is represented by g_1 is to respond to changes in h_1 by causing changes in f_1.

Yet, as I pointed out in discussing Beckner's application of feedback, there is an ambiguity as to which of the variables f_1, h_1, and g_1 is the regulated variable, and this is the first point at which the relevance of context must be recognized. Condition (vi) of my formal explication of feedback (Part III) stipulates that the intervention of the environment is limited to the variables $f_2, f_3, \ldots, f_n$. As I argued in that section, if the environment also perturbs some of the variables in the sets $\{h_i\}$ and $\{g_i\}$ the system cannot act as a regulator of any of its variables.

A second way in which the context helps to determine functionality is in providing a general setting in which feedback devices have uses. An isolated voltage regulator circuit, formed by a random collocation of atoms in, say, the Middle Ages, where there was no context in which voltages or their regulation had significance, would have been at most a *potentially* functional system. Some context of electronic design and use is required to justify calling this circuit by its functional designation. But I claim that, given this context, even a circuit produced by chance would qualify. Similarly, a coincidence of enormous improbability may (indeed, must if evolutionary theory is correct) have produced a single-loop chemical feedback system in an isolated pool of water somewhere on the pre-biotic earth. Again, without some sufficient context this system is only a potentially functional one.

Thirdly, the context determines the use to which a feedback device is put. The relations φ, ψ, and χ represent causal chains, and the analysis may be formulated in such a way as to telescope some of the details of these chains. Nothing in our formal explications of feedback forbids such an analytical ploy; in fact, the practice is both usual and necessary. But this unavoidable fact introduces an ambiguity which must be resolved by the context. To take a concrete example, consider an electronic circuit which regulates the voltage across a certain resistor. If its resistance is not subject to environmental perturbation, regulating the voltage across the resistor is equivalent to regulating the current passing through it. So the device may equally plausibly be called a current regulator. Now it may be the case that, owing to the details of construction of the device, it would make a better current regulator than a voltage regulator. Nevertheless, if it is used as a voltage regulator, a voltage regulator it is.

None of this means, however, that a reductionistic analysis of feedback and a cybernetic account of functionality are impossible. It may be the case, and frequently is with the products of human engineering, that some (at least) of the uses to which a feedback device is put are determined only by reference to elements in the consciousness of the designer and/or the user of the device. But many functional systems employing feedback are not the products of human design and yet are put to use; and their uses, I claim, are defined by the internal structure of the non-conscious systems in which they function.

But what sort of context is required? I have alluded earlier to the prevalence of cooperation among feedback loops both in living organisms and in simpler systems. In a typical complex of feedback systems, system A cannot act as a regulator of variable h_1 unless variable g_2 is protected from attack by the environment, and g_2 is protected because it is the regulated variable of system B. Moreover, the regulating ability of another system C in turn depends on the stability of h_1. Ashby's (1960) "homeostat" or "ultrastable system" displays an additional complexity. His device possesses a fixed repertoire of alternative strategies and a mechanism which switches from one of them to another when the protected variable is driven outside a certain range. When complexity of this sort becomes great enough — cooperating feedback loops, loops within loops — it becomes appropriate to apply the concept of *survival* (see Part I). The kind of stability achieved this way is active; by it the system as a whole preserves its structure against the onslaughts of the environment by actively compensating for these perturbations. Survival, or active self-maintenance, is qualitatively different from the merely passive endurance of less complex systems.

Thus feedback itself provides the context in which feedback becomes functional. Survival is not an alien or imported concept; to speak of survival is not to shift the focus of our discussion of function from cybernetics to something else, for survival is feedback 'fed back', feedback multiplied and involuted. Moreover, in this analysis functionality is a matter of degree. That this should be so is wholly consonant with an evolutionary theory which views life — and functionality — as gradually emerging from simpler, non-purposive precursors.

V. THE ROLES OF 'SELECTION' AND 'FEEDBACK' IN FUNCTIONAL EXPLANATIONS

What does a Functional Explanation Explain?

As Wimsatt points out, the concept of function is to be understood in the context of functional explanation — to see what function means we have to see how it is used in explaining something. Now, the standard form of explanation in the physical sciences is this: we explain a state of affairs by deriving it from an antecedent state of affairs by means of a 'covering law' which describes how any such system develops through time. So, too, with functional explanations, as Wimsatt ([1972], p. 70) discusses them: " ... a teleological explanation uses the results of a functional analysis to explain why the functional trait is there." In a similar vein, Campbell ([1974b], p. 181), writing of the jaw of the soldier termite, says, "We need the law of levers, *and organism level selection* ... to explain the [shape of the] jaw ... ." Finally, Wright ([1973], p. 154), discussing the example of a distributor cover in an automobile engine, states, "When we say that the distributor has that cover in order to keep the rain out, we are explaining *why* the distributor has that cover." Generalizing, Wright adds (p. 162), "In a functional explanation, the consequences of X's being there ... must be invoked to explain why X is there." According to this view, a functional explanation, like the ordinary explanations of physics, answers questions about how a given state of affairs came to be, but it does so in a special way. Why does this trout have a kidney? Several kinds of answer may be given to this question, but when we answer it by showing how the chance development of a proto-kidney in the ancestors of the trout helped them meet the criterion of natural selection, then we are making a functional explanation. I shall argue that this is possibly a misleading, and certainly a too narrow, view of what functional explanations do.

To see how a functional explanation of origins, based on a selection theory of function, is supposed to work, let us take a much simplified example of a selection[13] process: a group of model sailboat makers hold a regatta on the pond of a city park. Model boats of all types are present — fast ones, slow ones, some fitted with self-steering wind vanes, and others without. At a signal, all the boats are released on a course for a beach on the opposite side of the pond. During their passage several things happen: the fleet is so numerous that many collisions occur among the contestants; also, while most of the boats are only half-way through their voyage, the wind shifts to a new direction. There is, consequently, a selection process at work. Those boats which suffer collisions and are not fitted with self-steering equipment are deflected from their courses; those boats which are still far from the opposite side when the wind shift occurs and lack self-steering equipment are also deflected. The result is that only a subset of the fleet reaches the target beach; the rest come to shore at other points around the rim of the pond. If we look at the boats that 'succeed', we find that they are mainly of the self-steering type, with a few fast fixed-rudder boats among them. The latter attained the beach because of good luck in avoiding collisions and because they had reached shore before the wind shifted.

Now let us ask some questions about these boats. Picking one up, we observe a self-steering apparatus and we ask, why does this boat have a wind vane? Questions are, alas, rarely formulated clearly; so that the best indication of a questioner's real meaning is the sort of answer he deems acceptable. Conversely, the answer to a question often implicitly reformulates it; thus from an examination of the answer we may infer how the question should have been put so as to elicit just that kind of answer. At least three kinds of answer can be given in response to this question.

A functional explanation may be devised to answer the question why this boat has a self-steering apparatus, if the question is construed as an inquiry about "design function" (Wright [1973]). Under this interpretation the answer is that the designer of the model sailboat intended the boat to reach the target beach successfully, and that he believed that the wind vane would help the boat to cope with the vicissitudes of its prospective voyage. Note that this explanation does not refer to the voyage itself or to the actual beach — only to beliefs and intentions about them. To deal adequately with the question, the explanation need not refer beyond the beliefs and intentions to their objects; the wind vane would have been there even if the 'selection process' had, despite the expectations of the designer, never occurred. Clearly, it would be inappropriate to apply this interpretation to a question about the presence of a functional entity in a biological organism, for in a neo-Darwinian context design function is not an issue.

An ordinary, causal explanation of the wind vane's presence may also be produced, if the question is construed as a prosaic one about the causal sequence which led to the incorporation of the self-steering apparatus into the mechanism of the boat. The answer would be, roughly, that the vane is there because the builder put it there. This interpretation of the question, like the previous one, does not call for an answer in terms of the actual selection process through which the boat has passed. Unlike the previous answer, this one could be adapted to a biological system; it is a proper sort of answer both for the boat and for a trout. In the latter case one says that the part is there because it was specified by the DNA of the zygote that grew into this organism.

What sort of question, we may ask, *would* call for an explanation that refers to the actual voyage successfully negotiated by the boat? Here are two: 'why did the boat survive the voyage when others did not'? and, 'why is there a higher percentage of boats with self-steering devices among the survivors of the race than there were among the entire pool of entrants'? In these questions the continued existence of a particular boat, or the differential survival of a class of boats, is posed as a surprising event, as a problem to be explained. It is a puzzle precisely because the selection process (the race, with its collisions and shifting winds) is known to have occurred and its effect (the elimination of many boats from the race) is also known. If virtually any kind of boat could have, and in fact did, complete the race, then no appreciable selection would have resulted and the question about a particular boat's presence on the target beach would not arise. Clearly, the action of the selection process forms part of the background of the question and thus will not figure in the answer. So we explain the survival of the particular boat, or the class of boats, by referring to the self-steering device; that is, we construct a functional explanation which points out how the mechanism enables the boat to counteract the destructive influences which impinge upon it during the race. In short, we explain the survival by means of the wind vane, and not the wind vane by the survival.

To the question 'Why do trout have kidneys'? or 'Why does this boat have a self-steering device'? one might offer this third kind of explanation; but I think it is clear that when this type of answer is made, the question is being construed as an inquiry, not about origins or mere presence, but about survival. Not: 'Why is the kidney here'? but, 'Why is the kidney still here'? or, more exactly, 'Why did kidney-possessing fish survive the harsh conditions to which others succumbed?

Or, to take Wright's (1973) example, we may explain the presence of a distributor cover in an automobile engine by reference to the intention of its designer to keep rain out of the distributor, and to his belief (which need not be a true belief) that the cover would perform that function. Here the presence of a functioning part, that is, its membership in an organism, is explained by reference to beliefs about possible or potential functions, but (since the beliefs need not be correct) the explanation does not refer to the actual functioning of the device; hence the explanation is not a functional explanation in the sense we are considering here. Moreover, this type of explanation is not applicable to a neo-Darwinian account of biological systems, where there is no question of design function. Alternatively, we may explain the presence of the distributor cover in the mechanism of the automobile by referring to the process by which the engine was constructed. This type of answer is applicable to biological organisms (the analogous answer would be, roughly, that this trout possesses a kidney because certain mutations occurred in the DNA of its ancestors), but this explanation, too, makes no reference to the actual functioning of *this* kidney; hence it fails to conform to the general schema of Hempel or of the selection theorists. An etiological explanation may take as its starting point the newly formed zygote which grows into the organism in which we find this kidney, but in that case the story we tell is a simply causal one, since the concept of function plays no role in it. Or the etiological story may commence farther down the family tree of the present individual; in that case we must mention the health and survival of individual ancestral organisms, and our etiological explanation turns out to be parasitic on its rival. As we have seen, a truly function-referring answer to a question about the presence of the kidney may be constructed, but only at the price of radically re-interpreting the question. An inquiry that appears to be about the membership of a part in an organism must be treated as an indirect inquiry about survival: 'Why does the class of extant trout contain so many individuals with intact kidneys'? The answer comes that those in previous generations with defective kidneys did not survive long enough to reproduce.

Certainly the central position occupied by natural selection in modern biology lends impressive weight to attempts to explicate goals and functions in terms of selection. Even the soberly reductionistic George C. Williams (1966) asserts that the goal for which an organism is designed is the survival of its genes (p. 123). But this example shows that biologists work with two distinct concepts of survival: the active maintenance of the structural integrity of an individual organism in the face of exogenous threats, and the multiplying of copies of genes and gene clusters in the reproductive process. These two kinds of survival should be sharply distinguished, as should the two kinds of function ascriptions they support. The first is literal, and should be considered the primary sense of the term; the second is metaphorical and should be assigned a secondary place.

I have argued that we have good reasons for taking the survival of an individual as that organism's goal. The survival is accomplished by means of an objectively identifiable and rare kind of mechanism, namely interlocking feedback loops; and that mechanism closely resembles one of our paradigms of goal-direction, the action of the helmsman. We need not concern ourselves about the repeated copying of genes when we speculate about the function of the kidneys or of food gathering. But the concept sheds no light on other traits also called adaptations; from the point of view of the first and stricter sense of survival the suicidal behavior of spawning salmon, for example, simply makes no sense. For such adaptations we must invoke not just another goal but a target of a new kind; one, indeed, which may be incompatible with the first. Although some parts and behavior patterns serve the goal of individual survival, others, such as the anatomical and behavioral features associated with reproduction, promote the replication of certain patterns of bases in DNA, what Williams calls genic survival. What, if not this replication of molecules, are oviducts, nest building, and mating displays for? So genic survival may be called a goal, and Williams does so name it; but it has less impressive credentials for that title than does individual survival. The 'survival' of genes amounts to the winning out in the process of natural selection, an unremarkable process when compared with negative feedback, and one which bears but a thin resemblance to paradigms drawn from human goal-seeking. In natural selection there is no selecting agent at work, as Nagel reminds us ([1977], p. 286), and no distinguishable criterion informs the process by which the survivors separate themselves from the failures, as I have argued above.

Nor should we hesitate to admit that mechanisms which seek goals in the stronger, primary sense of the term have arisen in the service of the thinly metaphorical goal of genic survival, that the higher end 'serves' the lower. The emergence of greater sophistication from lesser epitomizes the evolutionary story.

Biological organisms have kinds of complexity not yet to be found in the products of human engineering. One of these is the mutual dependence of the component systems as a living organism; that is the phenomenon of self-maintainance on a large scale. Without the proper functioning of the other cybernetic mechanisms in the organism, most functional parts would cease to exist even as objects (skeletal structures being rare exceptions to this rule); but even in the machines of man dissolution of the global functional organization of the machine marks the end of a part at least logically *qua* functional unit, and sometimes physically as well. What was once a self-steering device becomes just a jumble of metal plates, discs, and string when the sailboat strikes a rock. So, even for mechanical systems, and especially for living organisms, the continued existence of a functioning part depends upon the survival of the organism or machine in which it functions. In this sense — and only in this sense — is it correct to say that a functional explanation accounts for the existence of a functional part. The explanation accounts for the survival of the organism, on which depends the continued presence of an intact, functioning part; hence the functioning of a part partially accounts for its own continued existence. Wright ([1973], p. 164), in another characterization of functional explanation, says:

When we explain the presence or existence of X by appeal to a consequence Z, the overriding consideration is that Z must be or create conditions conducive to the survival or maintainance of X.

Wright is certainly correct in using the concept of survival here, but, if the preceding arguments have been cogent, this statement would be rendered more precise by altering it to read: When we explain the continued presence or existence of X in the face of structure-destroying influences by appealing to a consequence Z, the overriding consideration must be that Z must be or create conditions conducive to the survival or maintainance of the organism in which X participates and thus of X itself.

I conclude that, both for natural and for man-made organisms, explanations of the origin or mere presence of a functioning part are not functional explanations at all, or if 'design function' is involved, do not refer to the actual functioning of the part in question; and true functional explanations are directed at other puzzles, one of which is the survival of the organism which possesses the functioning part.

Yet much good sense remains in our calling reproduction – i.e., the replication of genes – the function of the cellular machinery of meiosis, of the reproductive machinery of the organism, and of the mating behavior of sexual species. Here we pick out the replication of genes as an arresting and puzzling fact, a result achieved by truly outlandish machinery and behavior patterns in many species; and we flag for special attention those structures that contribute to that result. Even though the process of replication bears no resemblance to our paradigms of purposive action, yet 'goal' does not seem a prohibitively fanciful term to apply to the result, and 'function' at least a pardonable label for those actions and causal steps which lead to the replication of genes. We can also speak of a reproductive function as contributing to the current presence of the reproductive machinery in the weaker of the two senses we have distinguished; that is, although an element with a reproductive function does not contribute to its own survival and retention within the organism that contains it – as survival-promoting parts do – yet if we ask why intact reproductive machinery shows up with such near unanimity in a species at the present moment, the answer comes that members of previous generations who lacked the genes for such structures failed to contribute to the present population. Hence the functioning of the reproductive apparatus of earlier generations figures in the story we tell of how the present generation, displaying a preponderance of machinery very like that of its ancestors, came to be.

Functional explanations, in the primary sense for biology, are not explanations of the *origins* of phenotypic traits. Nor are they concerned exclusively with the evolutionary success of the traits, except insofar as that success depends upon the survival of individual organisms. I submit that there are many other questions, not having to do with origins or even selection processes, the answers to which employ the concept of function. Here are some examples:

(i) Is this animal in good health?
(ii) Why is it good for tennis players to gulp salt tablets between sets?
(iii) Why is this room getting so hot?
(iv) Why does the Zhabotinsky reaction (see Appendix) oscillate faster when heated?
(v) Would the Zhabotinsky solution still oscillate if I replaced the ferroin with cerium?
(vi) Could a virus molecule substitute for a cell's own DNA?

Partial answers to these questions, all involving the concept of function, might be:

(i) Yes, all its feedback systems are operating within their proper ranges.
(ii) The salt is needed to maintain the regulatory mechanisms within their proper ranges.
(iii) Because the thermostat's switch is stuck.
(iv) Because reaction (4), which controls the period of oscillation, proceeds faster at elevated temperatures.
(v) Yes, cerium could perform the same function as ferroin.
(vi) Yes, a virus can perform the same function as a cell's own DNA.

Explaining is not limited to accounting for the presence of something by showing how it arose out of an earlier situation. Viewed more broadly, to explain is to 'make sense' of something, to fit it into an intelligible pattern. Accordingly, I suggest that to give a functional explanation of a system is to recommend a selective way of looking at it, a way in which some causal connections are emphasized and others ignored. It is to suggest which connections are important and which are not. It is to adopt a perspective based on a pattern – and the pattern, at least in the examples mentioned above, is feedback.

In an ordinary physical explanation of a system we show that its behavior fits the pattern of the basic laws of physics, which are statements about enduring relations among the basic physical variables: mass, momentum, force, etc. From these laws we can obtain specific statements about the system. In a functional explanation we show that the behavior of a system fits another kind of pattern — a 'circular' pattern of causal connections, which can be represented informally by a feedback diagram or more formally by the seven criteria of Part III. From this pattern, too, we can obtain specific propositions about the behavior of the system; for example, we can state quantitatively how the action of the feedback loop reduces the sensitivity of the system to external perturbations (cf. the discussion of 'promote' and *'ceteris paribus'* in Section IV).

Of course, the basic physical explanation stands by itself. It is a complete explanation, at least for the kind of question it defines as meaningful. A functional explanation of the same system, then, would be in a sense superfluous. Question (iii) about the overheated room might be answered in the language of basic physics, by tracing the forces and motions from point to point through the machinery. But how much more illuminating it is to describe those levers, plates, and bimetallic strips as a feedback system, to describe the combined thermometer and switch as a member of that system so that it can be called a thermostat, i.e., a device whose *function* it is to *regulate* the temperature of the room.

Is a Functional Explanation Reductionistic?

I have claimed that my explication of feedback is reductionistic because it is done in the language of ordinary physical analysis. The explication of Part III refers to the laws of physics (or the reducing theory) in a general way, as determining the behavior or evolution of the system; but the essence of the explication is in stating what sorts of boundary conditions determine a feedback system. An object is a feedback system because of the way it is put together; because of the constraints and connections among its parts and the way they influence each other: all of these are what we mean generally by 'boundary conditions'. But Polanyi (1968) has argued that living beings and inanimate machines "transcend the laws of physics and chemistry" precisely because their essence is to be found in their boundary conditions. In living organisms and in machines one finds, he claims, a "hierarchy" of organizing principles, of which none contradicts another but each exerts its own kind of "control" on the total behavior of the organism. This "dual control" is possible because of the indifference of the laws of physics and chemistry to particular boundary conditions. Given a set of initial and boundary conditions, the laws of physics and chemistry determine the future development of the system; but these laws would equally determine a different path of development given different conditions. The "laws of inanimate nature", which constitute the lowest of Polanyi's hierarchical levels, are compatible with any possible set of boundary conditions, but certain kinds of boundary conditions — those found in machines and living organisms — embody "irreducible" higher level principles which "rely on" the workings of the lower level while imposing their additional controls on the system.

The kind of explanation one can make, using only physics and chemistry, cannot explain these boundary conditions: they must be *assumed*; then the behavior of the object can be deduced. Even though some ordinary physical explanations *seem* to be explanations of boundary conditions (we can, for example, explain why a certain object is in a certain place) these explanations merely shift the point at which the assumption must be made; they may explain today's boundary conditions, but only be reference to those of yesterday, and *those* must then be "fed in" to the calculation as assumptions.

Although all systems, animate and inanimate, are constrained by boundary conditions, Polanyi does not argue that all are therefore irreducible to physics and chemistry. The vast bulk of boundary conditions, being unremarkable, call for no explanation; some, however, embodying "higher level principles", exercise "control" over the lower level processes. These need explaining; but, being boundary conditions, they cannot be derived from the lower level laws and are in that sense irreducible to them. Polanyi's anti-reductionistic argument is an audacious move for, by treating machines and living organisms together, it pushes the frontier between physical–chemical explanations and higher level ones farther in the direction of inanimate nature.

Popper (1972), defending the freedom of human reason and will, develops an argument which is in many ways parallel to Polanyi's. Popper discerns in organisms "a hierarchical system of plastic controls" in which the lower functions "are constrained and controlled by the higher ones" (p. 245). Campbell (1974b) is careful to distinguish his position from that of the anti-reductionists; yet, though he identifies himself as a reductionist, he warns us against a simplistic reductionism. He argues that a complete explanation of biological systems cannot be accomplished by physics and chemistry alone, but "will often require reference to laws at a higher level of organization as well." For Campbell the paradigm of higher level laws is "downward causation," by which he means "causation by a selective system which edits the products of direct physical causation."

A counter-argument to this line of reasoning may be built upon an optimistic extrapolation of the work of Prigogine and others on dissipative structures (Glansdorff and Prigogine [1971]). As they have shown, a mixture of chemicals, formed by pouring various solutions into a container without any additional constraints on initial and boundary conditions, may sort itself out and become a rudimentary feedback system (see the discussion of the Zhabotinsky reaction in the Appendix). By extension it is plausible to assume that one will eventually be able to understand how the complex networks of cybernetic systems that characterize life could arise out of the dissipative processes present in the pre-biotic earth; so that the extraordinary phenomenon of life will be seen to be the common consequence of a wide range of initial conditions which are themselves quite ordinary. If this reductionistic program should be successful, it would, of course, still be correct to say that a physical–chemical explanation of life is merely an explanation of today's boundary conditions by reference to yesterday's, which remain unexplained. Nevertheless, the anti-reductionistic thrust of that retort would be considerably weakened if today's boundary conditions, by their remarkable complexity, demand an explanation; and yesterday's, being unremarkable, call for none.

Popper, Polanyi, and Campbell, though their arguments differ in many respects, seem to agree that a physical–chemical explanation of life is incomplete *as a causal explanation*. 'Control', 'editing' and 'downward causation' all convey the flavor of causal factors which supplement the ordinary causation of physics and which must be adduced as additional explanatory principles if we are to give an adequate causal explanation of the phenomenon of life. But I think that attempts to criticize a physical–chemical explanation as an incomplete causal explanation are mistaken. Physics–chemistry is eminently successful at what it sets out to do – to trace the present, by means of laws of development, from the past. Prigogine's program has every prospect of a triumphant conclusion. If a physical–chemical explanation of cybernetic systems is incomplete (as I shall argue it is) it is so in some other respect – not because it fails to give an adequate causal account of things.

Physical–chemical explanation is incomplete; not because it misses some of the causes operating in cybernetic systems, but because there is more to explaining than merely tracing causes and effects. The scientific challenge posed by biological organisms is not directed at our arsenal of causal influences, at our list of forces and interactions; it is, rather, directed at our understanding of what it means to understand. We need not add a 'vital force' to the four fundamental interactions of physics but we do need to add a new kind of explanation to the causal explanations of physics.

The explanations of physics proceed by fitting the flow of events to a pattern, by showing that one thing leads to another in accordance with, say, Newton's second law or Schrödinger's equation. These laws are patterns of a special type: they relate the state of a system at one time to its state at another; they are laws of development. I claim that there is no reason to suppose that they, or their successors in the haphazard march of physical theorizing, will be fundamentally inadequate in what they set out to do. But we may discern other patterns embodied in material systems — patterns of other sorts, not involving relations of temporal succession. A trivial and scientifically irrelevant example of this other sort is the geometrical pattern embodied by the stars of *Ursa Major*. But, as we have seen in Part III, an object's being a feedback system is largely a matter of its internal arrangements and connections and its interactions with the environment; that is, of its boundary conditions. Thus, a by no means trivial example of another sort of pattern is the set of structural and causal relations we call feedback.

Let us say that to understand something completely is to be aware of all the (non-trivial) patterns it instantiates, and to explain it completely is to point out those patterns. From a complete physical theory of the Milky Way galaxy one would be able to derive the earth-centered coordinates of the stars of *Ursa Major* in the year 1975 from their positions (and the positions of the rest of the members of the galaxy) at an earlier time. But to derive their positions is not to *state* that their appearance traces the outline of a water dipper, nor can a statement of this pattern be derived from a list of their coordinates. A physical explanation cannot be faulted for failing to yield a statement about the 'shape' of *Ursa Major*, just because the shape is scientifically irrelevant. But the cybernetic patterns instantiated by living organisms and certain machines are scientifically important and statements that they are instantiated are equally underivable from the 'physical—chemical topography' of the system. Physics—chemistry allows us to deduce the topography from initial and boundary conditions; but the statement that any particular system instantiates a certain cybernetic pattern cannot be derived from the topography. If we are to produce a complete explanation of the organism or machine, that is, if we are to point out all the (non-trivial) patterns which the system embodies, we must add to the physical—chemical explanation (which shows how the system conforms to the pattern of quantum mechanics, etc.) other explanations which point out that it conforms to these other patterns. In this sense it is proper to say that a physical—chemical explanation is incomplete: not because it fails to account — in its own way — for the behavior of the system, but because it fails to point out the other (scientifically important) patterns that the system embodies. If we have said, concerning an organism or a machine, all that physics and chemistry can say, we have not said enough: our understanding will not be complete unless we are aware of the cybernetic patterns to which the system also conforms and our explanation will not be complete unless we have pointed them out.

I suppose I am advocating here a muted sort of epistemological emergentism, albeit a reductionistic emergentism. Because an observer would be able to discover the feedback pattern equally well by examining either the system itself or its physical-chemical theoretical model, we may say that the pattern is recognizable in the theoretical model and, in that sense, is explained by or is reducible to it; yet the statement that the pattern is present in the system is not deducible from the model in the way that ordinary assertions about the behavior of the system are deducible. From a theoretical model of a thermostat we may deduce that a switch will close when the temperature drops below T_1, as well as many other details of behavior. But the important insight that the system in question is a feedback system is obtained, not by deduction, but by a kind of induction, namely by pattern recognition.

One could, of course, contrive a theory from which it would be possible to deduce statements that particular systems are feedback systems. For example: 'Any system whose theoretical model conforms to the criteria (i)–(vii) of Part III of this paper is a feedback system'. If we add a minor premise of the form: 'The theoretical model of system A conforms to the seven criteria of Part III', we may conclude: 'System A is a feedback system'. But the minor premise is a statement about the theoretical model, not a statement within that theory. So a deductive theory able to yield statements about the instantiation of the feedback pattern in specific systems would be meta-theoretic with respect to their physical–chemical theoretical models.

But this result is not unprecedented. An analogous case may be found within the theoretical structure of Newtonian mechanics. One of the standard 'proofs' of elementary mechanics demonstrates that the center of mass of a system of particles moves like a single particle with a mass equal to the total mass of the particles and subject to the total external force acting on the particles. This statement about the world is obtained, via the usual rules of interpretation of classical mechanics, from the observation that the differential equation for the motion of the center of mass has the same form as the differential equation for a single particle, with the mass and force terms as described. But that observation is meta-theoretic with respect to Newtonian mechanics. Hence, the 'theorem' is not deducible from the theoretical structure of mechanics, although it is obtainable from it by another means, namely by pattern recognition.

It seems, then, that the major point made by Polanyi in his (1968) is that important explanatory insights into the nature of certain complex systems are to be found in the patterns that reside in their boundary conditions. Moreover, as we have seen, statements pointing out these patterns are not deducible in the ordinary way from the physical–chemical theoretical models of the systems. If we interpret the term 'reducible to' to mean 'deducible from', then these pattern-statements are not reducible to chemistry and physics. But on a slightly looser interpretation of 'reducible to', as meaning, for instance, 'discoverable in', the pattern-statements are reducible.

If there should be important explanatory patterns discoverable in a system itself but not in its physical model, and if that model should be otherwise adequate, then we would have a case of genuine, irreducible emergence, on anyone's understanding of the term. Although Polanyi has not offered evidence of this kind of irreducibility, he has raised the caution that the task of reducing the theories of complex organisms to the theories of inanimate nature may be less straightforward and more interesting than some reductionists have supposed.

SUMMARY AND CONCLUSIONS

My central theme has been the mechanism of goal-orientation in biology. Two accounts of functionality have been contrasted and tested against a variety of examples. The first half of this paper (Sections I–III) was an attempt to produce a reductionistic explication of the concept of feedback. After stating the problem, I reviewed recent attempts at an explication and the criticisms of these attempts, adding my criticisms to those of others. I then put forward a revised explication, claiming that it escapes the errors of previous attempts and, because it is couched in the language of ordinary physical analysis, is reductionistic in spirit. (This explication is applied, in the Appendix to this paper, to a variety of examples drawn from the physical, biological, and psychological sciences.) I concluded that, since the concept of feedback can be made to be clear, a cybernetic theory of functionality has been cleared of the suspicion of radical incoherence in its central concept; thus that it is at least a plausible option.

In Section IV I examined the rival claims of two theories of function; the one justifying ascriptions of function by appeal to the presence of selective processes, and the other doing so by appeal to the presence of cybernetic organization in the system. Against the selection theory, I argued that, although conscious selection may provide a basis for functional ascriptions, the process of natural selection is neither a necessary nor a sufficient condition of functionality. For the cybernetic theory, I argued that the presence of feedback in a complex system, one to which the concept of survival is applicable, does provide a sufficient condition of functionality.

In Section V, I argued that a functional explanation is not properly an explanation of the origin of a functioning entity, but that explanations employing the concept of function are used in other ways; for example, they are used to explain the survival of organisms when survival is rare and depends on the peculiar characteristics of the organisms which do survive, and they are used to elucidate the connections and mutual influences of the subsystems of a surviving organism or cybernetically organized machine. Finally, I claimed that functional descriptions of cybernetically organized systems are, or can be, reductionistic, because the language of the description may be explicated in the language of ordinary physical theories; but that such descriptions are not simplistically reductionistic, because they cannot be deduced from the physical model of a system, but must be 'induced' by the process of pattern recognition.

APPENDIX

In this appendix, I shall apply the explication of feedback, worked out in Part III, to various examples drawn from physics, chemistry, biology and psychology. Some turn out to be genuine cases of feedback, as defined in Part III, others do not.

A.1. *Electronic Voltage Regulator*

It seems obligatory to include at least one example from workaday electronics, where feedback is such a stock device. In order to avoid a detailed discussion of the inner workings of transistors and the like, I shall use as the heart of the circuit an operational amplifier, which can be treated as a 'black box' with simple input and output properties. The amplifier, represented by the triangle in Figure 14 accepts two input voltages, V_1 and V_2, and produces a single output voltage, V_o, whose magnitude is proportional to their difference: $V_o = K(V_2 - V_1)$; where the proportionality constant K is called the *gain* of the device. The circuit shown in the figure is an example of a Type I feedback system, with the customary latitude of choice as to which

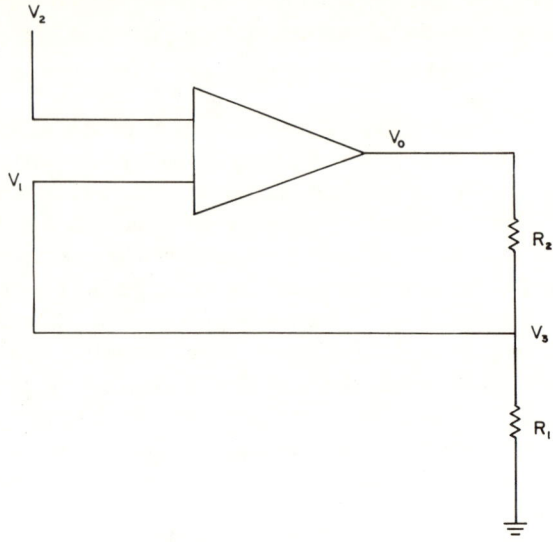

Fig. 14.

variable is to be regarded as the controlled variable. The standard block diagram of Figure 6 is applicable to this system. If we treat the voltage V_3 at the junction of the two resistors R_1 and R_2 as the h-variable, the circuit may be said to regulate V_3 with respect to perturbations of the values of R_1 or R_2. Qualitatively, what happens is this: if R_2 should increase, the current through it and R_1 (the same current passes through both) will immediately decrease, causing V_3 to drop. This causes (via the wire connecting the points labelled V_1 and V_3) a decrease in V_1, which makes V_o increase. As a result of that, the current increases, thus partially counteracting the original decrease. The three relations may be defined as follows:

φ: $h_1 =_{df} V_3 = V_o R_1/(R_1 + R_2)$

ψ: $g_1 =_{df} V_1 = V_3$ (because of the wire joining V_3 to the lower input to the amplifier)

χ: $f_1 =_{df} V_o = K(V_2 - V_1)$

It is easy to see that the relations are contingent, that χ is physically independent of ψ, and that both are physically independent of φ. Since all three relations are simple functions we can test whether the feedback is negative by forming the partial derivatives:

$$\frac{\partial \varphi}{\partial f_1} = \frac{\partial V_3}{\partial V_0} = \frac{R_1}{R_1 + R_2} \quad \text{(positive)}$$

$$\frac{\partial \psi}{\partial h_1} = \frac{\partial V_1}{\partial V_3} = 1 \quad \text{(positive)}$$

$$\frac{\partial \chi}{\partial g_1} = \frac{\partial V_0}{\partial V_1} = -K \quad \text{(negative)}$$

So their product is negative, as required. The change in V_3 produced by a change, say, in R_2 or in V_2 will be less by a factor $1 + KR_1/(R_1 + R_2)$ with the feedback connection in place than it would be if V_1 were unaffected by V_3; that is, if V_1 remained at an arbitrary constant value.

A.2. *Allosteric Enzymes*

Consider a chemical reaction one of whose steps is catalyzed by an allosteric enzyme. Reactions of this type are well known (Monod [1971], Chapter IV), but for the sake of clarity I prefer to invent a hypothetical and highly idealized example: The combination of A and B to produce C proceeds extremely slowly when left to itself, but when a very small quantity of enzyme E is added to the solution the reaction proceeds rapidly even though E is not itself consumed in the process.

The enzyme produces this effect by a sequence of steps illustrated in Figure 15. There are three active sites on the long molecule E (labelled 1, 2, and 3 in the figure) to which molecules of A, B, and C, respectively, can become attached by relatively weak bonds (Figure 15(a)). As a result of a chance encounter between an E molecule and an A molecule, A becomes temporarily attached at site 1 (Figure 15(b)). A bit later, before A has been shaken loose by thermal vibrations, a molecule of B wanders into the vicinity and becomes attached at point 2 (Figure 15(c)). The shape of E is such that these two attached molecules are held in just the right relative position to facilitate their joining together to become the product molecule, C (Figure 15(d)). This product is very weakly bonded to E, so it is soon shaken loose by the continual jostling it receives from neighboring molecules and it moves off into other regions of the solution (Figure 15(e)).

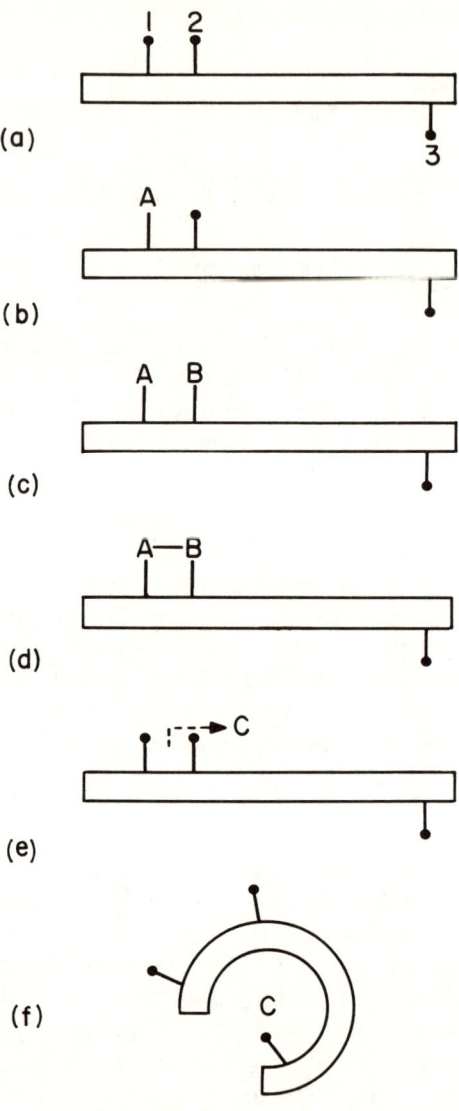

Fig. 15.

Initially there are so few molecules of C in the solution that the probability of one of them encountering an E molecule and becoming temporarily attached at site 3 is extremely small. As the reaction proceeds, however, the probability grows, so that an increasing fraction of the E molecules have C's attached. Now the enzyme molecules have the important property that, when C's are attached, they become bent out of their original shape (Figure 15(f)) so that they are no longer able to hold A and B molecules in the correct relative positions for chemical combination. This susceptibility to having the catalytic activity switched on and off is characteristic of the class of allosteric enzymes.

The unbent E molecules act as assembly devices for combining A and B to form C. Obviously the rate of production of C depends upon how many of these small 'factories' are wandering around ready for business. As more and more of the E's form temporary liaisons with C's and are thus put temporarily out of commission, the rate of production of C decreases. Eventually, if A and B are in plentiful supply, the production ceases altogether when a high enough concentration of C is achieved. Let us call this concentration c_0. Now suppose that there is a process $(C + D \rightarrow P)$ which consumes C at a rate which may vary, but is always less than a typical rate of production of C. As the concentration of C falls below c_0 the equilibrium

$$E + C \rightleftarrows E - C$$

begins to shift toward the left and molecules of unbent E become more numerous in the solution. As a result the production of C is resumed, thus compensating for the decrease caused by the process which consumes C.

Summarizing all this in more formal terms, we have this system of chemical reactions:

(1) $A + B + E \underset{k_{-1}}{\overset{k_1}{\rightleftarrows}} E - A - B \overset{k_2}{\rightarrow} C + E$

(2) $C + D \overset{k_3}{\rightarrow} P$

(3) $C + E \underset{k_{-4}}{\overset{k_4}{\rightleftarrows}} E - C$

A feedback diagram for this system can be formed as shown in Figure 16. Here it is assumed that the amount of C tied up in $E - C$ is always a negligible fraction of the total amount of C. The variable f_1 is defined as the rate of production of C by the combination of A and B, f_2 is the rate of consumption of C by its combination with D, φ stands for the relation

$$C(t) = C(0) + \int_0^t [f_1(t) + f_2(t)] \, dt,$$

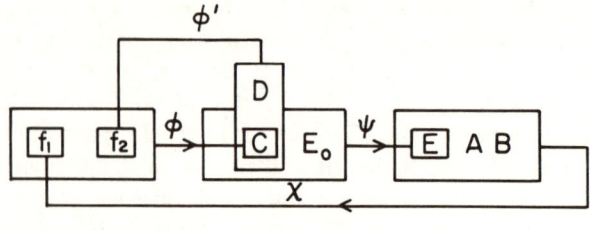

Fig. 16.

ψ stands for the equilibrium established via reaction (3):

$$E = \frac{K_4 E_0}{K_4 + C}, \text{ where } E_0 \text{ is the original amount of } E,$$

χ represents the dependence of the rate of reaction (1) on the concentrations of E, A and B, and φ' represents the dependence of the rate of reaction (2) on the concentrations of C and D.

The relations φ, ψ, χ and φ' are all synthetic and causally asymmetric. It is evident from the diagram that a reaction controlled by an allosteric enzyme with inhibitory feedback is an analog of a proportional thermostat.

A.3. *Oscillating Chemical Reactions*

Negative feedback is also involved in systems of chemical reactions which exhibit periodic variations in the concentrations of several of the ingredients. These 'oscillating' chemical reactions have attracted considerable attention from theoretical biologists both because they may be prototypes of the 'clocks' that govern biological rhythms and because they are examples of what Glansdorff and Prigogine (1971) have called "dissipative structures"; that is, they illustrate how ordered structures (ordered both temporally and spatially) can arise spontaneously from undifferentiated systems far from equilibrium. Thus they seem to provide some hints about the first stages of evolution.

One such system investigated by Zaikin and Zhabotinsky (1970) and by Winfree (1972), consists of these (simplified) chemical reactions:

(1) Bromide (Br^-) + Malonic acid (M) $\rightarrow$ Bromomalonate (MB).
(2) Bromate (BrO_3^-) + M $\rightarrow$ MB.
(3) Ferrous phenanthroline ($PhFe^{++}$) + BrO_3 $\rightarrow$ Ferric Phenanthroline ($PhFe^{+++}$) + Br^-.
(4) MB + $PhFe^{+++}$ $\rightarrow$ Br^- + CO_2.

In reaction (3), PhFe^{++} has a reddish color and PhFe^{+++} is blue. Reaction (3) proceeds — under normal circumstances — extremely rapidly, so that after a very short time all the phenanthroline molecules are in the ferric (blue) state. Even though reaction (4) steadily changes the blue ferric molecules to red ferrous ones, (3) restores them as fast as they are formed. However, reaction (3) is rarely allowed to have its way with the system because bromide, when present in the solution in amounts greater than a critical concentration, C_0, has the effect of inhibiting reaction (3), of turning it off.

With these few (and in fact greatly oversimplified) facts we can understand the 'mechanism' of this oscillator. At one state in the cycle Br$^-$ is present in excess of the critical concentration, so PhFe^{++} is not being oxidized to PhFe^{+++}. MB is also present, however, so after a short time (4) will put most of the phenanthroline into the ferrous state and the solution will have a reddish color. Then, of course, reaction (4) ceases because the solution has been depleted of one of its reactants. Reaction (1) continues to produce MB, using up the Br$^-$ as it does so. Eventually the concentration of Br$^-$ falls below the critical value and suddenly (3) turns on, converting the solution from red to blue in a fraction of a second. MB is still being produced, now primarily by (2), and since the low Br$^-$ level has allowed (3) to replenish the supply of PhFe^{+++}, reaction (4) swings into action again, producing Br$^-$ faster than (1) can consume it. Consequently, the concentration of Br$^-$ eventually exceeds the critical value and (3) is turned off once more, thus bringing the cycle back to the point at which this paragraph began.

Even in this highly idealized account the interactions among the variables are fairly complex; hence the detailed form of the feedback diagram depends strongly on what approximations one makes when representing the causal connections. I shall suggest one such choice of approximations, designed to produce as simple a diagram as possible. If we ignore the very short time required to convert PhFe^{++} to PhFe^{+++} in reaction (3) and if we assume that BrO$_3^-$ and MB are present in the solution in essentially steady-state amounts, then the relation between the concentrations of PhFe^{+++} and Br$^-$, PhFe$^{+++} = \psi$(Br$^-$), can be represented by the step-function shown in Figure 17. A more complete way of writing the relation would include the two ignorable variables thus:

$$\text{PhFe}^{+++} = \psi(\text{Br}^-, \text{BrO}_3^-, \text{MB}).$$

It is easy to show that the three variables on the right are all causally independent, so the relation is causally asymmetric.

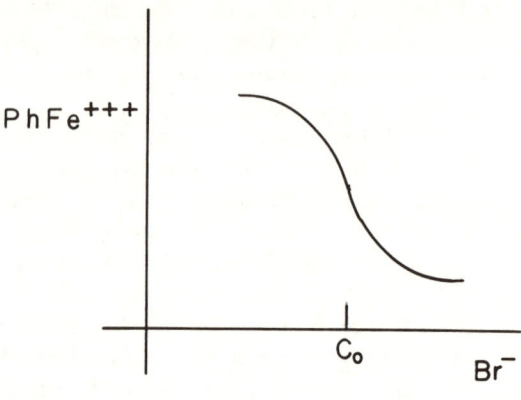

Fig. 17.

The consumption of Br⁻ is accomplished by two processes: primarily by reaction (1) but to a lesser degree by reaction (3) as well. Thus the rate of consumption of Br⁻ is affected slightly by the concentration of $PhFe^{++}$. If this minor effect is ignored, the feedback diagram takes the form shown in Figure 18(a). Here the variable f_1 is defined as the rate at which Br⁻ is detached from a carbon atom of M and f_2 is defined as the rate at which Br⁻ becomes connected to M. If the 'suppressed' variables, BrO_3^- and MB, are inserted, the diagram assumes this slightly more complicated appearance of Figure 18(b). Clearly, the diagram would acquire still greater complexity if the causal connections were represented more realistically. The three relations φ, ψ and χ are all causally asymmetric. The boundary conditions, B_φ, from which one may obtain the relation φ, consist in there being both Br⁻ and M in the solution. From these conditions it is not possible to conclude that Br⁻ is catalyzing the production of $PhFe^{+++}$ (relation ψ), nor is it possible to conclude that $PhFe^{+++}$ is being reduced by MB (relation χ), for the simple reason that B_φ does not specify that any PhFe is present. Consequently, the relations ψ and χ are physically independent of φ. Thus, the Zhabotinsky reaction is a feedback system. It is, however, a marginal case, as I pointed out in the discussion of complexity in Section IV.

A.4. *Temperature Control in Mammals*

According to the usual theory of how mammals control their body temperature, the regulated variable is the temperature of the blood. The sensing organ

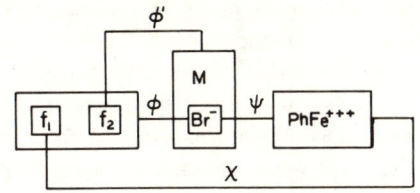

(a)

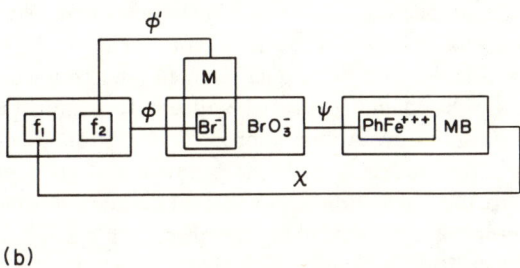

(b)

Fig. 18.

is the hypothalamus, which in turn is connected into the autonomous nervous system and can thereby cause constriction or dilation of the blood vessels, sweating, shivering, etc. The relation φ connects the temperature to rates of heat gain and loss via an integral over time; ψ relates the state of the hypothalamus to the temperature; and there are several χ relations, corresponding to the parallel and alternative mechanisms by which the nervous system affects the rates of heat transfer. Even in the absence of a mathematical model for this system it is easy to see that it conforms to the feedback pattern.

A.5. *Is Natural Selection a Feedback Process?*

Some philosophers of nature have professed to find a directed process (or at least a process with a discernible direction) immanent in nature. Now I do not wish to disparage the idea of such a process or the search for it, but it is important not to be looking in the wrong place. And I think that the process of natural selection is the wrong place. Of course, there is still a considerable amount of controversy among evolution theorists, paleobiologists and the like as to just what is involved in natural selection. There is, for example, the controversy over the question whether speciation or alleged cases of directed evolution or incidents of coincident extermination of species on a large scale can be understood purely with reference to a stochastic model — a model in which each species is treated equally and no directional tendencies are assumed (Kolata [1975]). A completely stochastic treatment of the evolutionary process is certainly at the extreme end of the spectrum of theoretical options, yet it is a tenable position and clearly is inimical to the idea of a controlling pattern operating in natural selection.

My arguments in Section IV have shown only that Wimsatt's view of natural selection as a purposive process is faulty. That is, I have shown that the separation process itself is not a goal-directed one, principally because it is not elicited by any 'problem' it might be alleged to solve. But there may be other ways to view the evolutionary process as goal-directed. Even if, as I claim, the separation process is itself the problem and not a problem-solving move directed to the alleviation of some other situation, there may still be some further response of the evolving organism to this separation mechanism which *can* be identified as a problem-solving move. Let us examine three variants of this hypothesis.

Some of those who claim to see some kind of organization in natural selection — at least something more orderly than the random model allows for — have suggested that it may be an example of a feedback process (Ruse [1973], Chapter 9 and Mackie [1974], Chapter II). Although all the observational returns are not yet in on how natural selection does operate, we can at least examine some of the current models of evolution to see whether they conform to the explication of feedback we have generated above.

At the outset, it is clear that the most obvious way in which natural selection *might* be a feedback process is ruled out by current neo-Darwinism. I refer to a kind of Lamarckian process whereby the species, or an individual member of the species, possesses a mechanism which senses a mismatch between the species and its environmental niche — a lack of adaptation — and this mechanism causes the organism to produce mutations. Preferably these mutations would be of the sort that lead to a closer adaptation to the niche (this would be a paradigmatic case of feedback), but even a mere increase in the overall rate of mutation (favorable and unfavorable mutations alike) would count as a response which tends to restore a favored state. But the evidence is that (a) the *kind* of mutation bears no relation to the adaptive 'problems' that the species faces, and (b) the *rate* of mutation is also uncorrelated with the adaptation or lack of it. Mutation goes on, in a given species, at essentially the same rate whether the organism is comfortably at home in its niche or on the brink of extinction. And the reason for this lack of correlation is that there is no causal mechanism connecting the problem of maladaptation faced by the organism to the actions it takes which lead to a solution. The 'problem-solving' actions are not elicited by the problem; hence they are not responses to it. Using the terminology of our analysis of feedback systems, there is no causal connection of the sort designated by the relation χ.

Undaunted by these facts, which they do not dispute, some theorists still see feedback in evolution. Michael Ruse (1973), for example, has constructed the following idealized but plausible example (see Williams [1966], p. 162). Consider a species of plover whose normal clutch is four eggs. The population has a stable, equilibrium size, with the birth rate equal to the total mortality rate from all causes; thus mating pairs that produce viable clutches of four chicks just manage to hold even in the equilibrium between birth and death rates. Given the environmental niche of this plover, five or more chicks place too great a strain on the nurturing capacities of the parent birds, with the result that the vitality of the chicks is reduced and there is a corresponding reduction in their survival rate. A clutch of fewer than four receives adequate attention from the parents, but since there is no significant increase in vitality of these chicks over those in a normal clutch, the total number surviving to the next generation will of course be smaller than the total from a clutch of four. Ruse invites us to consider the mechanism by which a normal clutch size of four is maintained in a population of plovers despite the fact that occasional mutations produce birds with clutch sizes greater or less than the optimum. Suppose, to be definite, that a mutant female produces a clutch of three chicks. There will be, says Ruse, a response to this event by a physically distinct part of the overall system (i.e., by the chicks which belong to normal clutches) − simply because the competitive pressure is slightly reduced by the failure of the mutant bird to produce a full complement of offspring. Thus, because of a primary variation (the deviant clutches) there is a response (increased survival of the chicks in normal clutches) which tends to restore the preferred value of the controlled variable (average clutch size), since the survival rate of normals will increase slightly.

There are three objections to this argument for the presence of feedback in natural selection. The first is that, even if Ruse has identified a feedback process here, it is a relatively minor addition to the primary process by which a normal clutch size is maintained — and the primary process does not involve feedback. Even if the survival rate of the birds in normal clutches remained constant, the mutant strains would be selected out, simply because of the inability of the smaller clutch to keep up with the mortality rate. Similarly, again with no reinforcing response from the normal birds, a mutant strain which produces clutches of five would also be reduced simply because of the impaired vitality of chicks raised in an overcrowded nest. The second objection is that the response of normal chicks to a primary variation toward *larger* clutch size is, by Ruse's model, in the wrong direction: if a mutant bird produces five chicks, the competitive pressure will increase and the survival rate of normal birds will decrease slightly. So if this is a feedback mechanism at all, it is a rather defective one, inasmuch as it responds properly only to decreases of the controlled variable and responds improperly to increases. The third objection is that, even in the case Ruse considers, the alleged response is not made by a separate, disconnectable part of the system. For *all* the chicks have survival rates which depend partly on the size of the total population (via competition for limited resources), even those that carry the genes for abnormal clutch sizes. In fact, on the plausible assumption that competition is keenest among members of the same clutch, the survival probability of the chicks in the small clutch will increase more than that of the normal birds. I conclude that the stability of the size of a clutch in this population of plovers is maintained by a far simpler process than feedback: it is merely the statistical outcome of the balance between competitive pressures and breeding rates.

Next let us examine a case of evolutionary *development*. Consider a population consisting of two phenotypes, distinguished by a trait which is governed by a single gene. There is a dominant allele, A, and a recessive allele, a. In the original environmental niche the phenotype associated with A is at a competitive disadvantage and the a phenotype has an advantage. Given enough time the A phenotype would eventually disappear, were it not for the fact that mutation from a to A occurs at a small but constant rate. As a result, a steady-state population is set up with a very small proportion possessing the A allele.

Suppose now that a change occurs in the environment such that the A phenotype acquires a competitive advantage and the a-type is at a disadvantage. Because of this change the adaptation of A becomes high, the adaptation of a becomes low and the average adaptation of the population, given the relative numbers of the two types at the time of the change, also drops. Let us see if we can fit the subsequent events into a problem-solving pattern of the feedback type, with adaptation as the controlled variable. Our strategy will be to try to divide the population into two parts: a regulated system and a separable, physically distinct part which responds in a contingent way to the problem faced by the controlled system; the response being such as to alleviate the problem. One way to partition the population is to divide it into an a-group and an A-group. The former group has a problem, namely, a low value of adaptation in the new environment; and the immediate effect of this problem is a decline in their numbers. As a result of decreased competition (among other things) the A-group now increases its numbers. So far, we have a system with a problem and another, disconnectable part, whose presence may be regarded as a contingent fact, which responds to the problem situation. But what kind of response is it? The result of the increase in size of the A-group is more trouble for the a-group. By increasing in numbers the A-group adds to the competitive pressure on the a-group, further increasing their maladaptation. So the response does not alleviate the problem – it reinforces it. Thus this analysis of the dynamics of the process also fails to conform to the feedback pattern.

I have not demonstrated that there is no other possible analysis which *might* conform to the feedback pattern; indeed, I do not know how such a proof of impossibility could be constructed. But there seems to be no other, more promising way of looking at the system. For example, if we take as our regulated system the entire population and let the regulated variable h be the *average* adaptation of the population, then as a result of the shift in numbers of the two types the value of h gradually moves from a low value immediately after the new environment becomes established to a high value as the balance of the population shifts toward the A-type. But on this analysis there is no other distinct part of the system which can be identified as the regulator, because we have included everything in the regulated system. Thus this analysis also fails to show feedback.

A.6. *Limitations on the Size of a Population*

If population growth were limited only by the food supply, one would expect wild populations to increase to the point where the death rate by starvation equals the birth rate. Thus populations in equilibrium with their food supply would normally be in a state of semi-starvation. In fact, in many species chronic starvation is not the case; let us speculate about a mechanism which might account for the relatively good health of a population which is supported by a limited food supply.

Suppose that for some species there is a hormonal mechanism which relates the size of a litter to the nutritional level of the adult animal. If this mechanism does operate in the individuals of a breeding population, there will be relation ψ between the level g of the litter-determining hormone and the average level h of nutrition in the population. But the nutrition level depends upon the amount f_1 of food and the size f_2 of the population by a relation we may designate by φ. Finally the size of the population is related, via the average size of litter to the hormone level by a relation χ. These variables and their relations are represented by the standard diagram of Figure 6. The effect of this regulatory process is that the average nutritional level (hence the average health) of the population is less sensitive to fluctuations of the food supply than it would otherwise be. There is a lag or response time of approximately one generation involved in relation χ, so regulation can occur only with respect to fluctuations whose characteristic time is longer than a generation or two.

The relation φ is causally asymmetric and the three relations φ, ψ, and χ are physically independent of each other. It is easy to see that the rest of the conditions for a feedback system are also fulfilled, so this example is a genuine case of feedback as we have defined it.

Another factor that may keep a population at a level below that at which general starvation would prevail is predation. Let us consider whether the balance between predator and prey is also a negative feedback mechanism which reduces the dependence of the prey population on external influences. Suppose that the predator population is regulated by an internal feedback process of the type just described, whereby the fertility depends upon the availability of food (in this case, the population density of the prey). Because of this assumed process, the size g of the predator population is related to the size h of the prey population by a relation ψ. This relation involves a time lag of the order of a few generations of the predator population. The size of the prey population depends, by a relation φ, on several variables, including the death rate due to predation, f_2. This variable, in turn, is related to the size of the predator population by a relation χ. Figure 6 again represents these variables and their relations. But, unlike our previous example, in this system the relation χ is not physically independent of the boundary conditions which determine the relation ψ whereby predator population depends on prey population. For both relations express aspects of the same process, namely the eating of prey by predators. The boundary conditions which determine the dependence of the predator population on the availability of prey are sufficient to determine the dependence of the death rate of the prey on the activities of the predators. To determine that the nutrition level (and thus the fertility) of the predators depends on the numbers of prey we must specify that the predators are present and that they eat the prey. But from that fact we can deduce that the death rate of the prey depends on the prevalence of predators. As a result, condition (v) is not fulfilled and this process is not an example of feedback.

A.7. *Learning by 'Variation and Selection'*

Donald T. Campbell has emphasized the analogy between natural selection and learning theory (Campbell [1960], [1974c], etc.). His analysis of learning involves three steps: (1) random variation of emitted behavior (trial solutions), (2) selective survival of certain variations, and (3) retention and duplication (via reinforcement) of the surviving (successful) variations.

Let us try to fit this analysis to the feedback pattern. Let the problem situation be characterized by variables $f_1, f_2, \ldots, f_i$. Let the trial solutions be represented by g_1. These activities will have some effect on some aspect of the problem situation, say on a variable f_1. Let h represent the problematical aspect of the problem situation, produced by the f_i and relevant to the well-being of the problem-solver. φ relates h_1 to the f_i; χ relates f_1 to g_1. The question is, can we find a relation ψ connecting g_1 to h_1? If so, the three relations required by our analysis of feedback (Figure 6) are present, and the various other criteria of feedback will be satisfied as well. Here is the point at which the analogy between theories of learning, habit formation and purposive problem solving on the one hand, and natural selection on the other, breaks down. And fruitful and suggestive though the analogy may be, the disanalogy is, I think, crucial. According to the theory of natural selection, as we have seen, the relation ψ does not exist; for this reason I have argued that it is a mistake to regard that process as instantiating feedback or even as goal-directed in some more general sense. In the theory of learning, however, there is a connection between the trial solutions and the situation they eventually correct. Campbell emphasizes that the trials are 'blind' in the sense that (1) "The variation emitted [are] independent of the environmental conditions of the occasion of their occurrence; (2) " . . . specific correct trials are no more likely to occur at any one point in a sequence of trials than another, nor than specific incorrect trials", and (3) it is not the case that "a variation subsequent to an incorrect trial is a 'correction' of the previous trial or makes use of the direction of error of the previous one" (Campbell [1960], p. 381).

Now as far as this analysis goes the analogy to natural selection is correct and Campbell would be justified in rejecting the label 'cybernetic' for the theory; however, it seems he has overlooked the fact that his "variations" or "trials" are related to the problem situation in at least two ways: (1) The rate of variation in a learning situation, unlike the rate of mutation in evolution, is correlated with the presence or absence of a problem; it seems fair to say that the organism's perception of the problem causes the trials to be 'emitted'. (2) Although the direction of the trial variation is unrelated to the detailed nature of the problem (presumably this means that the organism is not predisposed to try a correct response), yet the category of response is determined by the (perceived) nature of the problem. For example, in a situation where flight from a danger zone is the 'correct' response, the organism will try various types of locomotion (in random directions) but is unlikely to try, for example, ingestive activities, even though these are part of its repertoire of responses. For these reasons, it is appropriate to connect the trial variation variable g_1 to the problem variable h_1 by a relation ψ. Thus our analysis (so far) conforms to the feedback model.

But does it satisfy the other criteria of feedback? In particular, since there is an essential 'random' element in ψ, can the feedback be seen as negative? Again, I think the answer is yes, because of a further disanalogy with natural selection. In natural selection the organism produces all the variations that are physically possible; no constraint, other than those of the laws of chemistry, replication errors, thermal fluctuations, etc., is applied to the kind of variation that may occur; in particular, there is no auxiliary structure, built up in response to past successes and failures, that acts as an additional, contingent 'selector' on the mutation process. The situation is quite different in learning theory. As the learning process proceeds, some kind of inhibiting structure is built up which prevents the organism from making responses that are physically possible, but inappropriate. Moreover, this mechanism is contingent and its direction is related to and causally determined by the success and failure of past trials. This is the 'vicarious selection' of Campbell's analysis, and it makes all the difference between a goal-directed process and a truly blind one.

NOTES

* My thanks are due to the Department of Philosophy and the Department of Physics of Boston University for their hospitality during the early stages of the work reported here, and to Lake Forest College for providing a sabbatical leave of absence. I am especially grateful to Professor Abner Shimony of Boston University both for inspiring this project and for providing encouragement and helpful criticism along the way.

A preliminary version of a portion of this paper was read to The Boston Colloquium for the Philosophy of Science in March, 1975.

[1] See, for example, Taylor ([1950] and [1969]), Ashby (1960), and Manier (1971).

[2] The analogy to the action of a helmsman seems sufficiently close to justify the term goal-directed. I agree with Nagel's conclusion ([1977], p. 266) that Woodfield's criticism does not weaken the force of this paradigm.

[3] This word is used as Carnap ([1950], p. 3) defines it: "The task of *explication* consists in transforming a given more or less inexact concept into an exact one or, rather, in replacing the first by the second."

[4] The theoretical model which supports a counterfactual conditional statement may, for any particular statement, be very incompletely known. But the assertion implies the (at least potential) existence of the model. A particular assertion may imply only a low level generalization, but these implied regularities are the sort which, when fleshed out and systematized, form a full-blown theoretical model. In the statement 'if *B* had occurred instead of *A* then, *ceteris paribus, D* would have ensued instead of *C*', the *ceteris paribus* clause includes an implicit reference to the natural laws which govern the behavior of the system. A possible world, containing other occurrences than those found in the actual one, is implied; but not just any logically possible world. The statement assumes, not only that the same system is present, but also that the same natural laws are instantiated in the contrary-to-fact world. Altered circumstances are imagined, but not different laws of development. If this were not the case, the force of the counterfactual statement would be lost. Moreover, the statement is *not* equivalent to 'if *B* had occurred then, under the circumstances, *D might* have occurred'. A necessity is implied — not a logical, but a nomological or physical necessity.

[5] The reductionistic approach of this paper might fail if, for example, the concept of feedback turned out to involve an implicit and uneliminable appeal to the intentions of the designer of the device, or to some other element of consciousness. In that event any attempt to reduce feedback would be dependent on the success of the more ambitious program of reducing consciousness. It is my aim to discuss feedback, functionality and goal-directedness without assuming anything about the eventual success of physicalist theories of consciousness.

[6] Beckner's words are "teleological system," but he makes it clear that this is intended as a definition of feedback.

[7] In fact, so long as the relation $i = E/R$ remains valid there is no way to change i except by the causal mediation of E or R; if a varying magnetic field is employed, the circuit equation must be altered to read $i = E/R + (d\varphi/dt)/R$.

[8] To say 'logically infer' would be to restrict unduly the freedom of the physicist to invent variables and make approximations in the best traditions of his craft.

[9] The standard application of functional language in biology is to breeding populations, so that one speaks, for example, of the function of the liver in stoats. This is to be expected, for biological explanations, like those in any other science, are constructed to be as general as possible. Yet the concept of function must also be applicable to the organs of individual members of a species, and it is so applied in medical contexts; if livers in general have a function within stoats in general, then it must also be correct to assign that function to a particular liver in a particular stoat.

[10] We may, if we wish, restrict 'purposive' to situations where the action is consciously willed; but I am using the term in the looser sense which does not imply consciousness; hence a reflex action, involving a stimulus—response loop to the spinal cord or to the lower centers of the brain, but not involving conscious decision-making centers, would still count as purposive.

[11] If this analysis is correct, the term 'selection' must be regarded as tendentious and misleading, since it implies (incorrectly) that the separation itself serves some end. 'Natural separation' would, I suppose, be more acceptable, but in deference to entrenched usage I shall continue to refer to the process as selection.

[12] If there is one. See the next section.

[13] More exactly, a separation process. See footnote 11.

REFERENCES

Achinstein, P. 1977. 'Function Statements', *Philosophy of Science* 44 341–367.
Ashby, W. R. 1960. *Design for a Brain*. 2nd revised edition. New York: Wiley.
Ayala, F. J. 1970. 'Teleological Explanations in Evolutionary Biology', *Philosophy of Science* 37 1–15.
Beckner, M. 1959. *The Biological Way of Thought*. New York: Columbia University Press.
Beckner, M. 1967. 'Teleology', *s.v.* in P. Edwards (ed.): *The Encyclopedia of Philosophy*. New York: Macmillan.
Beckner, M. 1969. 'Function and Teleology', *Journal of the History of Biology* 2 151–161.
Boorse, C. 1976. 'Wright on Functions', *The Philosophical Review* 85 70–86.
Campbell, D. T. 1960. 'Blind Variation and Selective Retention in Creative Thought as in Other Knowledge Processes', *Psychological Review* 67 380–400.
Campbell, D. T. 1974a. 'Unjustified Variation and Selective Retention in Scientific Discovery', in F. J. Ayala and T. Dobzhansky (eds.): *Studies in the Philosophy of Biology*, pp. 139–161. Berkeley: University of California Press.
Campbell, D. T. 1974b. ' "Downward Causation" in Hierarchically Organized Biological Systems', in F. J. Ayala and T. Dobzhansky (eds.): *Studies in the Philosophy of Biology*, pp. 179–186. Berkeley: University of California Press.
Campbell, D. T. 1974c. 'Evolutionary Epistemology', in P. A. Schilpp (ed.): *The Philosophy of Karl Popper*, pp. 413–463. LaSalle, Illinois: Open Court.
Carnap, R. 1950. *Logical Foundations of Probability*. Chicago: University of Chicago Press.
Cummins, R. 1975. 'Functional Analysis', *The Journal of Philosophy* 72 741–765.
Edelman, G. M. 1974. 'The Problem of Molecular Recognition by a Selective System', in F. J. Ayala and T. Dobzhansky (eds.): *Studies in the Philosophy of Biology*, pp. 45–56. Berkeley: University of California Press.
Enc, B. 1979. 'Function Attributions and Functional Explanations', *Philosophy of Science* 46 343–365.
Glansdorff, P. and Prigogine, I. 1971. *Thermodynamic Theory of Structure, Stability, and Fluctuations*. New York: Wiley-Interscience.
Hempel, C. G. 1965. 'The Logic of Functional Analysis', in *Aspects of Scientific Explanation*. New York: Free Press.
Kolata, G. B. 1975. 'Paleobiology: Random Events over Geological Time', *Science* 189 625.

Mackie, J. L. 1974. *The Cement of the Universe: A Study of Causation*. Oxford: Clarendon Press.
Manier, E. 1971. 'Functionalism and the Negative Feedback Model in Biology', in R. C. Buck and R. S. Cohen (eds.): *Boston Studies in the Philosophy of Science* Vol. VIII, pp. 225–240. Dordrecht: Reidel.
Monod, J. 1971. *Chance and Necessity*. New York: Alfred A. Knopf.
Nagel, E. 1961. *The Structure of Science*. New York: Harcourt, Brace, and World.
Nagel, E. 1977. 'Teleology Revisited', *The Journal of Philosophy* 84 261–301.
Polanyi, M. 1968. 'Life's Irreducible Structure', *Science* 160 1308–1312.
Popper, K. R. 1972. 'Of Clouds and Clocks', in *Objective Knowledge*. Oxford: Clarendon Press.
Powers, W. T. 1973. *Behavior: The Control of Perception*. Chicago: Aldine.
Rosenblueth, A., Wiener, N., and Bigelow, J. 1943. 'Behavior, Purpose, and Teleology'. *Philosophy of Science* 10 18–24.
Rosenblueth, A. and Wiener, N. 1950. 'Purposeful and Non-Purposeful Behavior', *Philosophy of Science* 17 318–326.
Ruse, M. 1973. *The Philosophy of Biology*. London: Hutchinson.
Schaffner, K. 1967. 'Approaches to Reduction', *Philosophy of Science* 34 137–147.
Shimony, A. 1975. Personal communication.
Simon, H. A. and Rescher, N. 1966. 'Cause and Counterfactual', *Philosophy of Science* 33 323–340.
Taylor, R. 1950a. 'Comments on a Mechanistic Conception of Purposefulness', *Philosophy of Science* 17 310–317.
Taylor, R. 1950b. 'Purposeful and Non-Purposeful Behavior: A Rejoinder', *Philosophy of Science* 17 327–332.
Wimsatt, W. C. 1971. 'Some Problems with the Concept of Feedback', in R. C. Buck and R. S. Cohen (eds.): *Boston Studies in the Philosophy of Science* Vol. VIII, pp. 241–256. Dordrecht: Reidel.
Wimsatt, W. C. 1972. 'Teleology and the Logical Structure of Function Statements', *Studies in the History and Philosophy of Science* 3 1–80.
Wimsatt, W. C. 1974. 'Complexity and Organization', in K. Schaffner and R. S. Cohen (eds.): *Boston Studies in the Philosophy of Science* Vol. XX, pp. 67–86. Dordrecht: Reidel.
Winfree, A. T. 1972. 'Spiral Waves of Chemical Activity', *Science* 175 634–635.
Woodfield, A. 1976. *Teleology*. Cambridge: Cambridge University Press.
Wright, L. 1973. 'Functions', *Philosophical Review* 82 139–168.
Wright, L. 1976. *Teleological Explanations*. Berkeley: University of California Press.
Zaikin, A. and Zhabotinsky, A. 1970. 'Concentration Wave Propagation in Two-Dimensional Liquid Phase Self-Oscillating Systems', *Nature* 225 535–537.

Professor Roger Faber
Dept. of Physics
Lake Forest College
Sheridan and College Roads
Lake Forest, Ill. 60045
U.S.A.

PAUL K. FEYERABEND

PHILOSOPHY OF SCIENCE 2001*

One of the most important and least discussed tasks of a democracy is to protect its members from the ideologies it contains. All ideologies must be seen in perspective. One must not take them too seriously. One must read them like fairy tales which have lots of interesting things to say but which may also contain sizeable errors and wicked lies. Even when acting in accordance with a certain point of view, one's attitude should be that of an undercover agent who, in order to succeed in a strange country, adopts its beliefs and follows them in the most minute detail, but without ever being fully committed. This applies to all beliefs. It applies to the belief in the existence of a Christian god, it applies to the belief that all men are essentially equal, and it applies to science.

Now, is this not a strange and ridiculous attitude? Science, surely, was always in the forefront of the fight against authoritarianism and superstition. It is to science that we owe our increased intellectual freedom vis-a-vis religious beliefs; it is to science that we owe the liberation of mankind from ancient and rigid forms of thought. Today these forms of thought are nothing but bad dreams — and this we learned from science. Science and enlightenment are one and the same thing — even the most radical critics of society believe this. Kropotkin wants to overthrow all traditional institutions and forms of life, with the sole exception of science. Ibsen criticises the most intimate ramifications of 19th century bourgeois ideology, but he leaves science untouched. Lévi-Strauss has made us realise that Western thought is not the lonely peak of human achievement it was once believed to be, but he excludes science from his relativization of ideologies. Marx and Engels were convinced that science would aid the workers in their struggle for mental and social liberation. Are all these people deceived? Are they all mistaken about the role of science? Are they all the victims of a chimera? To these questions my answer is a firm: *yes and no*. Now, let me explain my answer.

My explanation consists of two parts; one more general, one more specific. The general explanation is simple. Any ideology that breaks the hold of a comprehensive system of thought contributes to the liberation of man. Any ideology that makes man question inherited beliefs is an aid to enlightenment.

A truth that reigns without checks and balances is a tyrant who must be overthrown and any falsehood that can aid us in the overthrow of this tyrant is to be welcomed. It follows that 17th and 18th century science was indeed an instrument of liberation and enlightenment. It does not follow that science is bound to *remain* such an instrument. There is nothing inherent in science or in any other ideology that makes it *essentially* liberating. Ideologies can deteriorate and become stupid religions. Look at Marxism. And that the science of today is very different from the science of 1650 is evident to the most superficial glance.

For example, consider the role science now plays in education. Scientific 'facts' are taught at a very early age and in the very same manner in which religious 'facts' were taught only a century ago. There is no attempt to waken the critical abilities of the pupil except *within* the confines of a certain intellectual attitude. There is no attempt to make him see this attitude itself in perspective. At the universities the situation is even worse, for indoctrination is here carried out in a much more systematic manner. Criticism is not entirely absent. Society, for example, and its institutions, are criticised most severely and often most unfairly, and this already at the elementary school level. But science is excepted from the criticism. In society at large the judgement of scientists is received with the same reverence as the judgement of bishops and cardinals was accepted not too long ago. The move towards a 'demythologization' of Christianity, for example, is largely motivated by the wish to avoid a clash between Christianity and scientific ideas. If such a clash occurs, then science is certainly right, and Christianity certainly wrong. So one wants to be on the safe side. (There are only a few courageous individuals who think differently.) Pursue this investigation further and you will see that science has now become as oppressive as the ideologies it had once to fight. Do not be misled by the fact that today hardly anyone gets killed for joining a scientific heresy. This has nothing to do with science. It has something to do with the general quality of our civilization. Heretics in science are still made to suffer from the *most severe* sanctions this relatively tolerant civilization has to offer.

However, is this description not utterly unfair? Have I not presented the matter in a very distorted light by using tendentious and emotional terminology? Must we not describe the situation in very different terms? I have said that science has become *rigid*, that it has ceased to be an instrument of *change* and *liberation* without adding that it has found the *truth*, or a large part of it. Considering this additional fact, so the objection goes, we realise that the rigidity of science is not due to human wilfullness. It lies in the nature of things. For once we have discovered the truth — what else can we do but follow it?

This reply is anything but original. It is used whenever an ideology wants to reinforce the faith of its followers. 'Truth' is such a nicely familiar word. Nobody would deny that it is commendable to speak the truth and wicked to tell lies. Nobody would deny that — and yet nobody knows what such an attitude amounts to. So it is easy to twist matters and to change allegiance to truth in one's everyday affairs into allegiance to The Truth of an ideology which is nothing but the dogmatic defence of that ideology. And it is of course *not* true that we *have* to follow The Truth. Human life is guided by many ideas. Truth is but one of them. Freedom and mental independence are others. If Truth, as conceived by ideologists, conflicts with freedom, then we have a *choice*. We may abandon freedom. But we may also abandon Truth. (Alternatively, we may adopt a more sophisticated idea of truth that no longer contradicts freedom; that was Hegel's solution.) My criticism of modern science is that it inhibits freedom of thought. If the reason is that it has found The Truth and now follows it, then I would say that there are better things than first finding, and then following such a monster.

I agree that science has produced some quite extraordinary individuals and some quite extraordinary results. There are parts of science that are as imaginative as a fairy tale and as free as a poet's dream. But these are exceptions. The bulk of science has become a rigid *business* institution that frowns on deviations and tries to maintain the status quo. This finishes the general part of my explanation.

There exist more specific arguments defending the exceptional position science has in society today. Put in a nutshell the arguments say (1) that science has finally found the correct *method* for achieving results and (2) that there are *results* to prove the excellence of the method. The arguments are mistaken — but most attempts to show this lead to a dead end. Methodology has by now become so crowded with empty sophistication that it is extremely difficult to perceive the simple errors at the basis. It is like fighting the hydra — cut off one ugly head, and eight formalizations take its place. In this situation the only answer is superficiality: when sophistication loses content then the only way of keeping in touch with reality is to be crude and superficial. This is what I intend to be.

There is a method, says part (1) of the argument. What is it? How does it work? One answer which is no longer as popular as it used to be is that science works by collecting facts and inferring theories from them. The answer is unsatisfactory as theories never *follow* from facts in the strict logical sense. To say that they may yet be *supported* by facts assumes a notion of support that (a) does not show this defect and (b) is sufficiently sophisticated to permit us to say to what extent, say, the theory of relativity is supported by the facts. No such notion exists today, nor is it likely that it will ever be found. This was realised by conventionalists and transcendental idealists who pointed out that theories *shape* and *order* facts and can therefore be retained come what may. They can be retained because the human mind either consciously or unconsciously carries out its ordering function. The trouble with these views is that they assume for the mind what they want to explain for the world, *viz.* that it works in a regular fashion. There is only one view that overcomes all these difficulties. It was invented twice in the 19th century, by Mill, in his immortal essay *On Liberty* and by some Darwinists (Boltzmann, Spencer, etc.) who extended Darwinism to the battle of ideas. This view takes the bull by the horns: theories cannot be justified and their excellence cannot be shown without reference to other theories. We may explain the *success* of a theory by reference to a more comprehensive theory (we may explain the success of Newton's theory by using the general theory of relativity); and we may explain our *preference* for it by comparing it with other theories. Such a comparison does not establish the intrinsic excellence of the theory we have chosen. As a matter of fact, the theory we have chosen may be pretty lousy. It may contain contradictions, it may conflict with well-known facts, it may be cumbersome, unclear, *ad hoc* in decisive places, and so on. But it may still be better than any other theory that is available at the time. It may be the best lousy theory there is. Nor are the standards of judgement chosen in an absolute manner. Our sophistication increases with every choice we make, and so do our standards. Standards compete just as theories compete and we choose the standards most appropriate to the historical situation in which the choice occurs. The rejected alternatives (theories, standards, etc.) are not eliminated. They serve as correctives (after all, we may have made the wrong choice) and they also explain the content of the preferred views (we understand relativity better when we understand the structure of its competitors; we know the full meaning of freedom only when we have an idea of life in a totalitarian state, of its advantages — and there are many advantages — as well as of its disadvantages). Knowledge so conceived is an *ocean of alternatives* channelled and subdivided by an ocean of standards. It forces our mind to make imaginative choices and thus makes it grow. It makes our mind capable of choosing, imagining, criticising.

Today this view is often connected with the name of Karl Popper. But there are decisive differences between Popper and Mill. To start with, Popper developed his view to solve a special problem of epistemology — he wanted to solve 'Hume's problem'. Mill, on the other hand, is interested in conditions favourable to human growth. His epistemology is the result of a certain theory of man, and not the other way around. Also, Popper, being influenced by the Vienna Circle, improves on the logical form of a theory before discussing it while Mill uses every theory in the form in which it occurs in science. Thirdly, Popper's standards eliminate competitors once and for all: theories that are either not falsifiable, or falsifiable and falsified, have no place in science. Finally, Popper's standards are fixed while Mill's standards are permitted to change with the historical situation. Of course, Popper's criteria are clear, unambiguous, precisely formulated while Mill's criteria are not. This would be an advantage if science itself were clear, unambiguous, and precisely formulated. Fortunately it is not.

To start with, no new and revolutionary scientific theory is ever formulated in a manner that permits us to say under what circumstances we must regard it as endangered. Falsifiable versions do exist, but they are hardly ever in agreement with accepted basic statements: every moderately interesting theory is falsified. Moreover, theories have formal flaws, many of them contain contradictions, ad hoc adjustments and so on. Applied resolutely, Popperian criteria would eliminate science without replacing it by anything comparable. They are useless as an aid to science.

In the past decade this has been realised by various thinkers, Kuhn and Lakatos among them. Kuhn's ideas are interesting but, alas, they are much too vague to give rise to anything but lots of hot air. If you don't believe me, look at the literature. Never before has the literature on the philosophy of science been invaded by so many creeps and incompetents. Kuhn encourages people who have no idea why a stone falls to the ground to talk with assurance about scientific method. I do not mean to say that *The Structure of Scientific Revolutions* is devoid of ideas. Quite the contrary; it is one of the most fascinating and informative books of the 20th century, but its fascination has turned out to be a double-edged sword. Nor do I have any objection to incompetence *per se* but I *do* object when incompetence is accompanied by boredom and self-righteousness. And this is exactly what happens. We do not get interesting false ideas, or interesting uninformed ideas, we get boring ideas, or words connected with no ideas at all. Secondly, whenever one tries to make Kuhn's ideas more definite one finds that they are false. Was there ever a period of normal science in the history of thought? No — and I challenge anyone to prove the contrary.

Lakatos is much more sophisticated than Kuhn. Instead of theories he considers research programmes which are sequences of theories connected by methods of modification, so-called heuristics. Each theory in the sequence may be full of faults. It may be beset by anomalies, contradictions, ambiguities. What counts is not the shape of the single theories, but the tendency exhibited by the sequence. We judge historical developments, achievements over a period of time rather than the situation at a particular time. History and methodology are combined into a single enterprise, history receives structure, methodology content. A research programme is said to progress if the sequence of its theories leads to novel predictions. It is said to degenerate if it is reduced to absorbing facts that have been discovered without its help. A decisive feature of Lakatos' methodology is that such evaluations are no longer tied to methodological rules which tell the scientist either to retain or to abandon his research programme. Scientists may stick to a degenerating programme, they may even succeed in making the programme overtake its rivals and they therefore proceed rationally whatever they are doing (provided they continue *calling* degenerating programmes degenerating and progressive programmes progressive). It is as if thieves get applauded by everyone and can continue stealing if only they say, whenever asked, 'oh yes, I am a thief': Lakatos offers *words* which *sound* like the elements of a methodology, he does not offer a methodology. There is no scientific method according to the most advanced methodology in existence today. This finishes my reply to part (1) of the specific argument.

According to part (2) science deserves a special position because it has produced *results*. This is an argument only if it can be taken for granted that nothing else has produced results. Now it is to be admitted that almost everyone who discusses the matter makes such an assumption. It must also be admitted that it is not easy to show that the assumption is false. Forms of life different from science have either disappeared, or they have degenerated to an extent that makes a fair comparison impossible. Still, the situation today is not as hopeless as it was only a decade ago. We have become acquainted with methods of medical diagnosis and therapy which are effective and which are yet based on an ideology that is radically different from the ideology of Western science (some of these methods successfully deal with circulatory disturbances not accessible to treatment by 'scientific' methods). We have discovered phenomena such as telepathy and telekinesis which are obliterated by a strictly scientific approach and which could be used to do research in a novel way (earlier thinkers such as Agrippa of Nettesheim, John Dee, and even Bacon were aware of the phenomena; today they are examined at numerous institutes all over the world, teams in the Soviet Union playing a major part). And then — is it not the case that the Church saves souls while science does no such thing? Of course, nobody now believes in the ontology that underlies this judgement. Why? Because of ideological pressures identical with those which today make us listen to science to the exclusion of everything else. It is also true that phenomena such as telekinesis and acupuncture may eventually be absorbed into the body of science and may therefore be called 'scientific'. But note that this happens only *after* a long period of resistance during which a science not yet containing the phenomena wants to get the upper hand over forms of life that contain them. And this leads to a further objection against part (2) of the specific argument. The fact that science has results counts in its favour only if these results were achieved by science *alone*, and without any outside help. A look at history shows that science hardly ever gets its results in this way. When Copernicus introduced a new view of the universe he did not only consult *scientific* predecessors, he also consulted a crazy Pythagorean such as Philolaus. He adopted his ideas and he maintained them in the face of all sound rules of scientific method. Mechanics and optics owes a lot to artisans, medicine to midwives and witches. The decision of 'scientists' to disregard all the phenomena recorded in the *Malleus Maleficarum* and in similar compilations had a strong retarding influence on the development of a useful and comprehensive theory of man. In our own day we have seen how the interference of the state can advance science: when the Chinese communists refused to be intimidated by the judgement of 'experts' and ordered traditional medicine back into universities and hospitals there was an outcry all over the world that medicine would now be ruined in China. The very opposite occurred: Chinese science advanced, and Western science learned from it. Wherever we look we see that great scientific advances are due to outside interference which is made to prevail in the face of the most basic, the most 'rational', and the most 'scientific' rules of procedure. The lesson is plain: there does not exist a single argument that could be used to support the exceptional role science today assumes in society. Science has done many things, but so have other ideologies. Science often proceeds systematically, but so do other ideologies (just consult the records of the many doctrinal debates that took place in the Church) and, besides, there are no overriding rules which are adhered to under *any* circumstances. There is no 'scientific methodology' that could be used to separate science from the rest. *Science is just one of the many ideologies that propel society (or that retard it) and it should be treated as such*. (This statement applies even to the most progressive and most dialectical sections of science.) What consequences can we draw from this result?

The most important consequence is that *there must be a formal separation between state and science* just as there is now a formal separation between state and church. Science may influence society but only to the extent to which any political or other pressure group is permitted to influence society. Scientists may be consulted on important projects but the final judgement must be left to the democratically elected consulting bodies. These bodies consist mainly of laymen. Will the laymen be able to come to a correct judgement? Most certainly, for the competence, the complications and the successes of science are vastly exaggerated. One of the most exhilarating experiences is to see how a lawyer, who is a layman, can find holes in the testimony, the highly technical testimony, of the most advanced expert and so prepare the jury, who are again laymen, for its verdict. Science is not a closed book that is understood only after years of training. It is an intellectual discipline that can be examined and criticised by anyone who is interested and that looks difficult and profound only because of a systematic campaign of obfuscation carried out by many scientists (though not by all, and not in all sciences; here the social sciences in their yearning for respectability are much greater culprits than the natural sciences). Organs of the state should never hesitate to reject the judgement of scientists when they have reason for doing so. Such rejection, and the preparation that has to be spent on the case will educate the general public, it will make it more confident of its ability to contribute to all aspects of government, and it may even lead to improvement. Considering the sizeable chauvinism of the scientific establishment we can say: the more Lysenko affairs, the better (it is not the *interference* of the state that is objectionable, but the *totalitarian* interference that kills the opponent rather than just neglecting his advice). Three cheers for the fundamentalists in California who succeeded for a time in having a dogmatic formulation of the theory of evolution removed from the textbooks and an account of Genesis inserted (but I know of course that they would become as totalitarian and chauvinistic as scientists are today when given the chance to run society all by themselves; ideologies are marvellous when used in the company of other ideologies — they become boring and doctrinaire as soon as their merits lead to the removal of their opponents). The most important change, however, will have to occur in the field of *education*.

The *purpose* of education, almost everyone believes, is to introduce the young into life and that means: into the *society* where they are born and into the *physical universe* that surrounds this society. The *method* of education consists in the teaching of some basic myth. The myth is available in various versions. More advanced versions may be taught by initiation rites which firmly implant them into the mind. Knowing the myth the adult can explain almost everything (or else he knows where more detailed information can be obtained). He is the master of Nature and of society. He understands them both and he knows how to interact with them. However, *he is not the master of the myth that guides his understanding*.

Such further mastery was aimed at, and was partly achieved by the *Presocratics*. The Presocratics not only tried to understand the *world*. They also tried to understand, and thus to become the masters of, the *means of understanding the world*. Instead of being content with a single myth they developed many and so diminished the power which a well-told story has over the minds of men. We cannot say that reducing this power was their *intention*. Their intention was the same as the intention of the older mythmakers they replaced. But the way in which they realised their intention created an atmosphere in which the power of myth was certainly reduced. The first thinkers (in the West) who did this reducing quite consciously and with a good deal of destructive glee were the *Sophists*. The achievements of these thinkers were not appreciated, and they certainly are not understood today. When teaching a myth we want to increase the chance that it will be *understood* (teaching is to remove puzzlement), *believed*, and *accepted*. Considering that the myth is supposed to guide us in our lives that seems to be the most sensible thing to do. What is overlooked is that people have always tried to change their surroundings, the ideas they possess included, and that such change has occasionally led to tremendous progress. After all, the surroundings to which we are to adapt ourselves include the effects of our actions and are therefore always in flux. Teaching must take care that our ideas remain in flux as well. Now a dogmatic method does not do any harm when a myth is counterbalanced by other myths: even the most dedicated (i.e. totalitarian) instructor in a certain version of Christianity cannot prevent his pupils from getting in touch with Buddhists, Jews, and other disreputable people. It is very different in the case of science, or of rationalism, where the field is almost completely dominated by the believers. In this case it is of paramount importance *to strengthen the minds of the young* and 'strengthening the minds of the young' means strengthening them against an easy acceptance of comprehensive views. What we need in such a case are *tough* minds who find problems everywhere, who never completely 'understand' and *not* mellow minds who acquire mental tumours at the drop of a hat. What we need is an education that makes people *contrary, countersuggestive*, without making them incapable of devoting themselves to the elaboration of a single view. How can this aim be achieved?

It can be achieved by protecting the tremendous imagination which children possess and by developing to the full the spirit of contradiction that exists in them. On the whole children are much more intelligent than their teachers. They succumb, they give up their intelligence because they are bullied, because their teachers get the better of them by emotional means. Children can learn, understand, keep separate two to three different languages (by 'children' I mean three- to five-year olds, *not* eight-year olds who were experimented upon quite recently in England and who did not come out too well; why? because they were already loused up by incompetent teaching at an earlier age). Of course, the languages must be introduced in a more interesting way than is usually done. There are marvellous writers in all languages who have told marvellous stories — let us begin our language teaching with them and not with 'mother has an umbrella' and similar inanities. Using stories we may of course also introduce scientific accounts, say, of the origin of the world and thus make the children acquainted with science as well. *But science must not be given a special position* except by saying that there are lots of people who believe in it and earn a living as a result of this belief. Later on the stories will be supplemented with 'reasons', that is, they will be *extended* in the manner customary in the tradition to which they belong. And, of course, there will also be contrary reasons. Both reasons and contrary reasons will be told by the experts in the field and so the young generation becomes acquainted with all kinds of sermons, all kind of songs, all types of wandering minstrels. It becomes acquainted with them, it becomes acquainted with their stories, it becomes acquainted with the fact that for every problem there are numerous answers, that there are people who believe in one answer only to the exclusion of everything else and that such people have a lot to say. By now everyone knows about the role of science in our society and so many will decide to become scientists. They will *become* scientists *without having been taken in by the ideology of science*, they will *be* scientists because they have made a free choice. But has not much time been wasted on unscientific subjects and will this not detract from their competence? Not at all! The progress of science, of good science (as opposed to hack-science) depends on novel ideas and on intellectual freedom: science has very often been advanced by outsiders (remember that Bohr and Einstein regarded themselves as outsiders). Will not many people make the wrong choice and finish in a dead end? Well, that depends on what you mean by a 'dead end'. Most scientists today are devoid of ideas, full of fear, intent on producing some paltry result so that they can add to the flood of empty papers that now constitutes 'scientific progress' in many areas. And, besides, what is more important? To lead a life one has chosen with open eyes, or to spend one's time in the nervous attempt to avoid what some not so intelligent people call 'dead ends'? Will not the number of scientists decrease so that in the end there will be no one to run our precious laboratories, sewer systems, and universities? I do not think so. Given the choice many people will choose science, for a science that is run by free agents looks much more attractive than the science of today which is run by slaves, slaves of institutions, slaves of 'reason'. And if there is a temporary shortage of scientists, the situation may always be remedied by incentives. Of course, scientists will not play any dominant role in the society which I imagine. They will be more than balanced by magicians, priests, astrologers. Such a situation seems quite unbearable for many people, young and old, right and left. Almost everywhere one finds the firm belief that at least *some* kind of truth has been found, that it must be preserved, and that the method of teaching I advocate and the form of society I defend will dilute it and make it finally disappear. Almost everyone has this belief; many people have reasons in addition to the belief. *But what one has to consider is that the absence of good contrary reasons is due to a historical accident; it does not lie in the nature of things*. Build up the kind of society I recommend, and the views that are now despised (without being known, to be sure) will return in such splendour that you will have to work hard to maintain science and you will perhaps be entirely unable to do so. You do not believe me? Then look at history. Scientific astronomy was firmly founded on Ptolemy and Aristotle, two of the greatest minds in the history of Western thought. Who upset their well argued, empirically adequate and precisely formulated system? Philolaus, the mad and antediluvian Pythagorean. How was it that Philolaus could stage such a comeback? Because he found an able defender: Copernicus. Who can say what will happen when Genesis finds an equally able defender? Of course, you may follow your intuitions as I am following mine. However, remember that your intuitions are the result of your 'scientific' training, and this training may mislead you. How can you find out whether you are being misled? By having a different reference system for the path you are travelling, i.e., by realising the kind of society I recommend; and with this I conclude my account of 'Philosophy of Science 2001'.

NOTE

* Written in 1974.

Prof. Paul Feyerabend
University of California, Berkeley, and
University of Zürich

JOST HALFMANN

THE DETHRONING OF THE PHILOSOPHY OF SCIENCE: IDEOLOGICAL AND TECHNICAL FUNCTIONS OF THE METASCIENCES*

1. SOCIAL INCORPORATION OF THE SCIENCES AND FUNCTIONAL CHANGE OF THE METASCIENCES

1.1. *Outline of the Argument*

In this paper I want to propose and elucidate the thesis that the metasciences have lost important elements of their functions with respect to the sciences. This evolution resulted from the social incorporation of the empirical sciences. Social incorporation of the sciences means growing subsumption of scientific research and development under industrial and political control. This process not only had, and still has, economic and political consequences for the scientific communities, it has also had considerable impact on the metasciences.

A second thesis which I want to develop in my argument is that within the metasciences — which comprise the philosophy, the historiography and the sociology of science — emphasis and functional weight have shifted from the philosophy of science towards the historiography and sociology of science. The perspective within which my theses are formulated is Marxist, although strictly speaking, the execution of some of my arguments is probably still proto-Marxist. The exposition of my theses may provoke some frowns. One might say 'Why should there be anything unclear about the function of the philosophy of science? It treats logical and methodological questions of the empirical sciences. Why should there be any reasoning about it from a sociological or even a Marxist perspective? Especially since philosophers of science in the socialist countries discuss philosophical questions in exactly the same way we do'. One might also feel that the notion of 'social incorporation of science' reminds one very much of externalism, that we have had enough of that after 20 years of debate over Kuhn and others, and that in times of 'back-to-business-as-usual', questions like mine are simply out of fashion.

Although I shall touch on questions of external and internal direction of scientific and metascientific development, and although my viewpoint certainly is 'externalist' in the sense that it is sociological, my main concern will be to contrast some features of the metascience of the late 18th century with those emerging in the 19th century and coming of age in the 20th century.

This paper is divided into four parts. I start with some general remarks about the historically changing interaction of science and technology on the one hand and the metasciences on the other hand. The second part consists of a description of — what I call — the ideological and technical functions of the metasciences. In the *third part*, I try to outline the relationship between metascience and science in early industrial capitalism; I chose Kant as the outstanding philosophical representative of that epoch. In the *last part*, I look at the same problem in developed (present) industrial capitalism.

1.2. Historical Changes in the Relation of Philosophy to Science and Technology

Before I start with the comparative argument let me briefly explain why it is justified to talk about metascience from a 'meta' standpoint, from a sociological, or even a Marxist viewpoint. The reason is that the social incorporation of science has brought about the awareness that social forces influence the course of scientific and metascientific development even in the metasciences themselves. The crisis in modern metascientific discussion, most visibly expressed in the aftermath of Kuhn's work, is a reflection of the 'industrialization' and 'politicization' of scientific work (Kuhn [1970]). It is not by accident that the main topics in modern metascientific debates are 'autonomy of science' and 'continuity versus discontinuity of scientific development'. The philosophy of science used to be the discipline that provided the sciences with logical, epistemological, and methodological guidelines for the conduct of inquiry. Philosophical metascience formulated the logic of research. I contend that this guiding function — although it has indeed existed once — has been taken away from philosophical metascience by the practical impact of technological evolution which is paralleled by an enormous increase in industrial and political organization of science. Philosophy of science has reacted to this process with a theoretical crisis. I suggest that the outcome of this crisis will be a shift in the ideological and technical functions of the metasciences, from the active guidance of scientific development to resistance against external influence, and from conceptual–ideological to reconstructive and legitimating theorizing. The explanation of the meaning of these terms and of the underlying material changes will take up most of my following argument.

To make this change plausible let me briefly touch on the evolutionary change in the relations between science, on the one side, and industry and the state on the other. The great advances which have been made in scientific inventions with the slow establishment of the capitalist mode of production — the industrial stage beginning in the late 18th century — were due to the separate institutionalization of scientific work in universities and academic societies. With that came the introduction of professionalized careers in the sciences and the establishment of allocation funds for basic research as well as applied science.

The advances in scientific research were made possible, among other things, by the separate social organization of science in universities during the 19th century (Ben-David and Zloczower [1965]), by the increasing mechanization of industrial production (Marx [1974], p. 591), and by the merging of science and technology in the late 19th century (Layton [1977], Noble [1977]). The first Technical Institutes in Germany in the middle of the 19th century were founded to enlarge scientific knowledge in physics and chemistry; but although they were institutionally separated from industrial corporations, their research goals were already defined by the technical needs of iron and steel and, later, of chemical entrepreneurs. The growing mechanization of industrial production provided the basis for the penetration of science into industry, because only a high degree of mechanization renders it possible that technology can be improved by science (Rosenberg [1976]). During most of the 19th century science was a 'gratis product' (Marx [1974]). Patronage or public financing supported research institutes. Scientific output, which, according to academic ethics, circulated freely, could only be appropriated by single inventors or industrial enterprises. In theory, scientific knowledge was available to everyone interested in it, but in practice it was utilized only by owners of the means of production. Large corporations sought the kind of scientific knowledge that could be profitably transformed into technological innovation. This, in turn, intensified the need to influence the very processes through which scientific knowledge is produced, thus leading to the 'rationalization' of the process of scientific invention and development.

The process of 'rationalization' of scientific work furthered not only a more theoretical approach to industrial problems of materials and production techniques, it also liberated science from traditional worldviews. Marx described science in this early period of industrial capitalism as 'general knowledge', not only because it contained knowledge with a higher degree of abstraction than the kind of knowledge involved in manual labor, but also because it was no longer restricted by particularistic considerations as expressed in religion and traditional political and cultural beliefs.

The final result of the process of incorporating science into industry was the foundation of industrial laboratories in the last decade of the 19th century. Edison, Bell and Baeyer changed the industrial approach to science decisively and somehow irreversibly (Bernal [1953]). In the wake of the integration of science into the industrial system the organizational forms of scientific work changed: formal organization, team work and hierarchization emerged. Scientific work, executed in industrial research laboratories, became part of the firm's organization and had to submit to calculi of time and money. Modern management claims that the research strategy of science has changed significantly as a result of the 'industrialization of science' (Merton [1971]).

The incorporation of science into industry has finally united the world of money and science. Until the middle of the 20th century universities did not profit much from the flow of money into science. It was only during the Second World War that the state discovered science (Greenberg [1969]). It was at that time that some of the scientific research projects reached a level which would surpass single corporations' capacity for research investment; direct state funding filled the gap. From then on much scientific research has been dependent on economic *and* political priorities.

1.3. *The Reflection of Scientific and Technological Change in the Philosophy of Science*

What effect does the orientation of science towards technological and economic prerogatives have for the metasciences? If the metasciences' objects are the empirical sciences, then changes in the reality that sciences seek to describe and to explain must be reflected somehow in the metasciences, too. I shall later try to specify what kind of changes in the reality that sciences deal with have taken place. Briefly, it is as follows: natural processes are no longer the objectives of science; rather technological artifacts dominate the experimental process and thus mediate the direct relationship between the scientist and 'nature'. One should speak of 'reality' rather than of 'nature' as the object of the sciences. Since scientific development depends so systematically on the course of technological innovation it makes sense to speak of an incorporation of the sciences into the capitalist mode of production, especially on the basis of their material means (e.g., experimental equipment). The development of solid state physics in the 20th century is a good example of the responsiveness of science to an urgent industrial need for knowledge of special mechanical and electrical properties of certain solid state materials (Halfmann [1984]).

I want to propose the thesis that this change in the 'reality' of the sciences, the two-tiered dependence of the sciences (upon technological development and, through technology, upon capitalist economy) has been reflected in the metasciences in two ways: 'discontinuity of scientific development' and 'autonomy and rationality of science'.

The question of continuity versus discontinuity of scientific development has become one of the more complicated problems in the philosophy of science. Kuhn, among others, has suggested that there are leaps and gaps in scientific development which cast doubt on a purely internalist interpretation of scientific development. I want to argue that the so-called discontinuity gaps in scientific development — and there are basically only two under discussion: the rise of Newtonian and of Einsteinian physics — coincide with changes in the social organization of the sciences and their relationship to reality. I want to explore this thesis later in this paper with respect to the role the metasciences play in this process. It is my view that the metasciences do play a role in the mediation and diffusion of new worldviews and working regulations of the sciences, but with historically changing weight and influence.

The mediative task of metascience becomes clearer if we consider the second topic in the post-Kuhnian debate: the autonomy of science, which is generally linked to the upholding of its rationality. The rationality of science again is seen as depending on the capacity of science to direct its own development on purely scientific grounds. Discontinuity of scientific development threatens scientific autonomy. The internalist restriction of the analytical philosopher's point of view (clearly to be demonstrated in Popper's or Lakatos' position) forces him to link the rationality of scientific inquiry to its autonomy, not seeing that its rationality, defined as the ability to promote scientific progress, may be increased and not endangered by the social incorporation of science.

The reason why I point to these two topics of recent metascientific debate is because they shed light on the technical and ideological functions metasciences have with respect to the sciences, the scientists and the public. Today the ideological function consists of defending science against external threats to the consistency of its theoretical or empirical procedures. The technical functions of metascience serve the preservation of the cognitive standards — cognitive is used in a non-psychological sense — of scientific work. This attempts to temper the domination of scientific work through innovation-oriented priorities of private enterprises.

In the following I shall try to convince the reader of a substantial change in the ideological and technical functions of metascience in the course of the historic change from the period of early industrial capitalism to the present era of fully developed capitalism.

2. TWO FUNCTIONS OF THE METASCIENCES

The historical turning-point which I want to compare with the modern stage can be seen in the theoretical relationship between Kant and Newton. The previous fundamental directive and ideological role of philosophy for science is contrasted with a more receptive and defensive function of the metasciences today.

There are two main functions the metasciences have with respect to the sciences. One function I want to call the technical function, the other the ideological function.

Historically, the technical function consisted predominantly of directive capacities. This resulted from the relatively distant relationship between scientific and natural philosophy on the one side, and the material process of production on the other. In ancient times philosophy and science were part of a differentiated system of social control and did receive impulses from the need of the ruling class for improved knowledge in matters of administration or cultural rituals. The emergence of geometry in Egypt was a result of knowledge needed for ideological control (such as the calculation of the sun's position in the context of religious practices) as well as for political–economic control (as in measuring the land for proper tax calculation). But because this knowledge did not become generally accessible and thus available for practical use in production, metaphysical reflections restrained scientific knowledge within the boundaries of the dominant world view. Since Newton, however, science left the context of reasoning which generalized everyday knowledge about dealing with physical objects, as Aristotelian natural philosophy had done. Newton operates with scientific terms which were not derived from any physical process related to the activities of craftsmen or engineers. Zilsel suggests that the notion of a general law — as fully developed by Newton — does not stem from experimental knowledge, but from legal reasoning about the emerging state in early capitalism (Zilsel [1976a], p. 96). In the realm of a materialistic analysis the physical laws which Newton developed are not to be explained by directly referring to the contemporary problems of military technology, as Hessen tried to prove in his famous essay (Hessen [1974]). Rather the opposite is true: the temporary detachment from problems of material production have made possible the historical leap in abstraction which Newton's physics represent. The higher level of abstraction in science has been one of the preconditions for its integration into the emerging capitalist mode of production. In other words, during a revolutionary period in the 18th century, production of scientific knowledge emancipated itself from the tradition of Aristotelianism and thus became prone to integration into the capitalist mode of production. Industrial capitalism was the first social formation in history which applied scientific knowledge for production purposes although this knowledge did not stem from knowledge developed in the process of material production. The way to achieve this was, to put it briefly, objectivation of nature through theoretical ideals which made it possible to transform the protoexperimental manufactural approach to nature into a systematic control of the organic and inorganic world. This is to say: only after a certain level of conceptual abstraction (as in a system of laws of nature) was reached did the concept of experiment change from trial and error based on handicraft experience to trial and error based on systematic inquiry into the material world.

The metasciences had two different guiding functions during this historical period. *First*, they developed a methodology which became adequate to put this radical change in perspective; instructions were formulated for the construction and proving of hypotheses and theories which were neither generalizations from rules of everyday knowledge nor deductions from metaphysical axioms. Thus the scientist distinguished himself from the craftman's as well as from the clergyman's approach to nature. *Second*, philosophy of science provided the sciences with normative criteria for the assessment of scientific progress and growth, dimensions which did not previously play any role. Again, it was Zilsel who demonstrated that the contents of the notion of progress (accretion, infinitiveness, public usefulness of knowledge) gained a consistent meaning in early capitalism only, and did not exist before Descartes (Zilsel [1976b]). These two directive functions can be distilled from Kant's philosophy where they are fully developed. Besides the technical or directive functions, metascience also has ideological functions.

I want to distinguish between three ideological functions. *First*, the normative function which provides science with teleological orientation is incomplete without a cosmological perspective or a general worldview. For Kant the progress of the new worldview and of bourgeois society consisted in the fact that humanity was now capable of planning its own fate, and that planned change of the world served the increase of freedom and reason. Science, as opposed to the old metaphysics, was the means by which the world could be made more rational. (This is *ideological function No. 1:* 'synthesis', the coordination of scientific worldview and social cosmology.) *Second:* the other ideological function consisted in the implicit sanctioning of the separation of mental and manual labor, of science and non-science. Kant wrote from the standpoint of 'common sense' against the old elitist ideologies; but he wrote in a way working people could not understand. He did not address himself to the common people, but to his colleagues. Kant's gesture had a double implication. It served the separation of manual and mental labor as well as that of science and non-science. Only the critical philosopher and scientist could express how the common person uses humanity's ability of cognition. (This is *ideological function No. 2:* 'demarcation'.) *Third*: the third function was geared towards the assimilation of the social incorporation of science. The empirical sciences do not have the means to reflect the changes in the social appropriation of knowledge. The metasciences theorize about it under the heading: freedom and autonomy of science. This topic was dealt with in Kant's philosophy in terms of the critique of metaphysics, the traditional version of which still opposed an unlimited research into nature and the secularization of topics previously dominated by religion. (This is *ideological function No. 3*: 'reflection'.) (See Table I and II, left-hand columns.)

TABLE I
Technical functions of the metasciences

	Early industrial capitalism	Developed industrial capitalism
Cognitive	Construction (philosophical criticism)	Reconstruction (methodological criticism)
Normative	Progress (material) (growth of reason)	Progress (formal) (growth of knowledge)
	Constructive Conceptual	Reconstructive Interpretative

TABLE II
Ideological functions of the metasciences

	Early industrial capitalism	Developed industrial capitalism
Scientific worldview and social cosmology	Synthesis	–
Mental/manual labor science/non-science	Demarcation	Institutional resistance
Social incorporation of science	Reflection	Cognitive resistance
	Conceptual	Legitimational

3. THE CONCEPTUAL ROLE OF PHILOSOPHY IN EARLY INDUSTRIAL CAPITALISM

I now want to reconstruct the technical and ideological functions of Kant's theory in relation to science. In the course of this argument I shall rely on three authors. Toulmin delineated the change which was marked by Newton's explanation of natural laws in relation to Aristotle by using the concept of "ideals of natural order" which define "criteria of rationality and plausibility" in the explanation of nature (Toulmin [1961]). Stegmüller will help to clarify the methodological guiding function of Kant's critique of pure reason with respect to Newton's theory (Stegmüller [1970]). Lang has tried to decipher the politico-ideological content of Kant's critique of pure reason (Lang [1977]).

3.1. *The Newtonian Revolution*

As I wrote earlier, I shall not try to undertake a historical reconstruction of the concepts of science from the beginning of bourgeois society to the present day. The illustrative reference to the Kant—Newton era is intended as a clarification of the change in function of metascience which has taken place in the last 200 years.

Toulmin was concerned with the change of scientific worldviews when he compared the Aristotelian to the Newtonian concept of nature, and thus arrived at an important distinction. While Aristotle's notion of science was based on the generalization of everyday knowledge, Newton's natural laws were formulated independently of any day-to-day experience. Toulmin explains the difference by comparing the Aristotelian paradigm of motion to the Newtonian. Aristotle understood motion as an activity which is executed against resistance: he imagined raftsmen towing a boat or a horse dragging a cart. Newton, however, construed the law of gravitation from the idea of 'free-falling motion'. He developed the laws of mechanics from Kepler's planet kinematics. Since Newton's physics is not deducible from the established body of knowledge in the production process, Kant's philosophical assistance was of great importance in order to elucidate the revolutionary impact of Newton's physics for the bourgeois cosmology and the capitalist mode of production. Kant provided his contemporaries with an explanation of why the Newtonian mechanics were a revolution in the scientific and social worldview of early capitalism. It can be shown that the normative content of Kant's notion of progress borrows heavily from the upcoming social ideology of the bourgeois class. At the same time, Kant tried to deploy the second guiding function of metascience, which was to outline a methodology of modern physics on the basis of a critique of reason.

Kant held an exceptional position in the relation of metascience to science, not only because he marked a turning point in the development of science and worldview, but also because he made it a topic of his philosophy in a peculiar way. He worked on a methodology of science as well as on the principles of the bourgeois idea of social progress. The *terrain* on which he unfolded these two metascientific functions was pure reason. The ideological and methodological *concepts* he wanted to criticize and integrate into one framework were rationalism and empiricism. The *instruments* with which he wished to establish the new concept of philosophy of science and metaphysics were the synthetic propositions *a priori*.

3.2. *Kant's "Philosophy of Science"*

Stegmüller tries to analyze the guiding function of Kant's philosophy by dissecting the content of the 'synthetic propositions *a priori*'. In his interpretation of what he calls Kant's "metaphysics of experience" Stegmüller contends that it was a comprehensive attempt to prove the possibility and truth of Newton's physics. In order to do this Kant had to separate himself from the rationalism of a Leibniz and from the empiricism of a Hume, respectively so that he could reconcile their opposing views and assign their concepts to the proper places in his own system.

Leibniz believed that everything worth knowing could be deduced from mathematics, which expresses the divine plan of the world; empirical knowledge was a secondary (because transitory and contingent) element in knowledge. Leibniz only admitted analytical propositions (in Kantian terms: analytical *a priori*) to his system. Hume, however, considered rationalism a comfortable habit of thinking; but reliable knowledge was only to be gained from experience and observation. In Hume's concept of science synthetic propositions (in Kant's terms: synthetic *a posteriori*) played the main role.

Kant, finally, built his philosophical program upon the compromising formula of synthetic propositions *a priori*, assuming that rationalist and empiricist procedures have to be reconciled.

Thoughts without content are empty; intuitions without concepts are blind. It is, therefore, just as necessary to make our concepts sensible, that is to add the object to them in intuition, as to make our intuitions intelligible, that is to bring them under concepts (Kant [1968], p. 93 – A51/B75).

Kant's intention was to teach metaphysics how to walk on the safe road of science by following the model of physics; he wanted to promote the division of labor between reason, intellect, and perception by establishing clear lines of demarcation. Kant began his task by clearing the philosophical terrain of old concepts, and then combined the critical and the conceptual work in his philosophy. The guiding element of Kant's philosophy consists in the attempt to promote the progress of science by liberating philosophy from the speculative use of reason. In order to obtain an acknowledged area of work in the field of knowledge philosophy of science had to disclose its relationship to science. Kant explored this relationship by explaining the character of synthetic propositions *a priori*. He not only tried to substantiate the assertion that scientific truth and progress in knowledge were only obtainable through a science of the Newtonian type. He also wanted to show that a metascientific concept can accomplish its guiding function (which is: coordination of perception, intellect, and reason) only if it further develops the critique of pure reason and makes it a positive philosophy of science.

Stegmüller asserts that Kant's attempt to prove Newton's physics as the only true science and to present a positive philosophy of science was based on four principles. According to Stegmüller, Kant developed the *'principle of epistemological existence'* in order to be sure of the empirical and rational conditions for the validity of Newton's theory. The empirical part of this principle claims that Newton's theory has been confirmed through observational propositions; the rational part implies that synthetic propositions *a priori* concerning science do exist. From both propositions the conclusion can be drawn that a theory of the Newtonian type exists and is considered valid. The second principle, '*the principle of structural reduction*', serves as a means to *a priori* exclude all possible competing physical sciences. It is supposed to ensure that only theories of the Newtonian structure are *possible*. This means that only theories with the following characteristics are acceptable: they have to agree with the axioms of Euclidean geometry; they have to contain an 'objective time relation' (which is: coexistence of spatially separate objects and an 'earlier-than' relation built on the causal structure of the universe); they have to imply universal determinism (which is: lawfulness in the universe and causal relations among all events); they have to acknowledge the substance character of matter and they have to follow the thesis that 'all physical properties are representable by constant functions'. According to this second principle, all non-Euclidean, time-discontinuous, indeterminist sciences are to be eliminated. The third principle, the '*principle of undisturbed observation*', postulates that all factors which can disturb the process of observation can be successfully eliminated. Finally, the '*principle of extended confirmation*' starts with the assumption that the proof of validity of Newton's theory for the present time applies for all future time in the same way. Kant saw no reason why a present nomological relation between two events should not be the same at any future point in time. Stegmüller summarizes the result of his analysis as follows:

By the 'principle of structural reduction' the immense class of logically possible theories is confined to the class of *a priori* possible theories, that is theories of 'Newtonian structure'. The principle of 'epistemological existence' guarantees that this class is not empty, that it contains just one true element, that this element can be found by humans, and that it provides complete knowledge of the world. The 'principle of undisturbed observation' guarantees that this true theory can be found by empirical means. Finally, the 'principle of extended confirmation' leads to the result that the present validity of this theory will be safeguarded for all future times (Stegmüller [1970], p. 56).

Stegmüller is critical of Kant's attempt. Kant's *a priori* argument may exclude non-Newtonian theories, says Stegmüller, but the class of theories of Newtonian structure will remain indefinitely large. Kant has not succeeded in proving the exclusive validity of Newton's theory, among other reasons, because *a priori* arguments are not able to discriminate between empirical theories with the same *a priori* qualities. For us, however, it is more interesting to stress the conclusion that Kant indeed represents a philosophical position in the history of philosophy of science which has successfully removed competing non-Newtonian theories from the metascientific battlefield.

In doing so, it has accomplished the social function of philosophy of science, namely to prepare the ground for the integration of the new sciences into the capitalist mode of appropriation of nature. Kant's task was critical. It cleared the ground for recognition of Newtonian physics in scientific and social communities before the former could receive recognition on its own grounds (namely, through experimental and observational proof). Kant states this role of philosophy clearly:

It is in philosophy, not in mathematics that Newton has made his biggest conquest (Durch Philosophie also, nicht durch Mathematik hat Newton die wichtigste Eroberung gemacht) (Kant [1971], p. 513.)

Newton's theory did not emerge from experimentalism, as I have already said. It was eminently theoretical because of the fact that planetary evolutions are not reproducible in laboratories, and thus Newton's theory exceeded contemporary technological knowledge. Moreover, and most importantly, the level of abstraction in his theory contrasted it with previous engineering and professional approaches to natural inquiry.

3.3. *Kant's Ideological Contribution*

Kant outlined the *a priori* arguments strongly in order to facilitate the process of accumulating empirical knowledge on a new theoretical basis in the physical sciences. The institutionalization of the sciences in the 17th and 18th centuries, contrary to the situation in the 20th century, is to be considered as a prerequisite of the institutional and cognitive separation of technology and science. Kant himself had been aware of the socio-political implications of his theory. The novelty in Newton's physics — that it established a safe way to the extension of knowledge by accretion — and the novelty in Kant's philosophy — that metaphysics could be based on rational foundations — required an explicit scientific-political and socio-political position. Kant conceived of the ideological implications of the critique of pure reason in the sense that rationalization of metaphysics meant establishing the new bourgeois worldview. Nevertheless, Kant did not exactly speak this language.

He was preoccupied with the detachment from the old cosmology. Kant still had to seek the language that the rising bourgeoisie would speak in the following centuries. Kant supported the ideological work of the revolutionary bourgeoisie. But he did it in a German way, as Marx stated: the French really make revolution, while the Germans philosophize about it (Marx [1969], p. 80).

In order to elucidate Kant's ideological efforts in the critique of pure reason, I want to refer to Lang's book. He has argued plausibly that we should consider Kant as the active conceptual ideologist of the new scientific and social world view and that the *Critique of Pure Reason* is more than a treatise on the 'philosophy of science'. The bourgeoisie was looking for ideological weapons to deal with its feudal opponents and to integrate itself as a social class. If the bourgeoisie wanted to use a new language, old meanings had to be replaced by new ones. Kant did this work in the critique of rationalism and empiricism. The *methodological* critique had its *ideological* companion. Kant made the assertion that the capacity of apprehension is such that the world can be subject to methodical alteration. Kant said: "Give me matter and I shall build a world from it" ("Gebt mir nur Materie ich will euch eine Welt daraus bauen," Kant [1910], p. 229). This postulate was to be understood as opposed to Leibniz's elitism, whose theocratic utopia offered itself as an ideological rationale of feudal sovereignty, as well as to Hume's empiricist scepticism, the conventionalist components of which did not provide a principle for the domination of nature and society. The *formal* side of bourgeois ideology was the creation of new principles of knowledge, namely the synthetic propositions *a priori* which allowed the transformation of the new worldview into scientific research practice. Therefore, Kant's eminent achievement consisted in the construction of the *formal* side of the new bourgeois ideology. This task was important even though the revolutions of the bourgeois classes in England and France had announced the irresistable rise of the new society. But to the same degree as the political revolutions preceded the economic transformation, the philosophy of science anticipated a scientific treatment of reality which became a technical commonplace only half a century later in the spreading industrial technology. The sciences of the 18th century were not a reflection of a parallel level of development in industrial production. Science was relatively more developed and, therefore, had to appear as natural philosophy fighting with the universalistic claims of theology. In this situation Kant took sides with science (and therefore with bourgeois society) and against the old metaphysics (and therefore against feudal restauration). The synthetic propositions *a priori* were battle slogans against the old metaphysics which held the view that science's task was to take cognizance of a world built according to divine plans and to find a true understanding of God's intentions towards humanity. Kant claimed that there is no premundane order. Instead, the synthetic propositions *a priori*

DETHRONING THE PHILOSOPHY OF SCIENCE

are the means to establish cognitive order in a chaotic world of matter. They are the principles of knowledge according to which reality is questioned exactly as in a court trial.

The old metaphysicists had illusions [according to Kant]: they believed the structure of consciousness to be the structure of matter. If there is proof in the new sciences that the success in dealing with matter is related to consciousness, which is that the synthetic *a priori* in mathematics and pure science provides the basis for successful dealings with matter, one can convert the errors of old metaphysics into a triumph: it must be possible to delineate an area to which the structure of consciousness is confined, the area of pure reason; this area is empty, but puts everything in order. Thus we can reconstruct the central grammatical phrase of Kant's philosophy – 'The conditions of the possibility of experience in general are the conditions of the possibility of the objects of experience' (Kant [1968], p. 194 – A158/B197) – as: the structure of consciousness is the structure of the consciousness of matter and not the structure of matter itself (as the old rationalism believed). Strong correspondence prevails between both (and not just chance, as the old empiricism claimed) (Lang [1977], p. 354–345).

Parallel to the instrumental conceptual relationship between philosophy and reality Kant introduced the political implications of the critique of pure reason. Kant battled against the classical philosophy with the means of common sense. He claimed that there are no social differences with respect to the use of the capacity for knowledge (= which is the horizon of possible practical projects). Every individual participates in the use of reason insofar he or she is part of humanity. Kant said:

Do you really require that a mode of knowledge which concerns all men should transcend the common understanding, and should only be revealed to you by philosophers? Precisely what you find fault with is the best confirmation of the correctness of the above assertions. For we have thereby revealed to us, what would not at the start have been foreseen; namely, that in matters which concern all men without distinction nature is not guilty of any partial distribution of her gifts, and that in regard to the essential ends of human nature the highest philosophy cannot advance further than is possible under the guidance which nature has bestowed even upon the most ordinary understanding (Kant [1968], p. 651–652 – A831/B859).

The gesture with which pure reason presented its political claims was determined by the new equalizing logic of the jurisdictional code. The critique of feudal metaphysics was brought forward during the trial of pure reason in which the different claims for certified knowledge are judged according to the rules of pure reason. Again, Kant developed the two main ideological implications of his philosophy, methodical alteration of the world and the postulate of equality with respect to the capacity for knowledge, according to their counterpart in Newton's natural philosophy, the principle of 'one world' and the principle of 'universal regularity'.

4. THE MODERN DEVALUATION OF METASCIENTIFIC FUNCTIONS

4.1. *The Dethroning of Philosophy*

The unity of philosophy and science lasted only until the beginning of the 19th century. It had been established because science itself contained speculations in natural philosophy and philosophy backed and directed these 'background assumptions' by critical reflection.

The unity of philosophy and science resulted from a mutuality in tasks: the resistance against competing world views such as those expressed in theology and the establishment of the modern ideology. The coalition of philosophy and science broke up as soon as the philosophical function of stabilizing the modern world view became obsolete. As soon as the new societal formation had made its way and the industrial cycle started working by its own momentum, cosmological designs which constitute a unity of natural and moral philosophy were no longer possible. Habermas interprets the loss of conceptual competence in philosophy as its devolution into philosophy of science (Habermas [1971], p. 26). This process is accomplished by the breakdown of metaphysics as a separate discipline. Marx noted in his eleventh thesis on Feuerbach the exhaustion of the ideological function of philosophy and recognized the chance to break the political power of the ruling classes with the impending loss of their ideological dominance. "The philosophers have only interpreted the world in various ways; the point is to change it" (Marx [1975], p. 423). Philosophy has been reduced, as Habermas notes, not only with respect to its constitutive role in regard to the dominating world view but also with respect to its technical guidance function in relation to science. "Philosophy had to give up its claim to be a basic science with respect to physics once it developed and founded a cosmology dependent on results of scientific research only and not by virtue of its own competence" (Habermas [1971], p. 26). The devolution of philosophy to philosophical metascience has been accompanied by a series of critiques of metaphysics. To the degree that the capitalist mode of production has extended its area of influence by developing industrial production processes and by advancing modern technology a new reality emerges for the sciences. It is determined by the modern production technology which, from that point on, dominates the level of experimental equipment in the sciences, directs the selection of scientific problems and determines the concept of matter itself. Philosophy is no longer required to compensate for the discrepancy between the low level of experimental technique and the high level of theoretical reasoning in science. As I mentioned before, it was exactly philosophy's support of science which compensated for the lack of experimental proof; this made Kant's contribution to science so important and provided Newton's theory with technical and ideological legitimacy.

Now, production technology steps into the gap which philosophy has left. With the loss of reality-constituting function, the retreat of philosophy to philosophical metascience, to reconstructive functions, starts. Philosophy of science now makes suggestions about how to interpret correctly the conduct of inquiry; it has become a logic of research. Philosophy tends to become superfluous for science because of the new unity of science and technology. This unity is established on the basis of organizational division: applied research takes place in the capitalist enterprise, basic research in the university or private research organization, metascience on the fifth floors of the educational system.

4.2. *From Steering to Reconstructing*

As I am primarily interested in the effects the developments in the relation of science and technology have on the metasciences I want to briefly discuss the changes in the technical and ideological functions of metascience that have taken place since Kant's time. The reversal of the relationship of dependency between science and metascience has been debated by the representatives of the so-called New Philosophy of Science (Kuhn, Toulmin, Lakatos, and Feyerabend) with several decades of delay with respect to the changes in reality. The result of this new perception of the history of science was that sciences do not follow the guidelines of the logic of inquiry any more. It is rather their function to reconstruct the real history of the sciences which developed at least independently from philosophy. In philosophy of science *reconstruction* substitutes for guiding (= historical change in the *technical* function of metasciences).

The *ideological* functions of metascience have changed, too. Instead of constituting the scientific and social worldview, the metasciences mediate cognitive norms with respect to scientists and defend scientific 'autonomy' against excessive claims by state and industry (see Tables I and II, right-hand columns). I want to discuss the devaluation of the technical function first.

I assert that the sciences not only depend financially on external sources, that they not only depend materially on the level of production technology. There is also an 'epistemological' dependency. Scientific knowledge does not relate to nature as such; that is a commonplace. It relates to socially mediated 'nature'.

> Scientific knowledge does not relate to nature as such, but to nature which has been transformed by socially organized labor and by technology. Scientific knowledge relates to 'reality' (in distinction to nature) (Dombrowski [1978], p. 174).

Insofar as physics is empirical, its reality is defined by experimental equipment. The latter is fully dependent on a highly developed mechanization and automatization of the production process. Mechanization is a precondition for the production of the most banal elements of experimental equipment, such as wheels, screws, pendulums, lenses, condensers, springs, etc., the precision standards of which determine the depth and range of modern experimental research. This differs greatly from Kant's scientific world view, which I want to mention parenthetically. Contrary to Kant's implicit principle of undisturbed observation, which stems from scientific research with little, or little-developed technical instrumentation, modern experimental technology has rendered this principle obsolete.

> Physical experiments are human undertakings with products of human labor; they are not simply natural processes. Those experiments, however, artificial processes with technical instruments, create the reality to which physical laws refer (Dombrowski [1978], p. 173).

It is because of the material foundation of physical reality that physics is not a natural, but a technical science. Technology is not applied natural science; but science depends materially upon technology (and through this, upon the level and logic of industrial production). It is not only, as I mentioned before, that mechanization of production allows science to merge with technology, mechanization of production also brings about axiomatization in the sciences (note that in Newton's theory we find an axiomatization of natural *philosophy*). Axiomatization also marks the completed institutional separation of science from manual labor because, again, the axiomatic elements of science are not deducible from the colloquial frame of reference in everyday life. The complete fusion of science and technology creates the dependence of scientific upon capitalist development. Capitalist economy is characterized by cyclical development and critical interruptions of the accumulation process, not by steadiness and continuity. These characteristics must penetrate into scientific development to the degree to which the sciences are incorporated into the capitalist production process. The metasciences reflect the material and theoretical changes of science in the debates on autonomy and external direction of scientific development. The debate on the discontinuity of scientific development also indicates that external influences might be responsible for leaps in the succession of theoretical paradigms. The fact that this controversy has not been settled so far shows that the new state of affairs has not quite penetrated scientific and metascientific reasoning. Many arguments by representatives of the New Philosophy of Science against the Popperian philosophy of science are centered around the question of how philosophy of science can mobilize a conceptual framework to adequately *reconstruct* the history of science. Explicitly or implicitly the idea of a directive function of the philosophy of science has been given up; there are too many proofs that real sciences did not care about the methodological guidelines that philosophy of science wanted to provide for the sciences.

It also seems to be evident that the content of the second technical function of the philosophy of science, the provision of norms of progress, has significantly changed. The idea that philosophy furthers the progress of reason — a concept that had substantial ideological and political connotations — has been replaced by a more non-specific idea of progress as growth of knowledge. Popper is proof of this assertion. Popper claims that progress in science is achieved by a critique of faulty hypotheses, the elimination of which shall bring science closer to truth; the critique is the medium of progress. The conceptual philosophical criticism of Kant still had a significant content: to set limits to the claims of theology and to conquer the terrain of reason for common sense.

Popper, however, formally clings to the ardour of the old notion of critique and truth without being able to share its ideological meaning with traditional philosophy. The factual restriction of philosophy to methodological reasoning makes the search for truth aimless, in the strictest sense of the world. Scientific development is strongest with respect to its means (= rational proof of theories), but weak when trying to set goals for scientific research. Popper states, in a speech given at the occasion of his receiving a honorary doctoral degree from the university of Frankfurt in 1979:

Science is a critical activity. We test our hypotheses critically in order to find errors, hoping to eliminate these errors and thus to approach truth.

Here, he expresses what the social change of science has imposed on metascience. It is at its best in reconstructing and examining scientific results, it can find logical or methodological flaws in scientific hypotheses, but it is no longer competent to assess the purpose of scientific development. Thus, metascience has been reduced to reconstructing the development of an externally guided science.

4.3. *From Intervention to Resistance*

Philosophy of science has also lost its *constitutive ideological* function to a large degree. Conceptual ideological tasks are replaced by those which legitimize a continuing separation of manual and mental labor under conditions of organizational reintegration and by those which stabilize science's resistance against external influences.

Historiography and sociology of science have been able to restrict philosophy of science's sole representation in the metasciences because they tend to be more able to accomplish certain socialization and demarcation functions which a socially incorporated science needs. Sociology of science advances the understanding towards the expected professionalization within science which exceeds purely cognitive capacities.

A theoretical example may serve to illustrate my thesis. The special character of scientific work, namely general and abstract mental labor, experiences external influences which are similar to those of manual labor during the process of industrialization. But the subsumption of science under profit or power calculi and the abandoning of self-determined research will not promote disinterest in the scientist towards the content of his work (as it is imposed on the blue-collar worker). The orientation of scientists towards truth has to be sustained — although there is no clear idea of the content of truth any more — and may not be simply substituted by orientations towards money or reputation. Thus, truth becomes a medium of organizational integration rather than an aspiration of the scientist which is embedded in the general cultural belief system and contributes to it in a privileged way.

Philosophy of science no longer has the task of founding or steering world views. This becomes evident in the fact that truth is no longer a problem that transcends the borderlines of science, truth has become a specific attribute to the system of science ... Truth has become a medium of the transferrability of meaning and of the reinforcement of the selectivity of the scientific system (Luhmann [1970], pp. 233–234).

For the same reason, surrogates for autonomy are established within the scientific community. Autonomy appears to be granted by the institutional separation of science from manual labor, by generalized social prestige ('scientists have more professional reputation than lawyers or physicians') and by the preservation of a predominantly collegial system of mutual recognition ('only peers are supposed to assess someone's contribution to scientific progress').

Sociology of science, therefore, discusses mainly problems of institutional preservation of scientific institutions and of organizational resistance against external claims. Historiography of science, parallel to sociology of science, treats questions of cognitive resistance: How to distinguish between science and non-science under conditions of rational discontinuity? By which means can science preserve rationality in order to guarantee scientific progress?

If we look at the legitimational and reconstructive functions of metascience from a sociological viewpoint we realize that a metascience which was merely philosophical would be insufficient. Ideological and technical integration of sciences and scientists into the modern type of scientific system cannot be accomplished with purely philosophical means because they only provide cognitive orientations. With the penetration of extra-scientific standards into scientific production of knowledge, a reformulation of the values and norms of scientific labor becomes necessary. Historiography of science offers arguments to restabilize the notion of rationality in science by adopting elements of the real history of science into metascientific reasoning. Sociology of science provides scientists with knowledge about the state of the formal organization of science and also presents practical knowledge needed for adequate socialization of scientists into the normative and institutional system of science. Sociology of science in particular takes care to balance the ambiguity between the fact that the scientific system creates products which are not commodities and the fact that science serves a societal system which is built on the production and distribution of commodities.

4. RECAPITULATION

One of the underlying ideas of my paper is that even the very abstract activity of metascience somehow reflects and expresses changes in the socio-economic environment. The process of the methodically intensified use of science by industry and state is expressed in the metasciences as the growing integration of 'real' empirical elements into metascientific theorizing. This is especially true of the historiography of science which was thus able to compete with a philosophy of science which restricts itself to logical and methodological analysis. The integration of the real history of scientific development into the historiography of science had the effect that the image of a self-guided science was undermined in the metasciences.

Those who took part in the emerging debate could not quite decide whether the rationality gap was to be looked after in the sciences or in the metasciences. But the fact of the discontinuous development of science was 'discovered'. If the philosophy of science cannot bring about continuity by its own means because it now has only reconstructing power (which Lakatos largely overestimated) and if the functional differentiation of colloquial and scientific knowledge cannot be reversed by mere proclamation (as Feyerabend's scientific–political illusion suggests), then discontinuity must indeed be caused by something stronger than scientific reasoning. Sociology of science has collected some arguments about the suggestion that discontinuity cannot be dealt with on purely cognitive grounds (Halfmann [1980]). There are social causes for the leaps and gaps in science's rationality. I have hinted at some of the causes.

But the problem of how growing social incorporation of science coincides with continuing (or increasing?) scientific discontinuity still remains. Since I assume that the social incorporation of science means growing dependence on economic and political development, I would expect increased contamination of science with the logic of uneven economic and political development. Therefore, I would conclude that a better understanding of the logic of scientific and metascientific development could be reached by concentrating on the critical structures and processes of modern capitalist society.

NOTE

* The following article is based on a lecture which I gave at the Boston Colloquium for the Philosophy of Science in October 1979. The original text has only been slightly revised, some bibliographical references have been added.

The text is little more than an exposé of a line of argument which I find challenging and fruitful. I cannot claim to have proven any of the hypotheses presented in this paper. Due to other research obligations I have not yet been able to elaborate the problem outlined in my lecture.

I want to thank Robert S. Cohen and Marx W. Wartofsky for helpful advice.

REFERENCES

Ben-David, J. and Zloczower, A. 1965. 'Universities and Academic Systems in Modern Societies', in N. Kaplan (ed.): *Science and Society*. Chicago: Rand McNally.
Bernal, J. D. 1953. *Science and Industry in the 19th Century*. London: Routledge and Paul.

Dombrowski, H. D. 1978. 'Gegenstand und Methode der exakten Wissenschaften in ihrem inneren Zusammenhang', in H. D. Dombrowski et al. (eds.): *Symposium Warenform – Denkform*. Frankfurt/New York.
Greenberg, D. S. 1969. *The Politics of American Science*. Harmondsworth: Penguin.
Habermas, J. 1971. 'Wozu noch Philosophie?' in J. Habermas: *Philosophisch-politische Profile*. Frankfurt am Main: Suhrkamp.
Halfmann, J. 1980. *Innenansichten der Wissenschaft*. Frankfurt; New York: Campus-Verlag.
Halfmann, J. 1984. *Die Entstehung der Mikroelektronik*. Frankfurt; New York.
Hessen, B. 1971. *The Social and Economic Roots of Newton's "Principia"*. New York: Fertig.
Kant, I. 1910. 'Allgemeine Naturgeschichte und Theorie des Himmels oder Versuch von der Verfassung und dem mechanischen Ursprung des ganzen Weltgebäudes nach Newtonischen Grundsätzen abgehandelt', in: *Kants Gesammelte Schriften*. Akademieausgabe, Bd. I. Berlin: Reimer.
Kant, I. 1968. *Critique of Pure Reason*. Translated by N. K. Smith. London: Macmillan. New York.
Kant, I. 1971. 'Handschriftlicher Nachlaß', in: *Kants Gesammelte Schriften*. Akademieausgabe, Bd. XXII, 2. photomechanischer Nachdruck, Berlin.
Kuhn, T. S. 1970. *The Structure of Scientific Revolutions*, 2nd edition. Chicago: University of Chicago Press.
Lang, M. 1977. *Sprachtheorie und Philosophie*. Osnabrück: Universität Osnabrück.
Layton, E. 1977. 'Conditions of Technological Development', in I. Spiegel-Rösing and D. de Solla Price (eds.): *Science, Technology, and Society*. London; Beverly Hills: Sage.
Luhmann, N. 1970. 'Selbststeuerung der Wissenschaft', in N. Luhmann: *Soziologische Aufklärung*. Cologne: Westdeutscher Verlag.
Marx, K. 1969. 'Das historische Manifest der deutschen Rechtsschule', in: *Marx–Engels-Werke*, Bd. 1. Berlin.
Marx, K. 1974. *Grundrisse der Kritik der politischen Ökonomie*. Berlin.
Marx, K. 1975. 'Concerning Feuerbach', in Karl Marx: *Early Writings*. Introduced by L. Colletti. New York: Vintage Books.
Morton, J. A. 1971. *Organizing for Innovation*. New York: McGraw-Hill.
Noble, D. 1977. *America by Design*. New York: Knopf.
Popper, K. R. 1979. 'Wissen und Nicht-Wissen', Vortrag anläßlich der Ehrendoktorwürde durch die Universität Frankfurt, in: *Frankfurter Rundschau* 19, June 1979.
Rosenberg, N. 1976. 'Karl Marx on the Economic Role of Science', in N. Rosenberg: *Perspectives on Technology*. Cambridge: Cambridge University Press.
Stegmüller, W. 1970. 'Gedanken über eine mögliche rationale Rekonstruktion von Kants Metaphysik der Erfahrung', in W. Stegmüller: *Aufsätze zu Kant und Wittgenstein*. Darmstadt: Wissenschaftliche Buchgesellschaft.
Toulmin, S. 1961. *Foresight and Understanding*. Bloomington: Indiana University Press.
Zilsel, E. 1976a. 'Die Entstehung des Begriffs des physikalischen Gesetzes', in E. Zilsel: *Die sozialen Ursprünge der neuzeitlichen Wissenschaft*. Frankfurt am Main: Suhrkamp.
Zilsel, E. 1976b. 'Die Entstehung des Begriffs des wissenschaftlichen Fortschritts', in E. Zilsel: *Die sozialen Ursprünge der neuzeitlichen Wissenschaft*. Frankfurt am Main: Suhrkamp.

Professor Jost Halfmann,
Universität Osnabrück,
Postfach 4469, Albrechtstrasse 28,
45 Osnabrück, West Germany

ALLAN JANIK

COMMENTS ON JOST HALFMANN'S 'DETHRONING THE PHILOSOPHY OF SCIENCE: IDEOLOGICAL AND TECHNICAL FUNCTIONS OF THE METASCIENCES'

> Thou shalt not sit
> With statisticians nor commit
> A social science.
> W. H. Auden
> 'Under Which Lyre' (1946)

Once upon a time there was a little girl who refused to eat her supper. In her efforts to persuade the child, her mother had recourse to the old ploy: "Think of all the starving children in India!" To which the little girl responded, "Name one!". Far from being a mere exercise in pseudo-spontaneity this story contains everything that needs to be said about Halfmann's thesis *in nuce*. The task of commenting upon Halfmann's thesis is indeed onerous precisely because it rests upon interpretations of interpretations and, like the little girl's mother's argument, suffers from lack of examples. Because this is the case, the task of the commentator becomes one of questioning the validity of the interpretations upon which Halfmann's interpretations rest. In short, the rarefied level at which his paper is written invites broadside cannon fire rather than dainty parry with a foil. Halfmann's original sub-title was 'Capitalism and the Philosophy of Science', which was accurate and useful in discussion. Thus, it is necessary to begin to ask Halfmann to clarify his thesis by explaining precisely what he means when he asserts that capitalism has so altered the function of metascience that government and industry have supplanted philosophy of science. Is this a thesis about the factors which must be given primary consideration in writing a history of a sociology of science (i.e., a research proposal) or is it also an explanation of current developments in philosophy of science, which assumes that there is a crisis of rationality in contemporary science — and not merely among those who claim to be K. R. Popper's legitimate successors — or is it both of these? Briefly, is Halfmann presenting an heuristic or a manifesto. There are as many reasons for welcoming the former as there are for being repelled by the latter.

If I understand Halfmann's thesis correctly, he is arguing that capitalism has somehow made metascience obsolete. This is an intriguing idea, not because Halfmann has presented it persuasively, but because historians and sociologists of science, at least in the United States, have paid far too little attention to the role of government and industry in the development of scientific knowledge. One indication that this is a worthy topic of investigation is the very fact that the role of finance in research does not even appear as an explicit item in the program developed to integrate philosophical, historical, sociological, and psychological perspectives on science into a systematic unity developed by Ian Mitroff and Ralph Kilman.[1] Basically, Mitroff and Kilman tried to form the chief themes currently discussed within the four metasciences into an integrated theory of science. The resulting sketch is interesting as it bears upon Halfmann's thesis for it fails to make explicit mention of the role of finance in science. I suggest that we might learn quite a lot about, say, the development of accounting where there is *prima facie* evidence that a paradigm shift has occurred in the last 20 years;[2] by taking Halfmann's thesis as an heuristic we might learn a good deal about this neglected area. Thus, the model that it suggests for research into scientific development may well provide a welcome *complement* and *corrective* to existing approaches. However, it is quite another thing to assert that Halfmann's approach has exclusive validity. The Marxist approach to problems in intellectual history and sociology generally has the refreshing effect of redirecting our attention from axiomatic and rational reconstruction models of understanding our intellectual enterprises to the concrete details of their day-to-day workings. However, it is a notorious fact that Marxist approaches to anything-you-like are prone to confusing interesting proposals with the results of investigation. Halfmann's program would seem to take some of these over from contemporary Marxism as well as having its own specific weak points and obscurities. To clarify such issues it will be necessary to ask him the following questions:

(1) In the wake of certain objection to the notion of 'advanced' capitalism, how can the use of such a category be justified?

(2) In the light of divergent uses of the term 'science' on both sides of the English Channel, how ought we to understand that term as it figures in his exposition?

(3) In the light of the distinctions Halfmann has drawn from the Positivist tradition such as science/metascience, internal/external, is there not a residual Positivism present which makes his position possible but which may also come into conflict with his Marxist commitments and which, in fact prevents him from appreciating the *philosophical* criticisms of Positivism. If I may put my point slightly differently, it is not at all clear to me that Kant, who was a scientist of sorts, would be happy to hear Halfmann term the First Critique 'metascience'.

(4) Finally, the crucial question remains: does not the rarefied level of

discussion obscure the all-important questions about exactly *when* and *how* capitalism has produced this alleged transformation of the function of metascience?

More briefly, Halfmann's assertions may follow from the general viewpoint of critical theory or some such speculative Marxism and even be a valuable contribution to that theory, but my question is why those of us who are not of that persuasion should be convinced by the considerations he has mentioned in the absence of detailed documentation of his general assumptions. Let's look at some of these.

To begin with there are assumptions in Halfmann's argument about the rationality, autonomy and continuity of science which are Kuhn-like, if not explicitly Kuhnian, and explicitly ignore certain objections to the sort of theorizing Kuhn and Halfmann indulge in. Confronted with these assumptions it is necessary to remind Halfmann of the reaction of one distinguished historian of science, L. Pearce Williams, to the Kuhn thesis. "The history of science", Willaims wrote, "simply cannot bear such a load [as to support Kuhn's thesis] at this time."[3] It is possible to read a purist, churlish or even a paranoid sentiment into Williams' remark but I think that it is necessary to remind ourselves from time to time that he just *might* be correct when he criticizes Kuhn for engaging in sweeping discussions which are grounded upon interpretations of hundreds of years of scientific, let alone social, development. We would probably do better to think of the history of science, which is, after all more than merely the history of physics, less as capable of yielding macro-historical interpretations than as being composed of a plethora of micro-histories of differing scope, some of which are better known to us than others and whose exact interrelationships remain to be determined.[4] Thus, it seems clear that the burden of proof that history can indeed sustain such large-scale interpretations must fall on the Kuhns and the Halfmanns who must provide a nuanced account of how and when the developments they allege to have transpired actually occurred. The pressing question for the Halfmann thesis, then, is that of establishing clear-cut examples of the theoretical implications of the complex practical interaction between the sciences, government and industry. As it stands, Halfmann's abstract discussion of these topics is seriously flawed in two respects, the first concerns his Marxist notions about something called 'advanced' capitalist development, the second has to do with his understanding of the term 'science'.

I think it is fair to question Halfmann about whether or not his Marxism commits him to the thesis that 'advanced' capitalist development is *everywhere the same*. Should we expect to find identical or merely analogous relations between science and industry in, say, Sweden, Japan, the United Kingdom and the United States? Do we *actually* find similarities? If so, what is their precise nature? It is clear enough that we shall find similarities if we operate at a sufficiently general level, as Halfmann seems to do; what is not clear is the question of whether the rarefied level of generality will trivialize our claims. It would seem, on the face of it, that Sweden with its homogeneous, small-scale Socialist state would differ from larger more pluralistic ones. Moreover, it would seem that the relationship between science and government would be different in a state like the United Kingdom where some research projects cannot be discussed legally in public due to the Official Secrets Act. Again, the peculiarities of the abrupt industrialization of Japan would seem to indicate that its version of capitalism will be different from those countries where it developed gradually. Finally, what does the fact that we have no direct counterpart in the United States to German academic '*verbeamtete Wissenschaftler*' do for Halfmann's thesis? In general, considerations such as these serve to remind us that we ignore at our peril Dahrendorf's arguments against predicating 'industrial society' univocally without examining the specific details of industrial development which appear to be unique to each society.[5] This is less because Dahrendorf has an iron-clad case but because he is sufficiently persuasive to require anyone who would appeal to such 'uniform social structures' to support that claim with unambiguous empirical data. I am personally inclined to accept Halfmann's thesis as it applies to his native Germany where all higher education, as Dahrendorf himself has pointed out, is much more directly under government control than it is in the United States. In short, no where more so than in the social sciences must we be continually reminding ourselves of William James's dictum, "Every difference to be a difference must make a difference".

What I find worthy of reflection in Dahrendorf, incidentally, is not incompatible with a certain understanding of Marx's own analysis of capital. Shlomo Avineri has argued with some cogency that for Marx 'capital' does not describe a single socio-economic reality but a potentiality within society, i.e., that capital ought to be understood as a sort of 'ideal type' abstracted from concretely existing societies.[6] The failure of the predictions that Marx generated from that abstraction itself is *prima facie* grounds for taking a circumspect attitude to the assumptions about the relationship between theory and practice which follow from it without a thorough analysis of the situation we confront in 1979. This is not only compatible with Marx's personal practice of relentless self-criticism but also warranted by the fact that the 'materialist' perspective from which Marx viewed social science has to a great extent been absorbed by modern social science. If there is any truth at all in my last assertion, manifests and confrontations with 'bourgeois' social science can only prove to be counter-productive and reactionary, for the distinction may be contrived in the first place. Thus, if there is any truth in the view I have taken, the task of the Marxist social scientist is to demonstrate the radical superiority of his methods in the form of case studies which are *clearly superior* to those of his alleged 'bourgeois' counterpart.

My concern for the meaning which Halfmann attaches to the term 'science' stems from different considerations but curiously converges with the other points concerning which I find his thesis wanting. No end of confusion has arisen from the fact that 'science' means quite different things in the Anglo-Saxon world and on the Continent. The differing ways of construing science is always a potential stumbling block whenever Anglo-Saxon and Continental thinkers sit down to discussion. The German *Wissenschaft* is normally the result of any sort of *Forschung* (research), whether it be in the natural sciences, the social sciences, the humanities or even in theology; whereas in English and American usage it almost always designates physical science. Among Anglo-Saxon philosophers science has been so identified with physics that bright undergraduate students of, say, physical anthropology or geology are unable to identify the concepts and procedures they have been learning with, say, the notion of science that we find in Reichenbach, Carnap or the early Popper (which is, of course, the most telling critique of that view of science, i.e., its irrelevance for scientific praxis). "That may be true of physics," they typically say upon being introduced to the logical positivist notion of science, "but its not true of *science*."[7] Thus, it is necessary to ask Halfmann for a factual clarification about his understanding of science, i.e., whether or not he would include the natural and the human sciences under a single rubric, which is to say, whether or not he would distinguish the study of mute objects from that of those that can talk back to us, as Tom McCarthy once formulated the famous Diltheyan distinction. Halfmann's response to this question will put us in a better position to evaluate his allegation that science in the latter part of the twentieth century studies not Nature, but "Nature mediated by technology." Further, we shall want to approach the distinction between science and metascience differently, if we restrict ourselves to the natural sciences. My factual question may appear pedantic but beneath my pedantry lurks a dilemma upon which Halfmann seems to have impaled himself: he seems to undermine his own position regardless of his answer to my question. Either he reduces science to the physical sciences with the

Positivists, as he seems to do when he speaks of the period of unity between philosophy and science extending from Kant into the early nineteenth century, on which account his own sociology has no claim to be a science — or he admits the *Nature/Geisteswissenschaften* distinction and with it his whole story about Kant is part of a much larger whole which he has declined to give us. From this question it is necessary to turn to that of whether the residual positivism implicit in Halfmann's central distinctions, science/metascience, internal/external, in fact fore-ordains his conclusions.

I have mentioned that the classical neo-Positivist understanding of science was basically a reified view of physics, so its metasciences were fundamentally philosophy of physics and history of physics, or to be absolutely precise, one view of the history of physics. Now, it is important to notice that the crucial distinctions Halfmann's argument depends upon are drawn from this Positivist scenario. The picture these terms convey remains that of the philosophical policeman protecting scientific development from the encroachments of metaphysics and religion. Halfmann's assertion that capitalism has made this sort of relationship between science and philosophy unnecessary fails to take into account that this position was always philosophically unsound, if for no other reason than that it fatally confused descriptive and prescriptive accounts of science. In short, it seems that Halfmann's reification of metascience makes it altogether too easy for him to pass over to his 'external' perspective without paying any attention to those philosophical criticisms of positivist metascience which, I would maintain, have in actuality led to a radical transformation of the program of metascience into something resembling, but not necessarily identical with, the historical and sociological inquiries which he allges have replaced traditional metascience. I think we can see this more clearly if we examine first his allegations concerning the relationship between Newton and Kant and a sketch of an 'internal' account of the structural and functional transformation of metascience at the hands of philosophers in the last forty years.

Halfmann insists that there was a point in time when the development of physics — he says 'science' — was such that it required 'steering' from philosophy and that its position in 'civil society', as it were, required legitimation through the establishment of a new civil and secular court of appeals to replace the discredited feudal onto-theological conception of rationality. Kant's Critical Philosophy is alleged to have performed each of these tasks. Because both of these functions were part of a single philosophical program, Halfmann seems to think that they are part of one conceptual task. I am not so sure they are. He thinks that what was done for physics in the name of science generally was actually what it claimed to be. I think, on the contrary, that he fails to see that Kant's Copernican Revolution in philosophy was less a matter of *one thing*, 'science', coming of age than it was a matter of one science, physics, attaining full conceptual development, i.e., becoming a unified or 'compact' science, and being established as a well-defined sphere of investigation. I suggest that Halfmann fails to recognize that the development of biology, for example, required both conceptual clarification and legitimation, but that in the case of biology these two functions did not find a single genius like Kant to weave these two issues into a single fabric. Thus, the conceptual development of biological science extended from Darwin's articulation of the notion of natural selection in the *Origin of Species* in 1859 to Wilhelm Johanssen's wedding of natural selection to a concept of the gene in the first decade of this century. However, the controversy about the 'civil' status of biology lingered on from the Oxford Debates of 1860 through the Scopes monkey trial in the twenties and after.[8] Even today the teaching of evolution in certain American high schools is problematic because it is offensive to local religious beliefs. Thus, publishers of high-school level biology texts are disinclined to publish books with chapters on evolution. Are we to take this as constituting a religious limitation on American capitalism? My point, then, is that conceptual development and legitimation are hardly *one* project and, consequently, that Halfmann's discussion of Kant's relationship with Newton is as distorting as it is illuminating. This is one of the considerations which leads me to question the appropriateness of Halfmann's science/metascience distinction and why I would be happier were he to formulate his position in terms of the relationship between philosophy and science generally rather than to continue to employ a distinction which has become increasingly irrelevant to methodological discussion. Perhaps the sources of my misgivings will become clearer if I provide the caricature of those developments in the philosophy of science during the last fifty years to which I am most attached.

Historically, metascience in the twentieth century has referred to a formalistically oriented version of empiricism, a theory which stipulated the formal requirements for what could and what could not legitimately be called science. The latter was fundamentally a matter of forging observation statements into theories via links provided by formal logic. Through a process of axiomatization it would then be possible to lay bare the 'logical structure' of the world. Briefly, Popper undermined the logical positivist's faith in verification and Quine undermined the analytic/synthetic distinction that made a logically grounded empiricism possible; whereas Wittgenstein and Collingwood, each in his own way, from radically different perspectives, undermined the claims of monolithic Positivist scientism to 'value-free' objectivity, on the one hand, and articulated a notion of philosophy as a self-reflective *practice* of examining precisely how the very fact of theorizing places limitations upon the theorizing intellect. The result was to transform philosophy of science from logical analysis into conceptual history.[9] However, in so doing, they did not eliminate the need for philosophical analysis of science but radically transformed its character. They taught us to philosophize by looking at what we actually *do* when we use concepts to solve problems rather than to theorize. Thus, philosophy of science was transformed from axiomatization, the search for 'covering laws' and 'rational reconstructions' to an analysis of the order implicit in our concrete investigative procedures. It would be less accurate to assert that the philosophy of science abandoned logic than it would be to assert that its primary focus shifted from formal logic to practical reasoning, i.e., to the *informal logics* actually employed in various processes of inquiry. We became aware of the role of questions in laying down criteria for answers to those questions. In short, logic became pluralized and context-dependent.[10] If this sketch bears any relation to reality, metascience is still very much with us and has not been transformed into traditional sociology and history. It is still very much a *philosophical* analysis which now admits heavy reliance upon sociology and history as essential to the discovery of the *logics of the sciences*. In short, I am claiming that the relationship between philosophy and the sciences has become *dialectical* in the last forty years. An essential element in this dialectic is the *contingent* nature of what is and what is not part of an inquiry into the logics of the sciences. Before we actually investigate the issue it is just as plausible that government and industry will have a major role in the development of science than that they will not. It is not necessary to be a Marxist to say this and, I fear, it is virtually impossible for a 'Marxist' of Halfmann's type to grasp it.

I want to close with the following consideration, which, I think, will emphasize at least one point upon which Halfmann and I concur: the need for an awareness of the complexities of just what is involved in being scientific is pressed upon us more urgently today than perhaps ever in modern democratic societies. One of the characteristics of such a society is that its members decide how public funds are allocated. This also includes funds for research. Citizens in a republic have the right and the duty to disburse public monies responsibly. To that end it seems that we have a long way to go before we manage to raise public interest in the role of capitalism in science to the level of its awareness of the role of capitalism, say, in professional sports.

NOTES

[1] Ian Mitroff and Ralph Kilman, 'Systemic Knowledge: Toward an Integrated Theory of Science', *Theory and Society* 4 (1977), 103–129. Note that I am *not* claiming that we know nothing whatsoever about the ways in which research is financed but that we do not have a general theory or 'bold conjecture' that would allow us to make reasonably sound assertions about *the* finance of science as it relates to the development of scientific ideas.

[2] Conversations with Professor Michael Di Angelis of the Department of Accounting, La Salle College, Philadelphia, 1971–1973. Briefly, the subject matter, the role of the accountant and the teaching of the subject have changed radically. Whereas accounting was once a set of glorified bookkeeping techniques, today it is a matter of employing alternative *systems* of accounting. This requires that the accountant be a manager rather than merely another white-collar worker. Both factors together require that an accountant receive a different education from the one he would have received 20 or 25 years ago.

[3] L. Pearce Williams, 'Normal Science, Scientific Revolutions and the History of Science', in *Criticism and the Growth of Knowledge*, Imre Lakatos and Alan Musgrave (eds.) (Cambridge, England: Cambridge University Press, 1970), pp. 49–50.

[4] Owen Flanagan of Wellesley College suggests that the only way to do justice to developments in psychology which coexist within the broader context of disputes about the fundamental nature of the science is to construe psychology somewhat in the manner of economics, where we are reasonably certain that we have made progress at the micro-level despite its lack at the macro-level. Personal communication.

[5] Ralf Dahrendorf, *Society and Democracy in Germany* (Garden City: Doubleday, 1967), pp. 31–45.

[6] Shlomo Avineri, *The Social and Political Thought of Karl Marx* (Cambridge, England: Cambridge University Press, 1968), pp. 158–159.

[7] This in fact occurred in my philosophy of science class at Wellesley College in spring 1978.

[8] These stories are grippingly recounted in Loren Eiseley, *Darwin's Century: Evolution and the Man Who Discovered It* (Garden City: Doubleday, 1961).

[9] See Stephen Toulmin, 'From Logical Analysis to Conceptual History', in: *The Legacy of Logical Positivism*, Peter Achinstein and Stephen Barker (eds.) (Baltimore: Johns Hopkins, 1969).

[10] This concept of logic was first sketched in Stephen Toulmin, *The Uses of Argument* (Cambridge, England: Cambridge University Press, 1958). A fuller account of this view of logic is found in Stephen Toulmin, Richard Rieke and Allan Janik, *An Introduction to Reasoning* (New York: Macmillan, 1979).

Prof. Allan Janik
Institut für Philosophy
Karl-Franzens Universität Graz
A 8010 Graz
Austria

HYMAN HARTMAN

PHILOSOPHY OF SCIENCE AND THE ORIGIN OF LIFE

MYTHS OF BIRTH AND DEATH

Birth (Development)

Birth and death are the two certainties of a human life. Birth is due to the fertilization of an egg which then develops into a human being. When mankind was young the embryological paradigm was used to explain the birth of the universe. "In the beginning time created the silver egg of the cosmos. Out of this egg burst Phanes-Dionysus. For the Orphics he was the first god to appear, the first born, whence he early became known as Protogonos. He was bisexual and bore within him the seeds of all gods and men." [1]* It required a symmetry breaking event to convert this god into male and female. It was only a later discovery which implied the necessity for a coupling between a man and a woman to fertilize the egg. Thus a whole new set of cosmogenies were invented involving world parents. As male chauvinism became dominant, the world mother turned into matter; until the Judeo-Christian tradition allowed god the father to create the universe out of nothing.

For man, trapped in the middle of the journey from birth to death, it was important to escape from history to his origins. He thus invented myth. As M. Eliade has observed, "that myth is always related to a 'creation', it tells how something came into existence, or how a pattern of behaviour, an institution, a manner of working were established; this is why myths constitute the paradigms for all significant human acts." [2]

We, in this god-forsaken century, have invented a cosmological myth. It may rejuvenate us.

In the beginning the cosmic egg hatched. It took a universe to create the elementary particles: electrons, protons, etc. Modern quantum chromodynamics sees these creations as a series of symmetry breaking events. As the universe develops galaxies, gravitationally interacting clouds of hydrogen and helium gas separate out. It takes a galaxy to form a star; a star to bake the hydrogen and helium nuclei into the nuclei of the periodic table; a solar system to create an earth; an earth to form a biosphere; and, finally, a biosphere to create man.

---
* References to the main text of this paper appear on pp. 202–203.

What is remarkable about this developmental sequence is that it goes from whole to part.

> The physical world is thus seen as a macroscopic totality encapsulating within it microscopic totalities all constituted on similar principles of unified order. It is a complex system to which the constituent elements are integral and mutually formative. In light of this conception, any talk of atomic facts is wholly incongruous and the sort of logic based on mutually independent propositions is obviously inappropriate. [3]

In other words the world is an organism and the need is for an organic logic. It would thus be useful to trace the present-day thought about embryogenesis to its source in Greek philosophy.

The study of morphogenesis is as old as the history of man. Many morphogenetic concepts were developed in the cosmogenies of the ancient world. With Aristotle, the simultaneous founder of biology and embryology, we find the first detailed scientific observations of morphogenetic phenomena. Aristotle formulated many ideas in embryology; however, there is one which has had perhaps the longest and most fruitful history. He put it like this:

> For it is *not* a fact, when an animal is formed, at that same moment a human being or a horse or any other particular sort of animal is formed, because the end or completion is formed last of all, and what is peculiar to each thing is the end of its process of formation. [4]

This generalization from embryology did not enter into the mainstream of science or philosophy until von Baer in 1828 formulated his great Laws of Development:

> (1) That the general character of the big group to which the embryo belongs appear in development earlier than the special characters.
> (2) The less general structural relations are formed after the more general, and so on until the most special appear. [5]

von Baer thus stated that biological development is a true evolution of the special or heterogeneous from the general or homogeneous, and ended by applying this principle to the physical universe.

> It is this same thought that in cosmic space gathered together the scattered masses into spheres and bound them together in the solar system, the same that from the weathered dust on the surface of the metallic planets brought forth the forms of life. And this thought is nought else but life itself, and the words and syllables in which life expresses itself, are the varied forms of the living. [6]

In 1852 Herbert Spencer became acquainted with von Baer's Laws of Development; of them he said:

The great aid rendered by von Baer's formula arose from its higher generality, since only when organic transformations had been expressed in the most general terms was the way opened for seeing what they had in common with inorganic transformations. [7]

Spencer later developed these ideas into his famous formula for evolution and morphogenesis:

Evolution is an integration of matter, and a concomitant dissipation of motion during which the matter passes from an indefinite incoherent homogeneity to a definite coherent heterogeneity, and during which the retained motion undergoes a parallel transformation. [8]

In his fundamental paper of 1952, 'The Chemical Basis of Morphogenesis', [9] A. M. Turing suggests a mathematical model for the mechanisms of morphogenesis or pattern formation.

Turing's theory is that development is governed by the reaction and diffusion of a system of morphogens (in a manner intimately determined by genetic information). Development takes place when some change in the physical or chemical data precipitates instability in the system. This may be due to growth of the system, for example, or to a change in the diffusion coefficient, or the catalytic effect of one morphogen on another. The form taken by the system in resolving this instability is already determined by the initial conditions. This explains, for example, how a homogeneous or symmetrical stage in development may give way to a less homogeneous, more highly patterned stage in a coherent rather than a random fashion (the phenomenon of breaking of symmetry). Turing suggests that such regular biological patterns as the arrangements of tentacles on Hydra and whorling in leaves may be accounted for by this model. He also relates a spherical version of his model to the phenomenon of gastrulation.

This paper founded the modern theory of embryogenesis. We shall meet Turing again.

Logic and Mathematics (Automaton)

Death and its consequences are the domain of the Shaman in primitive societies. As M. Eliade put it:

It is as a further result of his ability to travel in the supernatural worlds and to see the superhuman beings (gods, demons, spirits of the dead, etc.) that the shaman has been able to contribute decisively to the *knowledge of death*. In all probability many features of 'funerary geography', as well as some themes of the mythology of death, are the result of the ecstatic experiences of shamans. The lands that the shaman sees and the personages that he meets during his ecstatic journeys in the beyond are minutely described by the shaman himself, during or after his trance. The unknown and terrifying world of death assumes form, is organized in accordance with particular patterns; finally it displays a structure and, in course of time, becomes familiar and acceptable. In turn, the supernatural inhabitants of the world of death become visible; they show a form, display a personality, even a biography. Little by little the world of the dead becomes knowable, and death itself is evaluated primarily as a rite of passage to a spiritual mode of being. [10]

From the world of death, the Orphics, Pythagoreans and Platonists brought back mathematics and mathematical physics. For it is easy to trace back to Pythagoras the notions of symmetry and harmony, which underlie the notions of group and group representation. In quantum mechanics and modern mathematics, both ideas are of primary importance.

Our spectral series, dominated as they are by integral quantum numbers correspond in a sense, to the ancient triad of the lyre, from which the Pythagoreans 2500 years ago inferred the harmony of the natural phenomena; and our quanta remind us of the role which Pythagorean doctrine seems to have ascribed to the integers, not merely as attributes, but as the real essence of physical phenomena. [11]

Plato transmitted this tradition in his theory of forms and in his cosmology (*Timaeus*). It was the soul, when not perturbed by the fluxes of the body, that saw these abstract mathematical forms clearly (in the mind's eye, so to speak). It is odd that it was this tradition which tried to 'save the phenomena' by mapping it into a mathematical model which led directly into science as we know it. Aristotle, who had honed the tools of logic to a fine edge, would have none of this nonsense. He believed in the primacy of individual substance (e.g., a particular man Socrates). His logic, however, dealt with secondary substances, for example, species or genera:

For logic is a study of thought, and that which the individual contains over and above its specific nature is due to the particular matter in which it is embodied, and thus eludes thought. [12]

This opened up a debate as to the nature of species which has not ended. Realism versus nominalism. Whether Aristotle's razor can cut Plato's beard.

The modern debate began when Russell attempted to reduce mathematics to logic in 1900. Buttressed by Frege's invention of the predicate calculus, Cantor's invention of the theory of sets and Peano's elegant notation, Russell (initially a Platonist) discovered a paradox in set theory which initiated a crisis in the foundations of mathematics. Russell reacted to this crisis by inventing the theory of types and finally by a no-class theory. The crisis was resolved in mathematics by Zermelo who axiomatized the idea of limiting the sizes of sets. But how was one to know that the axioms of the *Principia* and Zermelo would not lead to further paradoxes?

Almost simultaneously with the earthquake in mathematics, theoretical physics was reacting to the results of the Michelson–Morley experiment, which brought into question the existence of the ether and which led to Einstein's theory of special relativity. Physics was also shaken by the discovery of quanta by Planck.

Symbolic logic, which had undergone an enormous expansion due to Frege, Peano and Russell, was now applied to the crisis in physics. From there it extended to philosophy in general, and thus logical positivism was born. It needed but correspondence rules to observation statements to fit physics into its *Weltanschaung*. However, how was it to deal with statements like 'sugar is soluble'? And thus were born operational definitions: e.g., If sugar is added to water, then it will dissolve. But what about wave functions which are, in principle, not observable; the program collapsed.

For mathematics and logic were considered to be analytical (or tautologies) and observation statements were synthetic. When a statement is true in all possible worlds then it is analytic. An example of such a statement is, 'either it is raining or it is not raining'. A synthetic statement would be of the form: 'At ten a.m. on the morning of July 4, 1978, a match was set to a fire cracker on the corner of Commonwealth and Beacon Streets in Boston'.

The logical positivists, wielding a tool of such formidable magnitude, exorcised metaphysical rubbish. Biology however, was untouched but psychology was rendered operational.

Logic itself did not stand still. Hilbert decided to prove modern mathematics consistent. He differentiated between mathematics (object language) and meta-mathematics (meta-language). He was going to prove consistency by finitary means in the meta-language.

In the meta-mathematical attempt to prove the consistency of number theory, Gödel discovered the impossibility of formalizing arithmetic. He did this by clarifying the distinction between provability and truth.

His main technique was to map meta-mathematics into arithmetic by means of Gödel numbers.

Simple as this method of arithmetizing a formalism is, it leads to an insight of considerable philosophical interest — namely that the natural numbers with their arithmetic constitute a field so wide that any theory (once it is completely formalized) can be mapped into it. The amazing power of number, which Pythagoras and Plato recognized more or less clearly, and which Swift made fun of, was utilized by Gödel for the purpose of the meta-mathematical study of a given mathematical formalism. [13]

Turing used Gödel's results to invent a computing machine which could calculate all partial recursive functions. It was von Neumann who used the Turing machine to invent the stored-program computer. It is worthwhile to quote von Neumann at length on the importance of the universal Turing machine as we shall need it later for his construction of a self-reproducing automaton (see Appendix A). [14]

It is thus a beautiful fact that the nineteenth-century attempt to reduce mathematics to logic failed, but produced the computer instead.

BIOLOGY (DARWIN AND THE ORGANISM)

It is remarkable that whilst all these convulsions and crises in both mathematics and physics were the order of the day in these disciplines, biology went on in a calm, self-confident manner. The reason for this is that biology had undergone its trauma forty years previously in the Darwinian revolution.

For Darwin had destroyed the Aristotelian notion of fixed species and genera and had substituted the notion of natural selection. Teleology and the Creator's design of organisms were obsolete.

> These laws taken in the largest sense, being Growth with Reproduction; Inheritance which is almost implied by reproduction; Variability from the indirect and direct action of the conditions of life, and from use and disuse: A Ratio of increase so high as to lead to a Struggle for Life, and as a consequence to Natural Selection, entailing Divergence of Character and the Extinction of less-improved forms. [15]

The man who saw most clearly what Darwin had wrought was Peirce, who wrote in 1877:

> The Darwinian controversy is, in large part, a question of logic. Mr Darwin proposed to apply the statistical method to biology. The same thing has been done in a widely different branch of science, the theory of gases. Though unable to say what the movements of any particular molecule of gas would be on a certain hypothesis regarding the constitution of this class of bodies, Clausius and Maxwell were yet able, eight years before the publication of Darwin's immortal works, by the application of the doctrine of probabilities, to predict that in the long run such and such a proportion of the molecules would, under given circumstances acquire such and such velocities, that there would take place every second such and such a relative number of collisions etc; and from these propositions were able to deduce certain properties of gases, especially in regard to their heat relations. In like manner, Darwin, though unable to say what the operation of variation and natural selection in any individual case will be, demonstrates that in the long run they will, or would, adapt animals to their circumstances. Whether or not existing animal forms are due to such action, or what position the theory ought to take, forms the subject of a discussion in which questions of fact and questions of logic are curiously interlaced. [16]

For a start, Darwin believed in a blending theory of inheritance. For if one crossed a red flower with a white flower, one would get a pink flower. If one then crossed a pink flower with a white one, the result would be a dimmer pink. He, furthermore, believed in a Lamarckian theory which claimed that the use and disuse of organs would in some manner be transmitted to the progeny. He finally proposed a theory of Pangenesis whereby every organ of the body secreted particles into the blood stream which then collected in the germ cells. This peculiar theory was taken up by the Russian Michurin who taught it to Lysenko.

It was the Austrian abbot Mendel who properly formulated the theory of inheritance based on a particulate model of the gene which allowed the conservation of variance. As R. Fisher put it:

The particulate theory of inheritance resembles the kinetic theory of gases with its perfectly elastic collisions, whereas the blending theory resembles a theory of gases with inelastic collisions and in which some outside agency would be required to be continually at work to keep the particles astir. [17]

The issue of Lamarckian inheritance was settled by Weismann who cut off the tails of mice for many generations and found that the resulting progeny still had tails. This meant that there was a split in the organism between genotype and phenotype. In other words, the hen is the means by which an egg makes another egg. What happens to the hen is not transmitted to her progeny. Today it is reflected in the central dogma of molecular biology (i.e., DNA → RNA → Protein). Thus clearly Darwin had to be supplemented by a correct theory of inheritance.

Just as the kinetic theory of gases told one very little about molecules (it required quantum mechanics), so Darwinian natural selection told one very little about organisms and their world.

In the nineteenth century, as the microscope was perfected, the cell was discovered. All organisms, including Man, were composed of cells. The cell's main function was to grow and divide. By the use of dyes, various parts of the cell were recognized: nucleus, mitrochondria, chloroplasts and centrioles. It was a great age of observation. Pasteur and Koch found a new type of very small cells called bacteria, which were implicated in many human diseases.

It was left to the twentieth century to analyze the cell into its chemical components: proteins, nucleic acids, fats, and carbohydrates. Of these chemical species, proteins were soon discovered to be giant molecules made up of thousands of atoms. Although the proteins were large, the atoms found in proteins were only carbon, hydrogen, oxygen, nitrogen, and sulfur. If proteins were gently broken down they fell apart into twenty molecules called amino acids. A protein was later discovered to be a linear sequence of amino acids. It can be described as a string over an alphabet of twenty letters (the amino acids). It was also found that the nucleic acids (DNA and RNA) were strings over an alphabet of four letters (nucleotides). In the case of DNA, the letters (or molecules) are adenosine (A), guanosine (G), cytosine (C) and thymidine (T). In the case of RNA it is A, G, C, as above, but uridine (U) instead of thymidine (T). The number of combinatorial possibilities are astronomical. Let us say we have formed a string of amino acids 100 letters long, how many different ones can be formed? There are twenty different possibilities for the first one and twenty for the second and so on. Therefore there are 20^{100} different proteins. This is an enormous number. In a bacterium like *E. coli* there are approximately 2000 different proteins.

The DNA in *E. coli* usually is found in a single molecule which is about 1 millimeter long. There are about 3 million nucleotide base pairs in one molecule of DNA. The different protein strings are encoded in the sequence of bases in the double stranded molecule of DNA.

A protein molecule called RNA polymerase transcribes sub-sequences of the DNA molecule into RNA molecules, called messenger RNA. This RNA molecule enters a protein synthesizing machine, which is best compared to a tape recorder, in which the RNA molecule is read; the output is a sequence of amino acids. Because there are four bases (A, G, C, U) in the RNA alphabet, it requires three bases to code for one amino acid ($4^3 = 64$). The many to one mapping from the sequence of bases in the nucleic acid to the protein is called the genetic code. (See Figure 1.) The way the table is read, the first base in the triplet is listed in the first column; the second base is listed along the row; the third base is listed in the last column. The sixty-four triplets are thus related to the twenty amino acids. For example the triplet GGG stands for the amino acid glycine. Thus the central dogma can be stated DNA → RNA → Protein.

What is of interest is that von Neumann had foreseen the nature of the cell before molecular biology had worked out the details.

The cell has been compared to a computer or automata. This analogy was first made by J. von Neumann. He asked the question: How is it possible to design a computer which would replicate itself? The answer which he gave in 1948 was similar to that which we now have outlined. Freeman Dyson gave an admirable summary of von Neumann's work in his Vanuxem Lecture at Princeton:

	2nd				
1st	G	C	A	U	3rd
G	gly gly gly gly	ala ala ala ala	glu glu asp asp	val val val val	G A C U
C	arg arg arg arg	pro pro pro pro	gln gln his his	leu leu leu leu	G A C U
A	arg arg ser ser	thr thr thr thr	lys lys asn asn	met ile ile ile	C A C U
U	trp non cys cys	ser ser ser ser	non non tyr tyr	leu leu phe phe	G A C U

Fig. 1.

von Neumann did not live long enough to bring his theory of automata into existence. He did live long enough to see his insight into the functioning of living organisms brilliantly confirmed by the biologists. The main theme of his 1948 lecture is an abstract analysis of the structure of an automaton which is of sufficient complexity to have the power of reproducing itself. He shows that a self-reproducing automaton must have four separate components with the following functions. Component A is an automatic factory, an automaton which collects raw materials and processes them into an output specified by a written instruction which must be supplied from the outside. Component B is a duplicator, an automaton which takes a written instruction and copies it. Component C is a controller, an automaton which is hooked up to both A and B. When C is given an instruction, it first passes the instruction to B for duplication, then passes it to A for action, and finally supplies the copied instruction to the output of A while keeping the original for itself. Component D is a written instruction containing the complete specifications which cause A to manufacture the combined system A plus B plus C. Von Neumann's analysis showed that a structure of this kind was logically necessary and sufficient for a self-reproducing automaton, and he conjectured that it must also exist in living cells. Five years later Crick and Watson discovered this structure of DNA, and now every child learns in high school the biological identification of von Neumann's four components. D is the genetic material, DNA, A is the ribosomes, B is the enzymes RNA and DNA polymerase, and C is the represser and derepresser control molecules and other items whose functioning is still imperfectly understood. So far as we know, the basic design of every micro-organism larger than a virus is precisely as von Neumann said it should be. [18]

For von Neumann had used Turing's notion of a universal computer and broadened it to a universal constructor. He considered in detail two models of self-reproduction: (1) the kinematic model which

deals with the geometric-kinematic problems of movement, contact, positioning, fusing and cutting but ignores problems of force and energy. The primitive elements of the kinematical model are of the following kinds: logical (switch) and memory (delay) elements, which store and process information; girders which provide structural rigidity; sensing elements, which sense objects in the environment; kinematic (muscle-like) elements which move objects around, and joining (welding) and cutting elements which connect and disconnect elements. [19]

Model [2], the cellular model:

It was stimulated by S. M. Ulam who suggested during a discussion of the kinematic model that a cellular framework would be more amenable to logical and mathematical treatment than the framework of the kinematic model. . . . von Neumann chose, for detailed development, an infinite array of square cells. Each cell contains the same 29-state finite automaton. Each cell communicates directly with its four contiguous neighbors with a delay of at least one unit of time. [20]

Since von Neumann's construction, other workers have simplified the cellular model. Now one can construct a universal constructing automaton with a four-state finite automaton in each cell [21]. Finally, Conway used two-state automata to construct his game of life in which one could construct a universal Turing machine.

It is paradoxical that, when all is said and done, modern molecular biology has reduced the simplest living organism (the bacterial cell) to a logical automaton. It is one of the unstated hypotheses of this field that it is the logical and informational relationships between the elements of the automaton which matter and not the physical and chemical relationships. One can build the same automaton out of mechanical flip-flops or electrical diodes and triodes, or whatever.

We have so far concentrated on the logical machine and not on the language for describing the instructions which the machine obeys. For after all what a Turing machine does is to map one string of symbols onto another string of symbols. Thus, one can also look upon the automaton as either generating or accepting strings of symbols. Thus a grammar is a device for generating sentences. The relationship between language as strings of symbols and automata are intimate and interchangeable. Human language and some artificial languages separate into three main divisions: morphology, syntax, and semantics.

Roman Jakobson noted the similarity between the phoneme and morpheme distinction, and the use of a triplet of bases in a codon; for one base is meaningless but three form a meaningful unit. He then asked the pertinent question:

> How should one interpret all these salient homologies between the genetic code which appears to be essentially the same in all organisms and the architectonic model underlying the verbal codes of all human languages and, *nota bene*, shared by no semiotic systems other than natural language or its substitutes? The question of these isomorphic features becomes particularly instructive when we realize that they find no analogue in any system of animal communication.
>
> The genetic code, the primary manifestation of life, and, on the other hand, language (the universal endowment of humanity) and its momentous leap from genetics to civilization, are two fundamental stores of information transmitted from ancestry to progeny, the molecular heredity and the verbal legacy as a necessary prerequisite of cultural tradition. [22]

The analogy between language and molecular biology is also seen at the level of syntax. The DNA string could be interpreted as deep structure and the translation into the protein a surface structure. The complicated three-dimensional folding of the string of amino acids is a result of a hierarchical process. The formation of secondary structure, for example, helices, pleated sheets, and random coils is transformed into a complicated tertiary structure. The transformational grammar for protein folding has yet to be written.

The proteins are the phenotype of the cell. They perform three main functions: control, structural and (most important), catalytic. Of the proteins of a bacterium, the majority are catalysts.

What is a catalyst? It is a molecule which alters the rate of a chemical reaction without itself being changed. It does this by altering the pathway the reaction takes. It does not alter the end state but alters the means to get there. It is very reminiscent of William James's definition of mind. He gave the following example: suppose one inverts a jar full of water in a pail of water and blows a bubble of air into the jar; it will rise to the surface of the inverted jar and remain there forever. If one places a frog in the same position it will discover a path around the brim of the jar and ascend to the air. The end is fixed but the means are varied [23]. Thus it is with a catalyst, the end is fixed but the means are varied.

In fact Pattee has claimed that a quantum theory of measurement is needed to explain enzymes [24]. It is precisely the quantum theory of measurement which deals with the reduction of the wave packet. It was this which convinced Wigner of the need to modify quantum mechanics [25]. It is clear that a very close examination of the quantum mechanics of catalysis is in order.

Monod realized that

proteins, which channel the activity of the chemical machine, assure its coherent functioning and put it together. All these teleonomic performances rest, in the final analysis, upon the protein's so called 'stereo specific' properties, that is to say upon their ability to 'recognize' other molecules (including other proteins) by their shape, this shape being determined by their molecular structure. At work here is, quite literally, a microscopic discriminative (if not 'cognitive') faculty. [26]

Thus, the complicated folded string of amino acids is a cognitive tool. There are two such types of tools, perceptual (such as a microscope or a telescope) and operational (such as a hammer or a saw). We have here quite explicitly language as tool.

To invent a language could mean to invent an instrument for a particular purpose on the bases of the laws of nature (or consistently with them). [27]

What about semantics? Modern semantics cannot assign a meaning to a sentence. As Quine put it:

The totality of our so-called knowledge or beliefs, from the most casual matters of geography and history to the profoundest laws of atomic physics or even of pure mathematics and logic, is a man-made fabric which impinges on experience only along the edges. Or to change the figure, total science is like a field of force whose boundary conditions are experience. A conflict with experience at the periphery occasions readjustments in the interior of the field. Truth values have to be redistributed over some of our statements. Reevaluation of some statements entails reevaluation of others because of their logical interconnections – the logical laws being in turn simply certain further statements of the system, certain further elements of the field. Having reevaluated one statement we must reevaluate some others which may be statements logically connected with the first or may be the statements of logical connections themselves. But the total field is so underdetermined by its boundary conditions, experience, that there is much latitude of choice as to what statements to reevaluate in the light of any single contrary experience. No particular experiences are linked with any particular statements in the interior of the field, except indirectly through consideration of equilibrium affecting the field as a whole. [28]

Given the 2000 or so proteins of a bacterium which are an interconnected set of strings, the meaning of any single protein is determined by whether or not the whole interconnected set of strings will replicate or survive. It is the whole organism which is tested against the world. Meaning goes from whole to part. Semantics of a sentence or a protein does not make sense. The sentence is not the atom of meaning; it is a set of interconnected sentences (e.g., proteins) which has meaning.

THE ORGANISM AND ITS WORLD

But what about the relationship between the organism and its world. It certainly is not that discussed by Skinner whose impoverished notion of experience makes semantics difficult.

> Experience, then for Peirce, Mead, and Dewey is part of the cosmos, an occurrence within the cosmos, and since cosmology in this philosophy must be based on observation, the cosmos as experienced is the base upon which generalized cosmological categories and theories must be built and tested. Experience, writes Dewey, is not a verb that shuts man off from nature, it is a means of penetrating continually into the heart of nature. [29]

It would be interesting to explore the topic of how a rich notion of experience worked out by pragmatists such as James, Dewey, Mead, and Peirce ended up with a pigeon in a Skinner box and not with observing the pigeon in its natural habitat.

The world experienced by an organism is what biologists call the ecosphere or biosphere (see Appendix C). The discipline that studies the biosphere is called ecology.

> Ecology is the area of the biological sciences that is concerned with living systems in their environmental contexts. In practice the living systems studied by ecologists are those of the highest levels of organization; individual organisms, populations, societies (as organizations of one species), communities (as systems of populations usually of many species) and ecosystems. An ecosystem is 'A community and its environment treated together as a functional system of complementary relationships' (e.g., transfer or circulation of energy and matter). [30]

The biosphere breaks up into communities which are populations of organisms which occupy a given area. The major new idea of ecology is the notion of population.

For Darwin had changed the Aristotelian concept of species into that of a population of interacting organisms.

> A population may be defined as a group of things which interact with one another. A group of gaseous molecules in a single vessel or the populace of a country form units of

such nature. Families, the House of Lords, a hive of bees or a shoal of mussels – in short social entities – all constitute populations. The class of red books, thanatocoeneses, or all hermaphrodites do not. The best test of whether or not a group forms a population is to ask whether or not it is possible to affect one member of the group by acting on another. Removing a worker from a hive of bees for example should influence the number of eggs the queen may lay, and therefore both worker and queen are parts of the same population. [31]

A population is, in other words, what physicists call a many-body system. These systems are very difficult to study as has been humorously noted.

How many bodies are required before we have a problem? G. E. Brown points out that this can be answered by a look at history. In eighteenth century Newtonian mechanics, the three-body problem was insoluble. With the birth of general relativity around 1910 and quantum electrodynamics in 1930, the two- and one-body problems became insoluble. And within modern quantum field theory, the problem of zero bodies (vacuum) is insoluble. [32]

Of course the physicist has had recourse to various approximations. Starting in the nineteenth century, statistical mechanics allowed one to derive macroscopic thermodynamics from microscopic dynamics. The major triumph was in the area of ideal gases, i.e., where there are only elastic collisions but no other interactions between molecules. When molecules begin to interact other than by collision the theoretical difficulties become enormous. A very fashionable problem is phase transition. The transition from liquid water to ice (from disorder to order) is such a phase transition. In recent years researchers have tried to push statistical mechanical reasoning from equilibrium states to those far from equilibrium. A recent attempt has been made by Haken in his program of synergetics.

The systems with which we are concerned possess an enormous number of subsystems (or degrees of freedom). The determination of the detailed behaviour of any individual subsystem is, in general, hopeless – but fortunately this is not needed. . . . Thus it is our task to select the relevant parameters and to do away with all unnecessary information. To achieve this goal, the concept of the order-parameter, well known in phase transition theory, has turned out to be a very useful tool. For example, consider the mean field theory of the ferromagnet. In this theory direct interaction (caused by the coulomb exchange interaction) between the spins is replaced by a two-step procedure: first, a macroscopic quantity (the magnetization) is constructed, generated by the different spins; then this magnetization acts on each individual spin to tell each how to behave. The magnetization acts as an *order parameter* in two respects. It gives orders to the subsystems and it also describes the degree of order (it is zero in the disordered state and acquires a maximum value in the ordered state). The order parameter (or a set of such parameters) represents the behaviour of the system on a macroscopic scale, and is thus a macroscopic variable.

This concept of the order parameter also sheds new light on the problem of self-organization: the subsystems themselves create fictitious or real quantities which, via feedback loops, organize the behaviour of the subsystems. [33]

What the order parameter notion states is that the whole is less than the sum of the parts. For the degrees of freedom of such an interacting population is less than the degrees of freedom of the organisms which compose it. In other words a system which admits an order parameter is a super-organism.

This idea of a system of interacting parts which gives rise to macroscopic constraints is one which haunts not only biology but economics (microeconomics *vs.* macroeconomics), anthropology (whether culture is a super-organic phenomena), and sociology (society as organism). It is a problem which is as old as philosophy itself (the one—many problem). For the first time it can be stated within a fairly well-defined context. Can a statistical mechanics of interacting subsystems be formulated? It is a problem which also haunts population genetics.

Genes in populations do not exist in random combinations with other genes. The alleles at a locus are segregating in a context that includes a great deal of correlation with the segregation of other genes at nearby loci. The fitness at a single locus ripped from its interactive context is about as relevant to real problems of evolutionary genetics as the study of the psychology of individuals isolated from their social context is to an understanding of man's sociopolitical evolution. In both cases context and interaction are not simply second-order effects to be superimposed on a primary monadic analysis. Context and interaction are of the essence. [34]

Finally, a reconciliation between Aristotle and Darwin is possible. For Aristotle, the embryonic paradigm is dominant. Within this global differentiation of the super-organism the statistical mechanics of Darwin is meaningful. An example of this type of thinking goes back to W. Roux, the founder of modern experimental embryology:

Roux calls 'functional adaptation' within the organism: that every organ, even every cell, possess its given structure, which changes if the conditions of the organ's function are changed, so that in normal circumstances the life of the body runs its even course; if this is disturbed by interference from outside, cells and tissues adapt themselves as required to repair the damage. This fact Roux considers to be due to a 'struggle for existence' between the cells in the body and even between molecules in every cell, each of which strives to force its way forward at the expense of its neighbours, an effort that is controlled by the general requirements of the body, the weakest elements being thrust aside and destroyed. [35]

If we want an example of a super-organism, our body, made up of billions and billions of cells, is one. Within it, according to Jerne and Burnet,

Darwinian evolution is going on; for within our bodies there are cells whose purpose is to protect us from invasion of foreign or harmful cells. These cells (lymphocytes) are predetermined by mutation and selection to distinguish between self (body cells) and non-self (foreign or tumor cells). The use of Darwinian theory in immunology is now dominant. Thus, it is useful to explore the relationship of a Darwinian statistical mechanics within the overall super-organism (biosphere). The microcosm, our body, is a model for the macrocosm, the biosphere.

We are now ready to discuss Darwin again. For like Copernicus, he had wrought a revolution; it changed our view of the world but it was not science. It took Galileo, Kepler, Descartes, and (finally) Newton to transmute the Copernican revolution into science. Likewise, it took Mendel, the discoverers of the cell, Pauling, Watson-Crick, and (finally) von Neumann to initiate the long trek of transmuting Darwin into science. We are not yet there, for it is going to take a lot of careful analysis to convert the still nebulous notion of super-organism into scientific gold.

Today we have the following Darwinian paradigm. A random event, like a cosmic ray, causes a change in the DNA (mutation). In other words a record is made of a random event. This change in the record is read out into some tool (a protein), either perceptual or operational. This change is then generalized by replication. It is the whole organism which is then selected by the biosphere which we can model as a super-organism. For us to make Darwin scientific we must understand the following concepts: organism, tool, and super-organism.

MEMORY

The other major facet which it is necessary to clarify is the idea of history. A convenient definition of history is that it is a method for transmuting noise into meaning. For it is in biology that history begins to become dominant. As Bernal put it:

> I had a very interesting discussion on this point with Einstein in Princeton in 1946, from which it appeared to me that the essential clue was that life involved another element, logically different from those occurring in physics at that time, by no means a mystical one, but an element of history. The phenomena of biology must be, as we say, contingent on events. In consequence, the unity of life is part of the history of life and, consequently, is involved in its origin. [36]

For history one needs memory. In the nineteenth century the question of genetic memory was related to vibrations and wave motions. It was, in other words, a holographic theory.

The originator of the argument, the German physiologist Ewald Hering, advocated vibrations of some arcane sort. The nervous system, as a united entity pervades and interconnects the whole body. The vibrations of an external stimulus are transferred to the nervous system and hence to all other organs, especially the gametes. [37]

It was Haeckel who took over this idea and based his formula of ontogeny recapitulating phylogeny on it. It was his student Semon who based his studies of genetic memory on this holographic analogy. Schrödinger was deeply influenced by Semon. It would be an interesting historical study to trace out whether or not the holographic theory of memory of Hering influenced Schrödinger in his discovery of wave mechanics. For in his article 'Are There Quantum Jumps' he describes how

The previously admitted discontinuity was not abandoned but it shifted from the states to something else, which is most easily grasped by the simile of a vibrating string or drumhead or metal plate, or of a bell that is tolling. If such a body is struck, it is set vibrating, that is to say it is slightly deformed, and then runs in rapid succession through a continuous series of slight deformations again and again. [38]

However, it was this same Schrödinger who proposed the modern theory of the aperiodic crystal which carries the information locked in genetic memory. As he put it:

the most essential part of a living cell – the chromosome fiber – may be suitably called an aperiodic crystal. In physics we have dealt hitherto only with periodic crystals. To a humble physicist's mind, these are very interesting and complicated objects; they constitute one of the most fascinating and complex material structures by which inanimate nature puzzles his wits. Yet compared with the aperiodic crystal, they are rather plain and dull. The difference in structure is of the same kind as that between an ordinary wallpaper in which the same pattern is repeated again and again in regular periodicity and a masterpiece of embroidery, say a Raphael tapestry, which shows no dull repetition but an elaborate, coherent, meaningful design traced by the great master. [39]

The aperiodic crystal of the chromosome is today known as DNA. In other words, an aperiodic crystal is a solid solution. A binary alloy is another example of a solid solution. One can imagine a two dimensional object such as an infinite two-dimensional checkerboard. On each square either a black or a red checker is placed. Any one such distribution of checkers on this checkerboard is a configuration of a two-dimensional solid solution. In DNA we have a one-dimensional solid solution with four bases. If one is going to use an aperiodic crystal for memory, one is limited to one and two dimensions; as it is very difficult to read out the information stored in a three-dimensional solid solution.

The two-dimensional case with two types of checkers (black and red) is modeled by the Ising model. We have to assume that neighbouring checkers interact with each other. A red checker may interact more favourably with a neighbouring red checker than with a neighbouring black checker. The problem at equilibrium (at some temperature) is how do the checkers distribute themselves given a nearest-neighbour interaction? It is especially important whether at some critical temperature, a phase transition can take place. It was an amazing mathematical feat of Onsager than enabled one to calculate the possibility of a phase transition at a temperature greater than zero in the two-dimensional Ising model.

We can now begin to analyze the problem of the origin of life. What we need is a Darwinian system of Natural Selection operating within some superorganism.

THE ORIGIN

The organism is the earth itself. For it was Hutton, the founder of modern geology, who envisaged the earth as an organism.

We are thus led to see a circulation in the matter of the globe, and a system of beautiful economy in the works of nature. This earth like the body of an animal, is wasted at the same time that it is repaired. It has a state of growth and augmentation, it has another state which is that of diminution and decay. The world is thus destroyed in one part but it is renewed in another. [40]

As McIntyre put it:

Analogy of microcosm and macrocosm, analogy of celestial spheres and atmosphere, analogy of heart and sun, analogy of blood and rain, this is the heredity of Hutton's Theory — of our theory. And the heart of the theory (if I may use the analogy) is the concept of circulation of matter in the Macrocosm. [41]

The greatest such circulation is the rock cycle, which weathers rocks and turns them mainly into clay.

For the theory of the origin of life which I would like to expound is that the original living organisms were clays. The major idea is that:

There are three kingdoms in nature — the animal, the vegetable and the mineral. The animal kingdom displays terrestial life in its highest development, the vegetable kingdom manifests life of a lower type: the members of the mineral kingdom are also alive, although their life is on a still lower plane and, indeed, quite rudimentary in character.

The members of the two higher kingdoms have the power of propagating themselves, which they do by 'seeds' given off from their own bodies. Arguing by analogy, it must be believed that the species of the mineral kingdom, metals, ores and other stones propagate by seeds also, although these are so small that they cannot be seen. The several metals therefore have their 'metalline seeds' by means of which they reproduce their respective species. [42]

So thought some geologists of the 16th, 17th, and 18th centuries. Of course, today we would call the propagation of metal or stones *crystallization*. Clays are two-dimensional solid solutions. The major new claim is that like DNA, one two-dimensional crystal (clay) is a template for another. These clay particles, furthermore, have catalytic abilities. Thus, given replication, the possibility for mutation, read out into catalytic abilities, the possibilities for natural selection by the 'biosphere' become possible. To quote the first proponent of this theory, Cairns-Smith:

the primitive genographs were patterns of substitutions in colloidal clay crystallites. The theoretical informational density in such crystallites is comparable to that in DNA. Evolution proceeded through selective elaboration of substitutional genographs that had survival value for the clay crystallites that held them ... within a complex dynamic primitive environment. A gradual 'takeover' of the control machinery by organic molecules — a genetic metamorphosis — is then considered to have occurred. [43]

In summary, clays are solid solutions, with the main ions being silicon, magnesium, and aluminium. They can carry out a large number of chemical reactions. Furthermore, a proposal has been made that clays could replicate, like DNA, then mutate and hence evolve. Thus before the proto-cells were clays, one can perhaps reconstruct the evolution of the proto-cell within an evolving system of clays.

The major theoretical model for such a system is a Kinetic Ising Model in two dimensions; that is, a model which is far away from equilibrium. The analysis of such a system has just begun. It is very similar to a probabilistic game of life [44]. It may be possible to join the theory of these two abstract systems and thus perhaps be able to generate a logical automaton from a physical and chemical system. This is a theoretical project for the future.

Furthermore, in the evolution of a complex system from a simpler one which occurs in a system with memory,

The pattern ... is one of stages of increasing inner complexity, following one another in order of time, each one including in itself structures and processes evolved at the lower levels. [45]

The system thus grows more complex, like an onion, and it must be possible to strip back the layers of the onion to expose the simpler earlier stages. One can carry out such an analysis in the evolution of the genetic code (see Appendix B), metabolism [46], the membrane, and other complex structures of the cell.

A scenario for the origin of life would go as follows. When the Earth had differentiated into core, mantle, and crust, the atmosphere would be mainly carbon dioxide, nitrogen and water, with some minor constituents like hydrogen sulfide. This atmosphere would not give rise to a soup of monomers. The coupling between atmosphere, water and the crust of the earth would lead to the formation of clays. If the clays behaved as Cairns-Smith proposed, they began to fix carbon dioxide using ultraviolet light and hydrogen sulfide. This would lead to thioesters, the citric acid cycle and to fatty acids. In a later stage, the fixation of nitrogen occurred, and the formation of amino acids and nucleotides became possible. The evolving clays began to polymerize the nucleotides and amino acids. In this system, the nucleic acids became coupled to the polypeptides through a genetic code. From this complex system, there peeled off a system now based on nucleic acids and proteins which began an independent evolution of its own.

In this picture, the earth is an oyster and life is a pearl. From a grain of sand, slowly layer upon layer of history has given birth to the cell.

REFERENCES

[1] Long, C. H. 1963. *Alpha; The Myths of Creation* (New York: Collier Books), p. 127.
[2] Eliade, M. 1963. *Myth and Reality* (New York: Harper and Row), p. 18.
[3] Harris, E. E. 1965. *The Foundations of Metaphysics in Science* (London: George Allen and Unwin), p. 39.
[4] Aristotle. 1943. *Generation of Animals* (Cambridge: Harvard University Press), p. 167.
[5] Russell, E. S. 1916. *Form and Function* (London: John Murray), p. 125.
[6] *Ibid.*, p. 128.
[7] Spencer, H. 1880. *First Principles*, 4th ed. (Philadelphia: David McKay), p. 284.
[8] Durant, W. 1954. *The Story of Philosophy* (New York: Pocket Library), p. 367.
[9] Turing, A. M. *Phil. Trans. R. Soc. London Ser. B* **237** 37.
[10] Eliade, M. 1964. *Shamanism* (New York: Pantheon Books), p. 509.
[11] Weyl, H. 1949. *Philosophy of Mathematics and Natural Science* (Princeton: Princeton Univ. Press), p. 185.
[12] Ross, W. D. 1959. *Aristotle* (New York: Meridian Books), p. 28.
[13] Weyl, H. 1949. *Philosophy of Mathematics and Natural Science* (Princeton: Princeton Univ. Press), p. 227.
[14] von Neumann, J. 1951. 'The General and Logical Theory of Automata', in *Cerebral Mechanisms in Behavior; the Hixon Symposium*, ed. by L. A. Jeffress (New York: Wiley), pp. 26–27.

[15] Darwin, C. 1958. *The Origin of Species* (New York: Mentor Book from New American Library), p. 450.
[16] Peirce, C. S. 1955. *Philosophical Writings of Peirce*, ed. by J. Buchler (New York: Dover), p. 7.
[17] Fisher, R. A. 1958. *The Genetical Theory of Natural Selection*, 2nd rev. ed. (New York: Dover), p. 11.
[18] Dyson, F. Unpublished lecture.
[19] von Neumann, J. 1966. *Theory of Self-Reproducing Automata* (Urbana: University of Illinois Press).
[20] *Ibid.*, p. 94.
[21] Gardner, M. 1971. *Scientific American* **224** 112.
[22] Jakobson, R. 1973. *Main Trends in the Science of Language* (London: George Allen & Unwin), p. 52.
[23] James, W. 1950. *Principles of Psychology* (New York: Dover Press), Vol. I, p. 7.
[24] Pattee, H. H. 1967. *J. Theor. Biology* **17** 410–420.
[25] Wigner, E. 1967. *Symmetries and Reflections* (Bloomington: Indiana University Press), p. 183.
[26] Monod, J. 1972. *Chance and Necessity* (New York: Vintage Books), p. 46.
[27] Wittgenstein, L. 1953. *Philosophical Investigations* (New York: Macmillan), p. 137.
[28] Quine, W. V. O. 1961. *From a Logical Point of View* (New York: Harper Torchbooks), p. 42.
[29] Morris, C. 1970. *The Pragmatic Movement in American Philosophy* (New York: Braziller), p. 116.
[30] Whittaker, R. H. 1975. *Communities and Ecosystems* (New York: Macmillan), p. 4.
[31] Ghiselin, M. T. 1969. *The Triumph of the Darwinian Method* (Berkeley: University of California Press), p. 54.
[32] Mattuck, R. D. 1976. *A Guide to Feynman Diagrams in the Many-Body Problem* 2nd ed. (New York: McGraw-Hill), p. 1.
[33] Haken, H. *Reviews of Modern Physics* **47** 68–69.
[34] Lewontin, R. C. 1974. *The Genetic Basis of Evolutionary Change* (New York: Comumbia University Press), p. 318.
[35] Nordenskiöld, E. 1928. *The History of Biology* (New York: Tudor), p. 566.
[36] Bernal, J. D. 1967. *The Origin of Life* (Cleveland: The World Publishing Co.), p. XL.
[37] Gould, S. J. 1977. *Ontogeny and Phylogeny* (Cambridge: Belknap Press/Harvard University Press), p. 97.
[38] Schrödinger, E. 1956. *What is Life* (Garden City, New York: Doubleday Anchor Books), p. 13.
[39] *Ibid.*, p. 3.
[40] Hutton, J. 1963. In *The Fabric of Geology*, ed. by C. C. Albritton (Reading: Addition-Wesley Publishing Co.), p. 9.
[41] McIntyre, D. B. *Ibid.*, p. 10.
[42] Adams, F. D. 1954. *The Birth and Development of the Geological Sciences* (New York: Dover), p. 289.
[43] Cairns-Smith, A. G. 1965. *J. Theor. Biology* **10** 53–88.
[44] Schulman, L. S. and Seiden, P. E. 1978. *J. Stat. Physics* **19** 293.
[45] Bernal, J. D. 1960. In *Aspects of the Origin of Life*, ed. by M. Florkin (New York: Pergamon Press), p. 30.
[46] Hartman, H. 1975. *J. Mol. Evol.* **4** 359–370.

APPENDIX A (FROM VON NEUMANN)*

Turing's theory of computing automata. The English logician, Turing, about twelve years ago attacked the following problem.

He wanted to give a general definition of what is meant by a computing automaton. The formal definition came out as follows:

An automaton is a 'black box', which will not be described in detail but is expected to have the following attributes. It possesses a finite number of states, which need be *prima facie* characterized only by stating their number, say n, and by enumerating them accordingly: $1, 2, \ldots n$. The essential operating characteristic of the automaton consists of describing how it is caused to change its state, that is, to go over from a state i into a state j. This change requires some interaction with the outside world, which will be standardized in the following manner. As far as the machine is concerned, let the whole outside world consist of a long paper tape. Let this tape be, say, 1 inch wide, and let it be subdivided into fields (squares) 1 inch long. On each field of this strip we may or may not put a sign, say, a dot, and it is assumed that it is possible to erase as well as to write in such a dot. A field marked with a dot will be called a '1', a field unmarked with a dot will be called a '0'. (We might permit more ways of marking, but Turing showed that this is irrelevant and does not lead to any essential gain in generality.) 'In describing the position of the tape relative to the automaton it is assumed that one particular field of the tape is under direct inspection by the automaton, and that the automaton has the ability to move the tape forward and backward, say, by one field at a time. In specifying this, let the automaton be in the state i ($= 1 \ldots, n$), and let it see on the tape a c ($= 0, 1$). It will then go over into the state j ($= 0, 1, \ldots, n$), move the tape by p fields ($p = 0, +1, -1$; $+1$ is a move forward, -1 is a move backward), and inscribe into the new field that it sees f ($= 0, 1$; inscribing 0 means erasing; inscribing 1 means putting in a dot). Specifying j, p, f as functions of i, e is then the complete definition of the functioning of such an automaton.

Turing carried out a careful analysis of what mathematical processes can be effected by automata of this type. In this connection he proved various theorems concerning the classical 'decision problem' of logic, but I shall not go into these matters here. He did, however, also introduce and analyze the concept of a 'universal automaton', and this is part of the subject that is relevant in the present context.

* From *The General and Logical Theory of Automata*, pp. 25–28. J. Wiley, New York, 1951.

An infinite sequence of digits e (= 0, 1) is one of the basic entities in mathematics. Viewed as a binary expansion, it is essentially equivalent to the concept of a real number. Turing, therefore, based his consideration on these sequences.

He investigated the question as to which automata were able to construct which sequences. That is, given a definite law for the formation of such a sequence, he inquired as to which automata can be used to form the sequence based on that law. The process of 'forming' a sequence is interpreted in this manner. An automaton is able to 'form' a certain sequence if it is possible to specify a finite length of tape, appropriately marked, so that, if this tape is fed to the automaton in question, the automaton will thereupon write the sequence on the remaining (infinite) free portion of the tape. This process of writing the infinite sequence is, of course, an indefinitely continuing one. What is meant is that the automaton will keep running indefinitely and, given a sufficiently long time, will have inscribed any desired (but of course finite) part of the (infinite) sequence. The finite, premarked, piece of tape constitutes the 'instruction' of the automaton for this problem.

An automaton is 'universal' if any sequence that can be produced by any automaton at all can also be solved by this particular automaton. It will, of course, require in general a different instruction for this purpose.

The main result of the Turing theory. We might expect *a priori* that this is impossible. How can there be an automaton which is at least as effective as any conceivable automaton, including, for example, one of twice its size and complexity?

Turing, nevertheless, proved that this is possible. While this construction is rather involved, the underlying principle is nevertheless quite simple. Turing observed that a completely general description of any conceivable automaton can be (in the sense of the foregoing definition) given in a finite number of words. This description will contain certain empty passages — those referring to the functions mentioned earlier (j, p, f in terms of i, e), which specify the actual functioning of the automaton. When these empty passages are filled in, we deal with a specific automaton. As long as they are left empty, this schema represents the general definition of the general automaton. Now it becomes possible to describe an automaton which has the ability to interpret such a definition. In other words, which, when fed the functions that in the sense described above define a specific automaton, will thereupon function like the object described. The ability to do this is no more mysterious than the ability to read a dictionary and a grammar and to follow their instructions about the uses and principles of combinations of words. This automaton, which is constructed to read a description and to imitate the object described, is then the universal automaton in the sense of Turing. To make it duplicate any operation that any other automaton can perform, it suffices to furnish it with a description of the automaton in question and, in addition, with the instructions which that device would have required for the operation under consideration.

APPENDIX B: SPECULATIONS ON THE EVOLUTION OF THE GENETIC CODE (FROM HARTMAN)*

In considering the evolution of a complex system it is useful to adopt an idea expressed by Bernal: "In general the pattern ... is one of stages of increasing inner complexity, following one another in order of time, each one including in itself structures and processes evolved at the lower levels" (Bernal [1960]). I like to call this the 'onion heuristic'. Complex systems evolve by adding layers to a simple system. The problems of the origin and evolution of the code depend therefore on the complexity of the biosphere postulated when the coupling of nucleic acid and protein occurred.

If one adopts the view of Cairns-Smith (1965) that life evolved through natural selection from inorganic crystals in particular clays, then one would conclude that the genetic code is a later invention and that the origin of life must be separated from the origin and evolution of the code. In recent years the experiments by Paecht-Horowitz, Berger and Katchalsky (1970) have given evidence of the importance of clays in the polymerization of activated amino acids. The hypothesis that clays were the original living systems which evolved is the view which I adopt in this paper.

The protein synthesis 'read-out' system as we know it today involves four interdependent subsystems: (1) Enzymes for the polymerization of nucleotides; (2) Ribosomes and enzymes for the polymerization of the activated amino acids; (3) Transfer RNA; and (4) activating enzymes.

The search for inorganic catalysts which will polymerize either nucleotides or amino acids is a very active field of research. However, the combination of tRNA and activating enzymes implies that today there is no specific physical–chemical interaction between the tRNA and the amino acid. Two views of a more primitive system can be presented. The first view presumes that the present system evolved from one in which a specific interaction existed between a tRNA and an amino acid. In other words there was no need for an activating enzyme. The second theory assumes that the existence of a primitive activating system was a necessity at the beginning. In other words

---

* From *Origins of Life* 6 (1975) 423–427.

PHILOSOPHY OF SCIENCE AND THE ORIGIN OF LIFE   207

there was no specific physical—chemical interaction between the amino acid and the tRNA. I propose to adopt this second hypothesis and assume that a primitive activating enzyme was necessary.

If the original living systems were clays, then the introduction of nucleic acids and proteins might have been solely for the additional structure which they provided to the clays (Cairns-Smith [1971]). The origin of the code would then lie in the interaction between nucleic acids, polypeptides and clays. Since these interactions have as yet received only preliminary study, much experimental work is needed.

A pattern inferring the evolution of the code and compatible with the above assumptions is proposed here which suggests how the various amino acids came into the coding system and correlates this entry with their function.

		→ 2nd					
1st	↓	G	C	A	U	↓	3rd
	G	gly	ala	glu	val		G
		gly	ala	asp	val		C
		gly	ala	glu	val		A
		gly	ala	asp	val		U
	C	arg	pro	gln	leu		G
		arg	pro	his	leu		C
		arg	pro	gln	leu		A
		arg	pro	his	leu		U
	A	arg	thr	lys	met		G
		ser	thr	asn	ile		C
		arg	thr	lys	ile		A
		ser	thr	asn	ile		U
	U	trp	ser	non	leu		G
		cys	ser	tyr	phe		C
		non	ser	non	leu		A
		cys	ser	tyr	phe		U

Fig. 1. The conventions used are as in (Stent [1971]).

If the code table is written in the form of Figure 1, a pattern is discerned which can be summarized as follows:

(1) The code originated as the polymerization of glycine which was specified by polyguanine. (Figure 2)

 →2nd
 1st ↓ G
 G glycine

Fig. 2.

(2) The code evolved to a doublet code specifying four amino acids by the addition of cytosine. (Figure 3)

 →2nd
 1st ↓ G C
 G gly ala
 C arg pro

Fig. 3.

(3) The code evolved from a doublet code to a triplet code upon the addition of adenine. A purine wobble appears in the third position (Figure 4). The wobble hypothesis concerns the codon-anticodon pairing. "It is suggested that while the standard base pairs may be used rather strictly in the first two positions of the triplet, there may be some wobble in the pairing of the third base" (Crick [1966]).

		→2nd				
1st	↓	G	C	A	↓	3rd
	G	gly	ala	glu	G	
		gly	ala	asp	C	
		gly	ala	glu	A	
	C	arg	pro	gln	G	
		arg	pro	his	C	
		arg	pro	gln	A	
	A	arg	thr	lys	G	
		ser	thr	asn	C	
		arg	thr	lys	A	

Fig. 4.

(4) The addition of uracil was the last base in the evolutionary sequence. There is the appearance of a pyrimidine wobble in the third position.

(5) The purine wobble in the third position of two codons was eliminated upon the entry of methionine and tryptophane into the coding scheme.

(6) When a new base was added, all the previous amino acid coding assignments were conserved, even when the code changed from a doublet to a triplet code. This is a restatement of the onion heuristic.

The succession of code tables can be seen in Figures 1, 2, 3, and 4. The amino acids which entered the coding system with each base have definite functional roles.

The amino acid which originated the code with guanine was glycine (Figure 2). The folding rules and structures for polyglycine are simple. Polyglycine either forms a hexagonal array of helices, each one of which has a threefold screw axis, or it folds into a pleated sheet (Crick and Rich [1955a]). Hence polyglycine probably had a structural role rather than a catalytic one. The choice of guanine and glycine rather than cytidine and proline is made on the basis of the simplicity of the amino acid and the interaction of polyglycine with clays (Akabori [1960]).

The amino acids which entered the doublet genetic code with cytidine were arginine, proline, and alanine (Figure 3). Again these amino acids gave rise to polypeptides whose role was structural. For example, a well-studied structural protein is collagen. Its structure of three polypeptide chains in a coiled coil configuration is determined by the properties of glycine and proline. Collagen is also rich in alanine and arginine (Crick and Rich [1955b]; Sefter and Gallop [1966]). The example of collagen is not used to infer that it is a primitive protein, but rather that a protein with these amino acids has simple folding rules.

Upon the addition of adenine to the code (Figure 4) it is evident that in the third position adenine was equivalent to guanine. It would appear, therefore, that when the code evolved from a doublet to a triplet code a purine wobble appeared in the third position (Crick [1966]). The amino acids which joined the code with adenine are all polar amino acids. Serine, aspartic acid and histidine are found in the active sites of proteases (Matheja and Degens [1971]). It is possible, therefore, that proteases appeared in the evolution of proteins at this point. The tertiary structure of polypeptides at this point in evolution could not have been overly complicated, however, as there were no strongly hydrophobic groups present. The choice of adenine rather than uracil is based on the supposed late entry of methionine and tryptophane into the coding system (Crick [1968]).

The addition of uracil to the code brought the non-polar amino acids to the code and now the complex tertiary structure of proteins became possible. A pyrimidine wobble appeared in the third position where uracil is equivalent to cytidine. However, the purine wobble in the third position is broken in two places, the coding for methionine (AUG) and tryptophane (UGG). It has been postulated by Crick that these two amino acids are later additions to the code.

It is plausible that UG and UA were the original doublet nonsense codons. As the code evolved, UA became UAA and UAG, while UG became UGG and UGA. The codon UGG was later used to code for tryptophane.

In considering the above hypothesis, the question naturally arises of how it is possible to go from a doublet to a triplet code. The answer depends very strongly on the mechanism by which the nucleotides are polymerized. If the mechanism is by Watson-Crick base pairing and subsequent polymerization of one strand using the original strand as template, then it is very difficult to go from a doublet to a triplet code. One could perhaps postulate a doublet code in which the third base of the triplet is a spacer to be utilized as the code evolved.

There is, however, another possible method of polymerization known as block polymerization in which the longer polymer chains are built up out of shorter chains (Bernal [1967]). If the mechanism is by block polymerization with no template, then the transition from a doublet to a triplet code is very simple; one polymerizes triplets instead of doublets. Of course the proteins coded for by such nucleotide polymers would be very simple and would probably be structural. At this point one may speculate that when a base was used in the codon, its complementary base went into the anticodon. The ability to base pair which is the distinct property of nucleic acids was used originally for translation rather than for replication. It is only when proteins became sufficiently complex, that a polymerase formed so that base pairing was then used in replication.

The adaptor hypothesis invented by Crick (1960) implies that there is no direct relationship between the amino acid and the codon or anticodon. The adaptor and the amino acid must be brought together by a third component, an activating enzyme which is itself a protein. It is at this point that the greatest difficulty in the origin of the code appears for it is a complex protein which must be present before the system can start. It is therefore necessary to invent some plausible mechanism which is not a protein and which can distinguish between adaptor RNA molecules and amino acids. One possibility is that of a lipid bilayer, for Overton showed that the cell membrane as lipid bilayer can distinguish molecules on the basis of their hydrophobicity (Overton [1899]).

Alkylamines and alcohols can interact with clays to form stable lipid bilayers (Weiss [1969]). Since proteins have in general their hydrophobic groups on the inside and hydrophilic groups on the outside (Perutz [1969]), they could have evolved to take over the recognition function played by the lipid bilayer. It is the recognition system which is the real reason behind the onion-like evolution of the code. For once adaptor and amino acid have been linked, they must remain so even if new amino acids and bases are brought into the coding system.

The problem concerning the origin of the code is related to speculations on the complexity of the system which led to its inception. If the foregoing hypotheses are correct, then the pre-biotic system seemed to invent the code for structural reasons, and then before the system turned into gelatin, invented hydrolases. When the catalytic power of proteins became evident, the system underwent a metamorphosis into one where the central dogma becomes central.

REFERENCES TO APPENDIX B

Akabori, S. 1960, in M. Florkin (ed.), *Aspects of the Origin of Life*, Pergamon Press, New York, p. 116.
Bernal, J. D. 1960, in M. Florkin (ed.), *Aspects of the Origin of Life*, Pergamon Press, New York, p. 30.
Bernal, J. D. 1967, *The Origin of Life*, The World Publishing Co., Cleveland.
Cairns-Smith, A. G. 1965, *J. Theor. Biol.* 10, 53.
Cairns-Smith, A. G. 1971, *The Life Puzzle*, University of Toronto Press, Toronto.
Crick, F. H. C. and Rich, A. 1955a, *Nature (London)* 176, 780.
Crick, F. H. C. and Rich, A. 1955b, *Nature (London)* 176, 915.
Crick, F. H. C. as quoted in M. Hoagland. 1960. in E. Chargaff and J. Davidson (eds.), *The Nucleic Acids*. Vol. III. Academic Press, New York.
Crick, F. H. C. 1966, *J. Molec. Biol.* 19, 548.
Crick, F. H. C. 1968, *J. Molec. Biol.* 38, 367.
Matheja, J. and Degens, E. T. 1971, in F. F. Nord (ed.), *Advances in Enzymology*, Vol. 34, Academic Press, New York, p. 12.
Overton, E. 1899, *Vjschr. Naturforsch. Ges. Zürich* 44, 88.
Paecht-Horowitz, M., Berger, J., and Katchalsky, A. 1970, *Nature (London)* 228, 636.
Perutz, M. F. 1969, *Proc. Roy. Soc., Ser. B* 173, 113.
Sefter, S., and Gallop, P. M. 1966, in H. Neurath (ed.), *The Proteins*, Vol. 4, 2nd ed., Academic Press, New York, p. 249.
Stent, G. S. 1971, *Molecular Genetics*, W. H. Freeman and Co., San Francisco, p. 532.
Weiss, A. 1969, in G. Eglinton and M. T. J. Murphy (eds.), *Organic Geochemistry*, Springer-Verlag, New York, p. 75.

APPENDIX C

The organism lives in an habitat and functions in an ecological niche. Darwin believed that niches were fixed like boxes. Into these fixed boxes organisms adapted, this left the organism passive.

Everywhere he looked he saw the same economic places: a hypostatized table or organization into which nature must fit its creation. [1]

Today, we define a habitat of an organism as the place where it lives; this means the geographical location.

The ecological niche of an organism depends not only on where it lives but also on what it does (how it transforms energy, behaves, responds to and modifies its physical and biotic environment) and how it is constrained by other species. By analogy it may be said that the habitat is the organism's 'address' and the niche is its 'profession', biologically speaking. [2]

The major difficulty

with the specification of empty niches to which organisms adapt is that it leaves out of account the role of the organism itself in creating the niche. Organisms do not experience environments passively; they create and define the environment in which they live. [3]

This relationship between organism and environment was worked out in detail by J. Dewey.

In actual experience there is never any such isolated singular object or event; an object or event is always a special part, phase, or aspect, of an environing experienced world – a situation. The singular object stands out conspicuously because of its especially focal and crucial position at a given time in determination of some problem of use or enjoyment which the total complex environment presents. There is always a field in which observation of this or that object or event occurs. Observation of the latter is made for the sake of finding out what that field is with reference to some active adaptive response to be made in carrying forward a course of behaviour. One has only to recur to animal perception occurring by means of sense organs, to note that isolation of what is perceived from the course of life-behaviour would be not only futile, but obstructive, in many cases fatally so. [4]

The crucial idea of John Dewey was that situations led the organism to inquire.

Inquiry is the controlled or directed transformation of an indeterminate situation into one that is so determinate in its constituent distinctions and relations as to convert the elements of the original situation into a unified whole. [5]

This notion of inquiry did not please Bertrand Russell.

It is clear that 'inquiry' as conceived by Dewey is part of the general process of attempting to make the world more organic. 'Unified whole' are to be the outcome of inquiries. Dewey's love of what is organic is due partly to biology, partly to the lingering influence of Hegel ... The metaphysic of organism underlies Dewey's theories but I do not know how far he is aware of this fact. [6]

The ecologist would agree with Dewey that a metaphysic of organism is more than relevant to understanding how an organism adapts and creates its ecological niche and habitat.

REFERENCES TO APPENDIX C

[1] Worster, D. *Nature's Economy*, p. 157. Anchor Books.
[2] Odum, E. P. 1971. *Fundamentals of Ecology*. 3rd Ed. W. B. Saunders, p. 234.
[3] Lewontin, R. C. 1978. *Scientific American* 239 214.
[4] Dewey, J. 1938. *Logic, The Theory of Inquiry*, p. 66. Henry Holt & Company.
[5] *Ibid.*, pp. 104–105.
[6] Russell, B. 1967. *The Basic Writings of Bertrand Russell*. Ed. by R. E. Egner and L. E. Denom, pp. 210–211. Simon and Schuster.

Dr. H. Hartman
Dept. of Earth and Planetary Sciences
Massachusetts Institute of Technology
Cambridge, Ma. 02139
U.S.A.

ANTHONY LEEDS

SOCIOBIOLOGY, ANTI-SOCIOBIOLOGY, EPISTEMOLOGY, AND HUMAN NATURE[1]

I. WILSON, HUMAN NATURE, AND EPISTEMOLOGY

E. O. Wilson's *Sociobiology, the New Synthesis* has received ebullient critical acclaim as a breakthrough in science in relation to its own claim to be a "new synthesis" of various biological and social sciences. It has elicited equally ebullient critical attack for its immeasurable failure to understand human cultural and social organization, a failure based on a biologically and socio-culturally misconceived human nature, as well as on some basic misconceptions in biology itself. In what follows, I treat only the question of human nature which I conceive to be central to Wilson's book and to most of the issues raised by its attackers, for whom that question is also a central, unresolved problem.

The key issue with which both sides must deal is that innumerable universals of human life clearly indicate a biological basis at the same time that *all specific* exemplifications of these human universals are highly varied in form and content, most of them exchangeable among distant populations, not necessarily in contact with each other, and rapidly responsive to changes in situation. That is, the *specific* forms *cannot* be genetically determined. All human socio-cultural behavior is based, at least in part, upon postulation. *What* is postulated has no genetic foundation whatever. Further, postulation joins with a uniquely human reflexivity which permits Man to observe himself as object — to detach himself from his physical, biological, or cultural self, with profound consequences.

These observations suggest that a quite different model of the relationship between the genetic foundation of the species and its behavior is needed for humans than that afforded by any other species — a model dealing with genetic structures generating broad, formless universals, within which non-genetic generators of behavior define form, content, and meaning. The biological sciences are remote from dealing with such a problem, while the social sciences and humanities have made enormous advances in dealing with the non-genetic generation of behavior.

Wilson's book presents a theory and conception of human nature which can be set forth fairly straightforwardly despite elements of contradiction within the text and in Wilson's own later public statements.[2] The latter essentially eviscerate his position as stated in the book. Since Wilson's text conceives human nature to be biological, the book clearly articulates the goal of the ultimate reduction of the social sciences and humanities to the biological sciences. Wilson says this variedly and repeatedly and his title indicates the road.

Underlying this conception is that the *subject*-matter of the social sciences and humanities — human society and culture — constitute essentially biological phenomena. This conception appears in a multiplicity of ways, including the postulation (and postulation it is, since none have ever been identified) of genes for this or that supposed single attribute of human behavior, such as guile, homosexuality, spite, gullibility, etc. His conception of genes or how they operate is for the geneticists and biochemists to treat, but it does seem palpably contrary to, or incompatible with, what is known of the intracorporal operations of neurological structures[3] and biochemical environments of genes. In any case, numberless such passages indicate not only a specific conception of human nature, but also that the human nature so conceived is based in specific genes, produced through a Darwinian natural and sexual selection.

This argument is bolstered by some 500 pages of data and argumentation about various animals purporting to show the evolution of characteristics such as "altruism" and "self-interest" (based on the conceptual work of Robert Trivers [1971], [1972], [1974]). Again, this material cannot be treated here.[4]

Key aspects of Wilson's conception of human nature — *giving the framework for the argument of the entire book* — appear in his opening page. Curiously, page 1 is devoted to two short quotations, standing entirely alone, on destruction and slaying, taken from a conversation between Lord Krishna and Arjuna (presumably from the *Bhagavad Gita*; no source is given). No explanation or exegesis is given of these brief, stark texts; one must assume they provide signposts for Wilson's conception of human nature. More explicit and crucial is the following passage which opens the text of the book itself (page 3):

> Camus said that the only serious philosophical question is suicide. That is wrong even in the strict sense intended. The biologist, who is concerned with questions of physiology and evolutionary history, realizes that self-knowledge is constrained and shaped by the emotional control centers of the hypothalamus and limbic system of the brain. These centers flood our consciousness with all the emotions — hate, love, guilt, fear, and others — that are consulted by ethical philosophers who wish to intuit the standards

of good and evil. What, we are then compelled to ask, made the hypothalamus and limbic system? They evolved by natural selection. That simple biological statement must be pursued to explain ethics and ethical philosophers, if not epistemology and epistemologists, at all depths. Self-existence, or the suicide that terminates it, is not the central question of philosophy. The hypothalamic–limbic complex automatically denies such logical reduction by countering it with feelings of guilt and altruism.

We have, here, a traditional Western mind/body dualism, sanctified by apparently scientific backing, but rooted in the overwhelming dualism of Western thought. Although most cultures display dualities of some kind or another, they do not override all classifications as they tend to do in the West. Often they dichotomize things quite different from those that most interest us. This fact itself suggests that many, or most, dualisms are arbitrary, clearly not inherent in nature, probably not inherent in many aspects of human thinking, but, at best, specific to linguistic thought (where the possibilities of predication and negation provide basic dualities and two-value oppositions). Dualisms did *not* occur as a significant paradigm of cosmos, person, and knowledge among the Greeks (especially Plato and Aristotle) — major ancestors of Western thought, radically altered by much of Greek thought. In sum, Wilson's dualism is itself suspect as a culture-specific, ethnocentric paradigm of the world, here the biosocial world.

Respecting humans, Wilson's dualism antinomizes reason, rationality, mind, intellect, on the one side, and, on the other, unreason, irrationality, body, emotions. The emotions are irrational, irruptive, violent, bestial: they "constrain our knowledge" and "flood our consciousness" and prevent "logical reduction". Emotion is opposed to and even blocks knowledge. Emotions destroy the order of the world. Emotions are located in an evolutionary archaic system – "the hypothalamic–limbic complex." Suicide is an irrational, not a moral–philosophical act. Since the limbic system appears relatively early in the evolution of higher animals and since it occurs in non-humans, Wilson *assumes* — a major *a priori* which structures the entire book — that this "system" is bestial in character.

Not only is this an *a priori* about human nature (see also Washburn [1975]), but it is an assumption about knowledge and knowing — about epistemology — which underlies an implicit theory of knowledge, essentially dismissed in the quotation by reducing epistemologists to resultants of natural selection. Knowing, for Wilson, comes not from emotions (after all, are they not "irrational"?), but from "reason", exemplified entirely by predicative discourse, as in language and mathematics. Wilson either does not know or simply avoids mentioning or treating epistemologies which deny this entire mode of thought, based on dualistic antinomies. For example, Lancelot Law Whyte, a physicist and biologist (not listed in Wilson's immense bibliography or index), turns the entire conception upside down, seeing essentially *all* primary knowledge as coming from the unconscious and emotional life (Whyte [1974], Chapter 1; see below).

II. THE SOCIOBIOLOGY STUDY GROUP, HUMAN NATURE, AND EPISTEMOLOGY

Shortly after Wilson's book appeared, an outcry against its conception of human nature arose. The earliest and, for long, one of the main antagonists was the Boston-based Sociobiology Study Group (hereinafter 'SSG' or 'antagonists') which published the first broadside against *Sociobiology* (SSG [1975]). I have been a member of the SSG from shortly after its inception, except for a year and a half, to the present, a signatory of the first and other publications (e.g., SSG [1976], [1977]), and have published separately from it ([1977b], [1981/2]; Leeds and Dusek [1981/2a] and Leeds and Dusek [1981/2b]).

In essence, the SSG asserted, first, that Wilson's whole conception of human nature was one specific to our time and place in the Universe — as a capitalist, competitive, invidiously alienated people in a United States which is now the central world power. I think this criticism is substantially true, but not basic.

Second, the SSG asserted that the scientific basis for attributing genetic foundations to human socio-cultural characteristics and traits was entirely absent. No specific genes for specific attributes have been isolated (nor could they be since genes do not work that way). Nor can specific genes exist for human institutions and cultural expressions because the similarities of human genetic structures are far greater among human populations than the enormous diversity of socio-cultural manifestations in time and space. Further, institutions, unlike genetic structures, can change drastically at very rapid rates in a given population or can be diffused from population to population, or even be taught to *older* generations. It is unique among humans, and necessarily *not* biological, that the sins of the sons can be visited on the fathers! Both parts of this assertion by the SSG seem substantially true, though it seems possible that certain kinds of traits may be genetically determined in some complex way, e.g., musicality.

However, despite the truth of the assertion, it is entirely clear that recognition of some broad relation between genetically-structured species characteristics of *Homo sapiens* and specifiable socio-cultural domains cannot be avoided. The relation is clearly *not* Wilson's kind of simple-minded genetic determinism, but it is also not his antagonists's simple-minded environmentalism. This point is absolutely central in the controversy between Wilson, the other sociobiologists, and their antagonists. I return to it at length below.

Third, although the SSG did not explicate this in detail, the very attributes that Wilson chooses — guile, spite, etc. — are linguistic categories (without exact semantic equivalents in many languages), which he treats as if they were real objects in the natural world. That is, he reifies them. Having reified them, Wilson makes them into actors in the scenarios of evolution and behavior — all based on the model of the semantics of the English language. In short, Wilson built the Anglo-American world, especially its 20th-century version, into the very world he wishes to study (see in this connection Barash's irresponsible verbal analogizing [1977] p. 119; see also Beach's comments on sociobiologists' mistreatment of the term 'homosexuality' [1978] pp. 120—135). The SSG's assertion seems to me substantially true, here, too.

In contrast to Wilson, the SSG saw humans as, essentially, beings of cultural norms and the institutions derived from them. It saw culture and institutional organization as innovated independently of any biological foundations (a position justified by the reasoning discussed above). Culture and institutions are symbolic forms and symbols are arbitrary forms not based in biology. By this argument, of course, Wilson's science is *itself* an arbitrary cultural phenomenon not based in, or derived from, his own or general human biology. His science is subject not only to radical change — even perhaps oblivion, as has, occasionally, happened in this history of science — but also to his cultural surroundings. This proposition seems to me also substantially true. By the same token, however, it applies to the sciences and scientific activities of the SSG, although clearly it considers its science better than Wilson's and other sociobiologists. In passing, it is important to mention that the SSG did not address the problem of the biological basis for the universal human capacity to symbolize, a point expanded upon below.

The SSG's form of argument or paradigm is called environmentalism. By some it is even called, rightly in my view, 'extreme environmentalism' (see, also, various forms of behaviorist psychology). Environmentalism derives the characteristics of (human) behavior from environmental conditions and situations; its extreme version denies that any significant *human* behavior has biological foundations. In the SSG's initial attack against Wilson in *The New York Review of Books*, this statement was included (over my strong objections):

But we suspect that human biological universals are to be discovered more in the generalities of eating, excreting, and sleeping than in [highly selected human habits] (SSG 1975, p. 48).

These are, of course, animal functions shared with, say, the dog, viper, goose, ass, and other animals with which I prefer not to be identified.

The SSG, then, found itself in a very peculiar position. It had, in fact, stated that the only attributes common to human — regarding which *Homo sapiens* might be said to have a human nature — were essentially identical to those of other animals — the "bestial" ones. At the same time, all those things that make *Homo sapiens* distinctively human — culture, institutions, "rationality", contained in cultural orders such as science and language — were denied the status of human nature on the grounds of social and cultural relativism, that is, the striking range of socio-cultural differences and variabilities. By this stroke, the SSG effectively denied the goals and activities of the social and humanistic sciences searching for universals. Most particularly, it denied those of anthropology which not only looks for socio-cultural universals, but also their basis in *Homo sapiens*'s biological species characteristics. Because of its gross evasion of specifically human universals (discussed below) and the disciplines which study them, the SSG's position appears thoroughly anti-intellectual and vacuous. By the same stroke, the SSG adopted, in effect, the same basic epistemology as Wilson — one separating mind and body, opposing culturally-based "rationalities" to "bestialities" — thereby also denying a large body of contemporary thought offering alternative epistemologies.

III. THESIS, ANTITHESIS, AND NEW SYNTHESIS

This was, and continues to be, the dialectic thesis and antithesis between Wilson and one group of his criticis. But, as just shown, the antithesis is, at best, only a part-antithesis. In what follows, I present several problems for consideration which now one, now the other party to the dialectic, and sometimes both, cannot treat, given their philosophical and substantive positions. These problems, in part, come out of various sorts of inquiry in anthropology, though not restricted to it; in part, out of my own professional and other work and experience, including, recently, intensive photographic work. Each is an area in which major research should, and could be, carried out with significant enlightenment on the issues set forth below.

IV. HUMAN UNIVERSALS

The first problem is that of human universals. Historically, the idea of universals was not phrased in terms of a *set* of elements, but appeared in concepts such as the (ethnocentric) universality or catholicity of the Church, along with its explanatory theology, world view, and epistemology. A later idea is that of the "psychic unity" of mankind which has permeated the thought of anthropology, explicitly, and of other disciplines, implicitly, for a least a century. It asserts that all human populations are characterized by the same attributes and functions of mind — "mind" itself being a universal, reified entity. Thus, the basic perceptual and cognitive processes are identical for all. In effect, we can all understand each other fully; cultural and language differences are only local and epiphenomenal. In principle, total translatability is possible within the human species, while inter-species translatabilities scarcely exist. Our concepts of most, or all, major human socio-cultural processes — e.g., enculturation and diffusion — are built on this conception.

More recently, inquiry has been more analytic and empirical and not so much concerned about the species as such, but rather about behavioral domains. A few examples will do. First, linguists have long been concerned with "language universals." These refer to structural properties found in all languages, for example, negation predication, question formation.

Second, many disciplines have been concerned with the universal occurrence of metaphor — visual, aural, and (especially), linguistic. In a sense, all language is metaphoric, since all words are, at best, arbitrarily encoded allusions to selected attributes only (a strictly human phenomenon involving *postulation*), and *not* to the totalities of situations experienced. In a narrower, more usual sense of metaphor, images from varieties of settings and times and in various distortions are juxtaposed in a single expressive form, a possibility uniquely afforded by the abstracted and arbitrarily referential, postulational, and extra-somatic human symbol. Only for humans does the tyger burn bright in the forest of the night, or does one rage, rage at the dying of the light. Only for humans can so many meanings be carried as provided by the visual image of Uncle Sam.

A third example — so beloved of anthropologists — is the incest taboo, also uniquely human and quite unexplained. All societies have some form of incest taboo consisting of some set of normative statements concerning what relatives you may not have sexual relations with. These always include mothers and fathers, sons and daughters, and almost universally — but not quite — siblings, while other relatives indicated vary greatly from society to society. But note that the taboo is a *rule* or *norm* which does not necessarily correspond to the behavioral actualities of the sex act, for example, rates of parent—child and sibling intercourse appear to be increasing in our society at present.

Fourth, *all* humans operate in society only through roles, which themselves are arbitrary, relativistic, and situational concatenations of non-somatic rights, duties, obligations, prerogatives, and ascriptions, without known parallel among other species. All humans potentially can, and all frequently do, change roles by choice, including adopting drastically divers roles in brief compass, even in different societies. Conversely, the same role can be filled by genetically quite different individuals, for example, a given job by scabs during a strike. This possibility includes, for example, the role, 'father,' which is universal among humans and universally embraces the social recognition of paternity *whether or not it is also biological*. Paternity is a *normative* status unknown among animals. Sociobiologists should encounter insurmountable difficulties with this fact and its generalization that *no human kinship role is necessarily a biological one* (see Sahlins [1976] on kinship). Lastly, transfers and exchanges are the foundations of economic and political systems. Transfers and exchanges are largely entailed *normatively* not biologically. Properly speaking, *contra* Wilson, no animal or plant population has an economic or political system at all, although it may have a social order.

It would be interesting to speculate how sociobiologists — who base so much on the ideas of parental investment, genetic determinacy, group selection, altruism (as they define it), and other inherently biological connections among individuals — would treat an economic system which effectively has no *biological* ties whatever, namely our own, the price-making supply—demand market system. Even the early metaphor for it is curiously ambivalent about this: the *unseen* hand. In the economic *system* of exchange, there are, indeed, in a very real sense of the sociobiologists, "no hands." (See Samuelson [1964] on the fallacy of composition; Becker [1976] [1977]; Hirshleifer [1977]; Tullock [1977].)

Fifth, all human societies, at least for the past score or two of millennia, appear to have had music. All musics have rules of linear progression analogous to the linearity of language, but lacking several major attributes (e.g., reference, negation, predication, question-formation). At the same time, they all have multi-level structures not parallelled in language, for example, polyrhythmy, polytonality, polyphony, harmony, orchestration, and combinations of the preceding. Music is a highly structured form of expression directly connected with emotions, without meaning or character if *lacking* articulation with the emotions, and a major mode of interpersonal, even inter-cultural emotional expression. Thus, one of the most abstract, one of the formally most arbitrary, one of the most "rationally" rule-controlled, one of the stylistically most localized, one of the most diffusable, one of the most changeable expressions of human culture — i.e., in its *specific* forms, clearly one of the most ineluctably *cultural* aspects of being human — is also the art form perhaps most directly connected with Wilson's "irrational", irruptive, bestial, and violent "emotions." One suspects the problem is Wilson's conception of "emotion", not anything that exists "in nature." (See Langer [1942], [1953], especially p. 27ff.)

Finally, all humans exist in a universe in which structures and meanings are "known" to humans only on the basis of postulation. The capacity or "drive" to postulate may be biologically—genetically based, but *what* is postulated and *what* is known has no known genetic basis at all. Further, in all socio-cultural specifics of behavior, humans respond to *what* is postulated. When postulation goes, meaning goes too — including Wilson's or the SSG's sciences and epistemologies. Finally, postulation is intimately connected with human reflexivity, discussed below. Many other human universals might profitably be discussed — humor, art, suicide — each with uniquely human characteristics, as well as some universals that are also shared with certain other species.

For Wilson and his antagonists, the problems concerning universals are the following. First, many appear to be strictly human and occur in all human societies. *Logically*, therefore, these are *species* characteristics and must be dealt with as such by Wilson's antagonists as "human nature". These universals are no mere arbitrary abstract categories or reifications, but domains of common human experience descriptively established by the social sciences and the humanities which both antagonists denigrate (see also Barash [1977], especially p. 109). The proof of commonality lies in the fact that, *in principle*, anyone with time and effort can learn culturally different forms — as any anthropologist, or, say, art historian, knows. That is, there is universal translatability from culture to culture *within* domains and, in certain senses, from domain to domain within a culture (e.g., symbols of prestige can be translated into resources of power).

The *fact* of the universality of the various experiential domains requires an explanation in terms of inherent — ultimately genetic — human capacities, as Noam Chomsky has argued for language. That there should be such species characteristics seems to me wholly unobjectionable on genetic grounds. I cannot understand why some of the geneticists of the SSG reject it. The problem is to formulate the *characteristics* of this unique kind of genetic foundation and to specify how it constrains and shapes human behavior, not to reject it out of hand.

On the other hand, the cultural *expressions* of these universals vary immensely from population to population, move around among them, appear and disappear. This is an aspect that simply cannot be coped with at all in terms of Wilson's simplified genetic causality. Clearly, what genetic determinacy there is does not apply to any particular expression but only constrains all-possible-expressions. This aspect the sociobiologists — Trivers, Wilson, and Barash, in particular — treat wholly inadequately or not at all. In fact, Wilson beat a drastic retreat in his claims for a sociobiology which would account for human behavior and subsume the social sciences by saying that perhaps only 10% of human behavior is genetically determined (*New York Times*, Nov. 9, 1975). What a sad social science it would be that explained only 10% of its materials! Wilson's disregard for the fact that human sociocultural features can be created, diffused, learned, lost, and translated into each other is clearly seen in his distressing Chapter 27 on human beings. Note the following quotation (Wilson [1975], p. 575):

The transition from purely phenomenological to fundamental theory in sociology must await a full, neuronal explanation of the human brain. Only when the machinery can be torn down on paper at the level of the cell and put together again will the properties of emotion and ethical judgement come clear. Simulations can then be employed to estimate the full range of behavioral responses and the precision of their homeostatic controls. Stress will be evaluated in terms of the neurophysiological perturbations and their relaxation times. Cognition will be translated into circuitry.

As a final comment here, please note that what I have presented above is *not* an argument for cultural relativism, but one about the character of human mind: a mind structured for, or capable of, non-built-in, impermanent, arbitrarily-constructed, cultural expressions in enormous variety. *That* is the — or a — human species characteristic, i.e., a part of human nature.

V. RULE BREAKING AND REFLEXIVITY

All "normal" humans know that they can break rules: 'After forty everything one wants to do is illegal, immoral, or fattening' goes the saying. This sense of rule breaking is itself an important fact, related to points raised just above against Wilson. It is one any child can tell you about.

More recently, however, I have become aware, through the photographic work I mentioned, of a much more fundamental sense of rule breaking. I began to discover that, despite great variation, I have tended to compose pictures in characteristic ways: centering, balancing, heavily emphasizing diagonals, "moving" from a relatively empty left to a rather full right, etc. The last rule was identified in the context of a broader discussion that this sort of "motion" or "reading" was *universal* in stills, whether in painting, graphics, or photography. Stacy (1977) argued that the foundations for such left–right patterns were genetic. Though exceptions were, of course, immediately pointed out to him, the various forms of centeredness, balance, left–right patterning as Stacy specified them are not merely my rules, not merely rules of Western art, but appear widespread among cultures in time and space. But the exceptions are also widespread. If these "rules" have a genetic component, it is still the case that *recognition* or *discovery* of *any* pattern, rule, or norm allows humans to break "rules" and change behaviors. Reflexivity, recognition, and rule-breaking of this sort are all strikingly and probably strictly human.

Clearly, the relation between the "rules", some sort of genetic foundations, and rule breaking must be dealt with by both sociobiologists and their antagonists in some comprehensive theory. Sociobiology must cope with the fact that "rules" are, for humans, at best "tendencies" which can be dealt with only statistically and that they can always be broken. The antagonists must deal with the fact of the statistical tendency toward rulefulness as a human species characteristic and that humans never operate *without* rules. The strongest possible environmentalist position the antagonists can take, I believe, is that departure from the biologically-based human "rule" tendencies is a slow process of cultural evolutionary cumulation, itself an uncertain tendency.

A final note. Reflexivity allows persons to look at their acts, their bodies, or parts thereof, or their psycho-cultural configurations of self as external objects – certainly a uniquely human characteristic. Reflexivity is conceivably a genetically-based capacity, although how it would be so would have to be specified by the genetically-oriented biologists. Together with postulation, it permits rule breaking, including the human rule of postulation itself. Through reflexivity and the breakdown of postulation, humans arrive at suicide – a uniquely human phenomenon, surely known in all societies, but practiced in ways suggesting no significant genetic patterning, that is, in culturally patterned ways such as hara-kiri. Since suicide involves postulation (or its dissolution) and meaning problems, it remains, despite Wilson, a philosophical, a cultural, as well as a biological and social problem.

VI. EMOTIONS AND EPISTEMOLOGY

Whyte, cited above, has argued ([1974], Chapter 1) that fundamental knowledge comes from the inner intuitive world and is merely given justification and communicable form in language. I concur fully (see Leeds [1977a]) and hold that Wilson's epistemology is wholly untenable and very narrowly culture-bound. Aside from the quotation given above, he has explicitly contrasted (in a Comment at Boston University's Colloquium on the History and Philosophy of Science in 1976) the "rationality" of mind to the "irrationality" of emotions. As in all human knowledge, based on postulation as it is, the consequences of this position are built into the premises. This is, of course, as true for Wilson's thought as it is for the ethical philosophers he was attacking (because their position negates sociobiology) when he said on a "Nova" program (Spring, 1976) that *their* consequences derive from *their* premises. Our contrary positions will not be resolved here, but it is worth sketching out the issues.

On any reasonable theory of the emotions, human or animal (see Plutchik [1962]), emotions "are" clearly not unitary Things, but very complex processes involving sensory inputs, their cognitive assessments, and evaluation of both the inputs and assessments. The key, here, is that the emotions are always *directed at*, and are about, *externalities* (including, from some sensory locus in the body (see Jacobson [1967]), reactions about other parts of the body). In short, they are object-oriented, assessing external objects and their states of being, contexts, and dispositions. That is, the emotions are, as Whyte argues, the sources of *basic* knowledge, while language and all logical forms derived from it merely translate this knowledge into our major form of communication by abstracting and decontextualizing (with much loss of *objective* information). One consequence of this view is that most of the central core of the "subjective"/"objective" distinction falls to pieces (see Leeds [1974]). "Rationality" and "irrationality" display themselves equally in the worlds of the scientists' postulations such as 'the ether', 'phlogiston', 'quarks', or 'ant "slavery" and "queens",' and in all human beings' object-oriented emotional sorting out of meanings through sensory scansion of the external world.

Among humans, however, the emotions take on a special character which Wilson, from the perspective of non-human animals, seems thoroughly unaware of and with which, given his paradigm, he cannot deal. All known humans, at least for the last 50,000 to 100,000 years or more, have lived in cultural environments. I mean this in two fundamental senses.

First, all human beings shape their environments and do so by cultural means, especially technology, in terms of cultural, normative conceptions, formulated as goals and ends. The degree of shaping varies, of course, with the effectiveness of the technology and probably with the scope of the conceptions, but shaping, in some degree or another, takes place (as it does among innumerable animals and even among many plants — a point Wilson inadequately deals with) in some culturally, that is, *not* biologically determined way. It is quite obvious, of course, that tools made by humans are not genetically determined at the same time that these same tools are evolutionarily increasingly involved in the shaping of human behavior.

Second, all humans define their environment conceptually (postulation, again). All human action is directed *only* at objects so defined and given value. In effect, whatever is undefined conceptually does not exist, although, as external analysts, we may say that these cultural non-existents in fact affect the culture carriers. The point is, however, that from the point of view of all culture-carriers, they live exclusively in culturally conceptualized environments — including the very Nature that the Enlightenment — and the great French Sage, Claude Lévi-Strauss — set against Culture. For example, the concept that there is a "struggle" between "Man" and "Nature" is entirely a cultural artifact — a relatively recent Western one — implying an ontology and epistemology of the sort Wilson accepts as "natural": "Rational" Man *vs.* "Bestial" Nature. By his very acceptance of it as "natural," rather than a cultural artifact of this time and that place, it becomes ideology, as well as being philosophically naïve.

From the point of view of the development of the individual, human beings' chief mode of knowing, the emotions, are *necessarily* shaped from birth by externalities which, in a double sense, are all cultural or culturalized, and also encoded in language, itself a cultural form and the basic repository of "reason." The very form and content of the emotions are, therefore, necessarily *cultural*, not bestial. By the very nature of the human beast, emotions, the most fundamental basis of human knowing, encompass all the *logics, rationales, and rationalities* which Wilson denies them, including even Wilson's narrowly conceived "reason" itself. Indeed, the human beast, including even Wilson, lives in an almost (but not quite) tautological world: the means of knowing are shaped by what the means of knowing permit us to create, in an *almost* closed circle.

This view of the nature of human knowing is at sharpest variance both with Wilson's conception of human nature and with his epistemology. Since *Sociobiology: the New Synthesis* is built on these two conceptions, much of the logical structure and interpretation would collapse if they were untenable.

In particular, his views on the genetic basis of human behavior become still more ambiguous or entirely untenable if one disavows his conception of emotions in human nature and epistemology.

His antagonists, on the other hand, must cope with this characteristically *human* phenomenon, one which I see no worthy objection to calling an aspect of human nature and which must have its roots in the genetically-based biology of *Homo sapiens*.

VII. SEXUAL DIMORPHISM

Wilson's antagonists have accused him of "sexism" in his sociobiology: the pervasive reading into animal and human life of the *particular cultural norms* of relations between the sexes in American life. Although I think the assertion true, the SSG fails to deal with systematic cross-societal expressions of sexual differences which I discuss here as sexual dimorphism. On the other hand, the "rules" of sexual dimorphism can be, and are, broken, a fact the sociobiologists must cope with.

I was led to the subject of sexual dimorphism by interest in the use of the human body as a set of tools (levers, pincers, borers, containers, etc.), by observation of a number of pregnancies, post-partum conditions, and early childhood diseases, disabilities, and child "management" problems, and by reading a paper on osteoporosis (y'Edynak [1976]). The sexually lopsided − or dimorphic − distribution of various characteristics all at once struck me and suggested further research (begun in an exploratory way in 1975–76). This preliminary evidence is very strong: hopefully research in depth will follow.

The general proposition is as follows. Under primitive techno-social conditions, there is a sharply marked tendency for statistically relatively standardized cross-societal *forms* of the division of labor to occur, though necessarily varied in *content* because of divers technical and geographical variables. Hypothetically, the patterning of the division of labor is based on systemic aspects of the sexual differences between male and female humans. If, *a priori*, one denies that such differences have significant effect, as Harris (1974, p. 84) suggests, one tends not to look for them. But if one asks if it is possible that they have significant sociological consequences and pursue the question systematically, interesting things begin to emerge.

Two properties of these sexually dimorphic features strike one. One is that the characteristics involved pattern consistently. That they are distributed differentially by sex is demonstrable and the *directions* of these differences tend to fit each other. That is, they are patterned for the males and patterned for the females. The other property is that, with one exception, the dimorphic features are never exclusively male or female — the features occur in both sexes — but the distribution of their variations tends to be heavily lopsided sexually. The deduction from this second point is that the effects of these dimorphisms ought *sociologically* to appear as *tendencies*, not absolute dichotomies. Because the features occur in both sexes, men and women can both do the same work, and sometimes do, but because the features have very lopsided distributions, the division of labor is significantly differentiated by sex. The absence of exclusivity is demonstrably related to choice — that is, to reflexivity and "rule"-breaking. The one exception mentioned above is, of course the pregnancy—child—birth—lactation sequence. This exclusively female set of attributes is accompanied by *statistically* significant dimorphisms of body build, especially of the pelvic area and leg articulations, as well as of gait and possibly other motor behaviors such as squatting. Not clearly related to reproduction are accompanying, statistically significant dimorphisms in leg-bone proportions, foot structure, etc. All these clearly suggest differential male/female behaviors, in part linked to reproductive functions, selected for during the primate evolutionary process leading to *Homo sapiens*. "Man the hunter" and "woman the gatherer" (and baby-producer) surely have a major genetico-biological basis which is, in my view, folly to deny as long as one recognizes that the behavior observed is a variable statistical distribution, not a biologically absolute requirement (as queen bee performance is, for example). Women *can* hunt and men *can* gather; they both do so. But in the divisions of labor observed, they tend statistically not to. In my language, the rule, though it can be broken, tends to be observed. The cross-societal data leaves no doubt about this.

This argument can be extended by considering numerous distributions. I mention only a few here. (1) The dimorphic distribution of protein (male) and fat (female) deposition which involves differentiated patterns of calorie usage, consistent with differences in hunting and gathering behavior and with widely observed social patterns of food consumption (e.g., more meat going to men); (2) osteoporosis, often related to hormonal changes associated with menopause, occurs much more frequently among women and is associated with debilitating bone breakage, especially of the hip bone; (3) dyslexia occurs mostly among males and appears to be associated with certain cognitive functions of the right or "configurational" hemisphere of the brain which, in my view, are eminently adapted to the markedly configurational aural—visual cognizing of hunting as opposed to the rather linear visual cognizing in gathering; (4) the greater variability of body size of dispensable males under environmental stress which may function to stabilize food resources for the child-bearing and nurturing females.

Though aspects of such dimorphism are found for other animals, especially the higher primates, their human occurrences must be dealt with as part of human nature and their implications even for contemporary human life assessed, unpopular as this may ideologically be for many today. Clearly, the evolution of extra-somatic technology loosens the biological hold; clearly the principle of rule-breaking means that any of these tendencies can be disregarded — and increasingly, in the course of cultural evolution, tend to be. Yet all societies display sexual divisions of labor. Universally, the heaviest work, the long-distance running, and the like are standardly done by the males, while the equally important, relatively sedentary work, as well as the main work of child care, is done by the women. On the whole, this pattern is still evident in our society.

The problem of human sexual dimorphism is one that both Wilson and his antagonists must cope with. It requires major research. The sociobiologists, especially Trivers, Wilson, and Barash, must review their entire approach to the relationships between the sexes, especially where humans are concerned, particularly their highly invidious, individualistic conception of genetic competition between the sexes.

The antisociobiologists must deal with the possibility of systemic, rule-generating, distributions of male and female human characteristics seen in relation to socio-cultural evolution. The characteristics discussed occur in all human populations. That is, they may be seen as species characteristics (though some may overlap with other animals), hence as aspects of human nature rooted in biology.

VIII. CONCLUSIONS

In the controversy before us, the polar positions of the sociobiologists and the anti-sociobiologists have led to stalemate. Despite the opposed content, the antagonists were, and remain, in basic ways locked into the same rules of the game. The result is that crucial questions have not been posed because crucial aspects of human beings remain unexamined, namely, the virtually universal appearance of attributes, most of them uniquely human, in all kinds of domains; the great variability of the form and content in which these universals appear; the detachability, diffusability, learnability, and losability of these forms and content; and mankind's peculiar capacity to break practically any rule, including even, perhaps, those epistemological rules which are rooted in the biology of his sensory equipment by virtue of the cultural evolutionary capacity to create observational tools of an extra-somatic kind. In a generic sense, the variability within the species universals, the spatial and temporal unfixedness of the universals, and the rule-breaking, are core aspects of human nature.

Both protagonists and antagonists flee especially from those central capacities of humans — of being human — which underlie these core aspects, namely postulation and reflexivity. Their statements have indicated both scientific and philosophical inability to treat the non-biological, purely human dilemma so powerfully expressed by Benjamin Nelson:

> Culture always cries out to be regarded as symbolic form translating experience as dramatic design. Depending upon one's perspective, mood, or philosophic tradition, the design is either celebrated as the ultimate revelation underlying all appearance or exposed as sheer convention barely concealing the world of chance. On this view, culture in the sense of form is man's supreme albeit most ambiguous discovery. Were it not for the intervention of human concern, the flux of nature and time would seem without distinction or direction. Events intrinsically empty of meaning or at best agonizingly equivocal in implication achieve the status of a representative symbol; come, indeed to constitute a higher Truth through the human device of Postulation and the human production of consensus induced by postulation.[6]

For those who have agonized at the sheer edge of convention, where ambiguous culture and its postulational underpinnings erode and reflexive, rational, emotional assessment of self in a universe empty of meaning follows, Wilson's *Sociobiology*, for all its monumental amassing of data, will have little to say. They will find Truth, if little comfort, in Camus's view that the only serious philosophical problem is suicide, or in the view that suicide may be the ultimate of rational and moral acts.

NOTES

[1] Thanks to Lynne Hollingshead, Michael Lieber, and Gloria y'Edynak for thorough readings and criticism of a draft. Basic issues raised by y'Edynak — particularly the Western (and Wilson's) commitment to a characteristic view of causality, hence directionality and Progress — must be treated separately later.

An earlier, shorter, and rather simplified version of this paper was published in (ironically!) *The Wilson Quarterly*, 1977. That version omitted material presented here, to which considerable addition has been made. I have also made some clarifications and stylistic revisions. The relationship to Benjamin Nelson's thought was lost entirely in the 1976 publication, but is quite explicit here: it germinated in my earliest contact with him in 1956. An immeasurable indebtedness to Ben Nelson can be seen (I like to think) in much of my work. I always felt him to be one of the two or three greatest intellectual influences of my life – that kind of influence which sets problems and guidelines for a lifetime. Not the least of his being a cynosure for me to emulate was his style – perhaps, one day, I shall achieve writing as elegantly as he did. The quotation from Ben Nelson which closes this paper has, for me, been almost as influential stylistically as it has in my thinking about my subject matter and life in general these eighteen years.

[2] On one hand, Wilson says (*Sociobiology*, p. 550), "Variation in the rules among human cultures, however slight, might provide clues to underlying genetic differences, particularly when it is correlated with variations in behavioral traits known to be heritable." The whole section from which this is drawn attempts to geneticize human behavior. After the sharp attacks of the SSG, e.g., 1975, Wilson saw fit to be more cautious, one might even say weak: "I see maybe 10 percent of human behavior as genetic and 90 percent as environmental." (From an interview in *New York Times*, Nov. 9, 1975). The two positions are incompatible.

[3] Eichenbaum (1978) points out that similar neurological structures are often correlated with radically different behaviors, while dissimilar neurological structures are often correlated with practically identical behaviors. The similar neurological structures are presumably determined by similar genetic structures, while the dissimilar ones cannot be. In short, no direct inference from genes or genetic structure to behavior can be made even if we know a great deal more about genes, especially those of humans, than we do. It is essential to deal with the problem of levels and reductionism which Wilson addresses only with respect to the non-reducibility of biology to the lower level units of physics and chemistry, but obviously rejects as relevant to the problem of the social sciences and humanities (see quote on p. 224). See also various passages in Eisenberg, 1972.

[4] One key technical issues of this entire literature is the unstated presupposition of the striving of animals to reproduce their own genes (e.g., see Wilson's Chapter 1 on the "morality" of the gene). The assumption builds in an untenable, in fact undemonstrable, teleology to make the system work. It is not actually an empirically-derived element of the theory but a metaphysical axiom historically derived from the Aristotelian conception of a prime mover. If one considers the axiom false, as I do, the entire theory collapses. Striving necessarily implies some form of cognition, however much one hides it under the notion of 'programming' (Wilson, p. 4), here, cognition of its own genes and their action by an animal – a patent absurdity.

[5] In view of the loud and rather acerbic rhetorical noises that occurred after the SSG's opening fire on Wilson, it is important to say something about the membership of the SSG. For example Wade (1976) and others made essentially political responses to the SSG, painting them as firebrands and leftists (associated with Science for the People) who were making basically political comments. Aside from the fact that these responses misrepresented what the SSG actually said on scientific matters (see my text above for some of the assertions of a scientific order), it *was* also part of the SSG's point that the very *character* of sociobiology was political in nature, deeply influenced in its most central conceptions by the present world in which the text was written. It is important to re-examine the SSG documents in terms of their *scientific* criticism partly because Wade and others, by omission, did *not* make plain to their readers that these criticisms were coming from scientists *from a variety of disciplines*, fully as eminent and competent

as Wilson. Some of the responses to the SSG comments were presented as an individualized and personalized "battle between giants" (Lewontin and Wilson) completely failing to deal with the collective and multi-disciplinary criticism by fully qualified scientists that had been put forth. Among the very eminent scientists were Jon Beckwith, Stephen Chorover, Ruth Hubbard, Stephen Gould, and Richard Lewontin, to mention only a few. There were also reputable and equally competent, although perhaps less well-known younger scientists, e.g., H. Inouye, Robert Lange, Michael Lieber, Larry Miller, Herbert Schreier. The disciplines included anthropology, biology, genetics, micro-biology, medicine, physics, philosophy, psychology, psychiatry, political science.

[6] The quotation is from Nelson (1964, p. 142), a paper originally read at a symposium I organized at the Annual Meeting of the American Association for the Advancement of Science, Philadelphia, December 1962: 'The Structure of Meaning Systems'.

REFERENCES

Barash, David P. 1977. 'The New Synthesis'. *Wilson Quarterly* 1 108–120.
Beach, Frank A. 1978. 'Sociobiology and Interspecific Comparisons of Behavior'. In Michael S. Gregory, Anita Silvers, and Diane Sutch (eds.): *Sociobiology and Human Nature*. pp. 116–135. San Francisco: Jossey-Bass.
Becker, Gary S. 1976. 'Altruism, Egoism, and Genetic Fitness: Economics and Sociobiology'. *J. of Economic Literature,* 14 817–826.
Becker, Gary S. 1977. 'Reply to Hirshleifer and Tullock'. *J. of Economic Literature* 15 506–507.
Eichenbaum, Howard. 1978. 'Neurology and Sociobiology'. Paper delivered at the Symposium *Sociobiology: Man, Nature, and Culture.* Farmingham State College, October 16.
Eisenberg, Leon. 1972. 'The *Human* Nature of Human Nature'. *Science* 176 123–128.
Harris, Marvin. 1974. *Cows, Pigs, Wars, and Witches: The Riddles of Culture.* New York: Random House.
Hirshleifer, Jack. 1977. 'Shakespeare vs. Becker on Altruism: The Importance of Having the Last Word'. *J. of Economic Literature* 15 500–502.
Jacobsen, Edmund. 1967. *Biology of Emotions.* Springfield: Charles C. Thomas.
Langer, Susanne K. 1942. *Philosophy in a New Key.* Cambridge: Harvard University.
Langer, Susanne K. 1953. *Feeling and Form.* New York: Charles Scribner's.
Leeds, Anthony. 1974. ' "Subjective" and "Objective" in Social Anthropological Epistemology'. In R. J. Seeger and Robert S. Cohen (eds.): *Philosophical Foundations of Science . . . Boston Studies in the Philosophy of Science.* Vol. XI, pp. 349–361. Dordrecht and Boston: D. Reidel.
Leeds, Anthony. 1977a. 'Culture-Filledness, Complexity, and Co-occurrence of Emotions in Human Knowing.' In Symposium *The Epistemic Status of Human Emotions* (organized by A. Leads), Annual Meeting, American Association for the Advancement of Science, Denver.
Leeds, Anthony. 1977b. 'Sociobiology, Anti-Sociobiology, and Human Nature'. *The Wilson Quarterly* 1 127–139.
Leeds, Anthony. 1981/2. 'The Language of Sociobiology: Reduction, Emergence, History, Social Science, Normativeness.' *The Philosophical Forum* 13 161–206.
Leeds, Anthony and Valentine Dusek, ed. 1981/2a. 'Sociobiology: The Debate Evolves.' *The Philosophical Forum* (double issue) 13 (2–3) i–xxxv, 1–323.

Leeds, Anthony and Valentine Dusek, eds. 1981/2b. 'Sociobiology: A Paradigm's Unnatural Selection through Science, Philosophy, and Ideology.' *Ibid.*, pp. i–xxxv.
Nelson, Benjamin 1964. 'Actors, Directors, Roles, Cues, Meanings, Identities: Further Thoughts on "Anomie"'. *Psychoanalytic Review* 51 135–160.
Plutchik, Robert. 1962. *The Emotions: Facts, Theories, and a New Model*. New York: Random House.
Sahlins, Marshall D. 1976. *The Use and Abuse of Biology: An Anthropological Critique of Sociobiology*. Ann Arbor: University of Michigan Press.
Samuelson, Paul A. 1948. *Economics: An Introductory Analysis*. New York: McGraw-Hill.
Sociobiology Study Group (SSG). 1975. 'Against Sociobiology'. *New York Review of Books* 22 47–48.
SSG. 1976. 'Sociobiology – Another Biological Determinism'. *Bioscience,* 26 182, 184–86.
SSG. 1977. 'Sociobiology – A New Biological Determinism'. In The Ann Arbor Science for the People Editorial Collective (ed.) *Biology as a Social Weapon*, pp. 133–149. Minneapolis: Burgess.
Stacy, Paul. 1977. 'Response Manipulation in Movies'. Paper presented at the *Conference on Culture and Communication*. Temple Univ., Philadelphia.
Trivers, Robert L. 1971. 'The Evolution of Reciprocal Altruism'. *Quarterly Review of Biology* 46 35–57.
Trivers, Robert L. 1972. 'Parental Investment and Sexual Selection'. In B. Campbell (ed.): *Sexual Selection and the Descent of Man, 1871–1971*, pp. 136–179. Chicago: Aldine.
Trivers, Robert L. 1974. 'Parent–Offspring Conflict'. *American Zoologist* 14 249–264.
Tullock, Gordon. 1977. 'Economics and Sociobiology: A Comment'. *J. of Economic Literature* 15 502–505.
Wade, Nicholas. 1976. 'Sociobiology: Troubled Birth for New Discipline'. *Science,* 191 1151–1155. (March 19.)
Washburn, Sherwood L. and Elizabeth R. McCown. 1975. 'Man: A Product of the Evolutionary Process'. Paper presented at the Annual Meeting, American Anthropological Association, San Francisco, December.
Whyte, Lancelot Law. 1971. *The Universe of Experience: A World View Beyond Science and Religion*. New York: Harper, Torchbooks.
Wilson, Edward O. 1975a. *Sociobiology: The New Synthesis*. Cambridge: Harvard University Press.
Wilson, E. O. 1975b. Interview, *New York Times*, November 9.
y'Edynak, Gloria. 1976. 'Estimating Life Styles from Human Skeletal Material: A Medieval Jugoslavian Example'. In J. Friedlander and E. Giles (eds.): *Measures for Man*. Cambridge: Harvard University Press, Papers of the Peabody Museum.

Prof. Anthony Leeds
Dept. of Anthropology
Boston University
College of Liberal Arts
232 Bay State Road
Boston, Ma. 02215
U.S.A.

ELENA PANOVA

SUBSTANCE AND ITS LOGICAL SIGNIFICANCE

It may be said that philosophy came into being together with the problem of substance and even that it sprang from it. Substance, as a 'prime cause', or prime 'matter', embodied the ancient conceptions of the nature of the world as a whole; it was a notion dictated by the persistent necessity to relate both observations of experience and theoretical conclusions to one and the same reality. Yet it was only in classical Western philosophy that substance became the center of philosophical speculation. What is it that makes things real, that makes them belong to one and the same reality? This question is radical for every science because the answer to it demarcates science from religion and superstition alike, for, there is no science of non-existing things. Even during the Middle Ages no scholar took 'diabolology' seriously, because no '-ology' can be set up of something, the reality of which cannot be proved.

A careful analysis will inevitably lead us to the conclusion drawn by the ancient philosophers that the immediately given does not yet guarantee the reality of any object: we may be dreaming, hallucinating, or seeing mirages, yet remain at the same time firmly convinced that the object we see is real. For the subjective idealist *'esse est percipi'* (i.e., sense perception as a subjective process is the only case in which the existence of the objects of cognition is possible). Such is also the meaning and basic significance of 'ostensive definitions'. However, a materialist can never agree with a treatment of the reality of an object as dependent upon the perceiving human consciousness. Only if the object exists as a 'thing-in-itself' independently of perception, will a materialist call it real.

For modern man, and especially for the modern scientist, there can almost be no doubt that the best way to reveal the reality of something is to 'force' it to act, i.e., to make manifest its interaction with the surrounding reality. Modern physics judges the reality of the strangest properties of its imperceptible objects above all by means of the changes which these objects bring to their environment. "The world does not consist of finished and finite objects, but represents an agglomeration of processes", as Engels wrote in *Ludwig Feuerbach*.

What is it that makes things capable of interacting? This question then, is to a large measure equivalent to the question: what is it that makes an object real?

We may give the following answer: If the object does not have any properties — a certain size, structure, hardness, etc. — then it could not possibly enter into any contact with something else and produce any change or become subjected to any change.

But the fact that things come into interaction with each other as a result of their intrinsic properties does not yet explain the principle of interaction itself, because the properties are not the bearers or the subjects of any change. The principle of interaction, and consequently of the reality of the object, presupposes above all, the identity through time of the object with itself, which makes it 'this' definite object, capable of 'acting' and of 'suffering action', of undergoing change.

Could an object, then, be considered an absolute individual entity, identifiable by the properties inherent in it and peculiar only to it, properties which make it something exceptional if not unique? Indeed, it is impossible to find two objects fully identical, sharing all properties, because then they would not be two objects, but one. In this sense, Leibniz's principle of the identity of indiscernibles preserves both its truth and its significance. There can always be a difference in the color, form, hardness, place, etc., which mark off one object from another most closely similar to it. But in spite of this, if the object 'as such' is determined by the *absolute* specificity of its properties, it would not be able to retain its identity even in the course of two successive moments; it would be thus unable to endure even the most insignificant modification of its properties. For, any change would be the negation of some of these properties whose presence is the absolute necessity for this object's identity or presence. The matter becomes still more complicated if we follow up the whole 'biography' of an object, which includes past stages characterized by properties that have disappeared long ago.

Is it possible, then, to find some properties that are essential to, i.e., absolutely specific for, any given object? A property is always a combination of the individual and the universal. That is why, to express the specificity of any object means at the same time to exhibit features it shares with other objects.

The description of the individual and the universal differences and relations constitutes a necessary stage along the road of elucidating the relation of a given object to its environment. However, if the question is posed more widely, as a question concerning the principle of interaction in general, it is clear that we must come to the essence of all things, and that means to their substance.

It was René Descartes who gave us the classical definition of substance as an "essence which needs nothing else for its definition". All further substance philosophers started from this definition but they differ most importantly in their views as to 'what' is this 'essence' and how it can be known.

Engels is certainly right when he considers both Descartes and Spinoza to be "brilliant representatives of dialectics in the New Philosophy". For the assessment of a philosophical teaching with respect to its place in the development of the dialectical view of the world, mere dialectical phrases and formulations are of no great importance. Something much more essential is of importance — how far has this philosophy succeeded in revealing and explaining interaction as a basic principle in the existence of real things. Interaction is precisely the real object of dialectics.

In his endeavors to come to substance, to the essence of things, Descartes came upon dualism — for scientific, religious, and other reasons. He was unable to consider thinking to be a property of matter (much less a product of social life) and so he was forced to view it as an independent substance, side by side with spatial extension, not defined by anything else. He saw the world as divided into two kinds of substance; by definition, then, they cannot interact with each other, because this would mean to delimit them. And so dualism is the death knell to interaction. Descartes found himself unable to explain the obvious correspondence between physical and mental processes — a correspondence which finds its highest expression in human knowledge. And with the aid of some hypotheses about God he tried to abate his dualism: he viewed God as the supernatural being, and thus as a supreme substance, which embraces both other substances and thereby enables them to interact. From that, he only fell into a still deeper logical contradiction, because the two subordinate substances, the material and mental substances, proved to have originated from a substance higher than them, something by definition inadmissible for a substance.

Benedict Spinoza undertook a devastating criticism of the dualism of Descartes with a view to making the understanding of substance as a single essence logically incontrovertible.

By substance I mean that which is in itself and is conceived through itself; that is, that the conception of which does not require the conception of another thing from which it has to be formed (*Ethics*, Part I, Definition 3).

It was on this definition that Spinoza built the whole impressive edifice of his monistic system. On the strength of the principle '*omnis determinatio est negatio*', substance can have no cause outside itself, because this would mean that it was limited, and thereby it would deny itself as substance. For this reason substance can be only one. It also follows from this that substance is indivisible, infinite, eternal. It is all-embracing, nothing has the right to limit it, and it exists of necessity.

These characteristic features, stemming from his very notion of substance, force us, according to Spinoza, to accept substance as being identical with the whole of nature. Nothing exists outside of nature, as nothing exists outside of substance. And insofar as it is the all-embracing essence of everything, nature is God.

The strictly deductive method of Spinoza's exposition, the almost total absence of any explanations of the adopted definitions, make possible a very great diversity of interpretations of Spinoza, particularly this central conceptual substance. Kuno Fischer, for instance, treats Spinoza's substance as 'qualityless cause' of everything existing; whereas Windelband conceives it as 'metaphysical space'. Hegel went so far as to identify Spinoza's philosophy with absolute idealism, and Spinoza's identification of substance with nature meant to him only that the essence of nature is God, where God is concerned as a spiritual principle.

For us, the role of substance in Spinoza is thought to reveal the profound unity of all things. Spinoza has a deep understanding of the fact that if things, despite their infinite diversity, are not united into a single essence, they would not be in a position to interact, to be causes of each other, to exist. "Because of the unity which we see everywhere in Nature. If there were different beings in it then it would be impossible for them to unite with one another,"[1] "finally, if we take the analogy in the whole of nature, we can examine it as one essence."[2]

In spite of the brilliant dialectics which characterize Spinoza's philosophical conception of substance, it is not free from the basic flaws of the abstract metaphysical approach. They manifest themselves with particular force in the relations between creative and created nature, *Natura naturans* and *Natura naturata*, as he called them. *Natura naturans* logically precedes all its concrete creations: "Substance is by nature prior to its affections,"[3] as he says. Thus, the existence of Spinoza's substance is independent of the existence of its individual manifestations. And although substance in a necessary way produces everything else, its reality does not depend upon the reality of all 'created nature'.

Spinoza understood the danger of counterposing 'creative' to 'created' nature. And he strove to throw a bridge between absolute substance and concrete changing reality by introducing the concepts of 'attributes' and 'modes'.

As we have seen earlier, for a thing to retain its identity through change it has to have some essential properties, some non-accidental properties, they are called. And the distinction between essential and accidental properties or qualities, or between essence (or substance) and accidents is characteristic of almost all the preceding theories of substance, including those of Hobbes and Descartes, as well as of most later ones. Spinoza abandoned this differentiation between substance and accident, replacing it by the relation between substance and modes:

... existence is divided into substance and mode, and not into substance and accidents, because accident is only a mode of thinking, since it expresses only one side.[4]

By 'accident' philosophers understand separate qualities (color, smell, form, size, etc.), through which a body is known and characterized. That is why a body, as substance of its accidents, remains beyond and inaccessible to cognition.

Spinoza could not define the individual object as substance and accident, because this would be an admission of the existence of a plurality of substances (as many as are the objects in the world) — something inadmissible for Spinoza because of the definition of substance, as we saw. For Spinoza the individual thing is only a mode, a concrete manifestation of the one and only substance, but not a part of it (unlike bodies with respect to material substance in the philosophies of Hobbes or Descartes). The aggregate of modes constitutes the sphere of *Natura naturans*, but not also of *Natura naturata*. The individual object cannot be a substance of its accidents, because it does not create its own properties but is only a moment in 'created nature'.

On the other hand, the counterposing of substance to accidents means lending a relative independence to an individual quality with respect to the object it is a quality of. But for Spinoza, the accident has no existence or significance outside the object. It can be separated from the object only mentally and that is why he presents it as a 'mode of thinking'.

At places Spinoza also speaks of 'finite substances' bearing in mind the relative independence of individual things, but in principle modes do not exist by themselves. They have an existence only thanks to substance, in it, and through it.

By mode I mean the affections of substance; that is, that which is in something else and is conceived through something else.[5]

The link between substance and modes is effected through the attributes.

The attributes do not exist separately from the substance and the modes. They are the essential definitions of substance, and as such they are eternal, not caused, determined by their own conception and infinite 'in their species'. The attributes by definition are infinitely many (which follows from the infinity of substance), says Spinoza, but he speaks of only two attributes: extension and thought.

Extension and thought as essential characteristic features of substance are at the same time essential definitions of individual things. In this way, thanks to the attributes, the modes which have no essence of their own, receive essence from substance and become capable, though only partially, to express its nature.

Particular things are nothing but affections of the attributes of God; that is, modes wherein the attributes of God find expression in a definite and determinate way.[6]

The penetration of attributes into the world of modes constitutes the basis of Spinoza's conception of the relation between concrete existence and the universal, substantial nature of things. As individual, finite, determined and changing, individual things do not have essences of their own. They exist only thanks to substance and are subordinated to the iron necessity of nature. The modes of substance express the universal essence of nature through that attribute which is manifested in them as their essence. In this way Spinoza protects the modes from being transformed into simple phenomena, devoid of any essential significance. At the same time he avoids counterposing essence as something independent with respect to the concrete existence of modes.

The relative differentiation between existence and the essence of modes then explains the transitional character of the existence of the individual throughout the preservation of its genus. The individual perishes, because it is dependent not only upon the essence of substance, but also upon causes that are external to it.

... no thing whatever can come to naught except through external causes ...[7]

The determination of individual things not only by substance but also by 'external' causes, which other concrete things prove to be, enables Spinoza to avoid the extreme of fatalism. The existence and destruction of the concrete is not predetermined by substance. This relative counterposing of essence and existence, according to Spinoza, does not exist only with regard to God as substance. Essence depends only upon the 'eternal laws of nature', and existence:

upon the consequence and order of causes ... In God, however, essence and existence do not differ and that is why the necessity of his existence does not differ from the necessity of existence.[8]

With the aid of the notion of 'attribute' Spinoza strives as a matter of principle to overcome the dualism of Descartes. Thought and extension are no longer two independent substances, but only two attributes of the single substance, expressing its infinite essence. But, in spite of the underscored monism, this equalizing of extension and thinking as attributes of substance does not allow Spinoza to avoid certain fateful dualistic contradictions.

The counterposing of thought and extension as two attributes of substance makes the philosophical treatment of knowledge particularly difficult, because on this basis it is not much easier to explain the coordination between bodily and mental processes than with the dualism of Descartes. Spinoza seeks relief from these philosophical difficulties in the single substance. Because they express the essential nature of one and the same substance, "thinking substance and extended substance are one and the same substance, comprehended now under this attribute, now under that. So, too, a mode of Extension and the idea of that mode are one and the same thing, expressed in two ways."[9]

It is from here that Spinoza formulates his famous principle: *"Ordo, et connexio idearum idem est, ac ordo, et connexio rerum."*[10]

Although it is not founded on an understanding of knowledge as a reflection of external reality, but on the attributive explanation of thought and extension, Spinoza's thesis of the correspondence between the order of ideas and the order of things is of exceptionally great significance in solving most difficult philosophical problems. Both the ontological and the epistemological significance of all basic notions in the philosophical system of Spinoza follows directly from it. This circumstance which is usually considered by the non-dialectical minds as a great difficulty in elucidating the essence of Spinoza's philosophy, is one of its most interesting results. But, notwithstanding the dialectical profundity and logical consistency of the philosophy of Spinoza, it is built upon a profound logical contradiction which stems from the very essence of its conception of substance.

Spinoza's position cannot accommodate, even with the use of the attributes, the necessary relation of the substance to the world of modes. Substance remains in Spinoza's system as something transcendent towards 'created nature' which, owing to its non-substantial character does not have the right to be recognized as 'existing in itself'. Thus, the individual thing, in spite of some stipulations made by Spinoza, proves to be logically incapable of producing any change in its environment. This makes it a concrete thing not thanks to its concrete existence, but only in its capacity of a bearer of the universal, eternal, and infinite substantial nature. That is why *Natura naturata* can never acquire the creative monopoly of *Natura naturans*, and 'created nature' is always dependent upon creative substance for its own modification and development.

But how can 'creative nature' carry within it the principle of development and change, when it is absolutely indivisible and unchanging? Movement is not an attribute of substance, but is an 'eternal mode', inherent only to the world of modes. In Spinoza, substance has no 'needs' of change. Called to life to explain the possibility of the existence and interaction of things, Spinoza's substance, owing to its remoteness from concrete things, proves unable to fulfil this role to the very end.

Nevertheless, a careful study of Spinoza's theory contributes greatly to the comprehension of the modern meaning of substance. It reveals both the wealth and the necessity of this concept, as well as the metaphysical dangers stemming from the absence of a consistent dialectical approach as well as from the complexity and difficulty of the problem itself.

The dialectical materialistic conception of matter as substance has nothing in common with the old metaphysical conceptions of matter, since the old conception contrasts substance with the entire sensuous diversity of concrete things.

Material substance is obliged, by this conception, to be present both in the most necessary and in the most changeable properties of reality, for otherwise it would lose its universal significance. And insofar as an object manifests its properties only in its interrelations with its environment, its substance must be disclosed in this universal quality of things which links them with one another and enables them to interact. What, in the most general sense, characterizes and permits interaction among things is thus their true substance (the Spinozist *causa sui* excellently expresses interaction, wrote Engels in *Dialectics of Nature*).

There is no doubt that in the most general sense things interact on the strength of their independent existence — in themselves and outside of any consciousness. On this basis, regardless of all differences, they are objectively real and they belong to a single reality. That is why, for a materialist dialectician, the philosophical definition of matter, i.e., its definition as a universal essence, as a substance of the existing, is equal to the "universal property of things to be objectively real" (Lenin).

The identification of matter with objective reality safeguards us against contrasting it as substance with all of its different manifestations. The objectively real, i.e., the material, are precisely the concrete things, with their properties and relations. That is why there is no danger that matter as substance might be turned into a Spinozist or a Hegelian Absolute, into something preceding the particular things. Matter did not exist prior to its individual manifestations. In the literal sense of the word, it is embodied in the great, endless, order of things, in the infinite change and interaction of things. And this is not (Hegel's) 'bad infinity', because it is precisely through it that the true essence of matter is comprehended.

In this way the identity of the substantial significance of matter with the universal property of everything that exists becomes the objective reality, and achieves that position towards which Hegel in his *Logic* wanted reality to arrive, i.e., to reveal what is unified as an immanent essence of everything concrete, the unified which manifests itself and develops in the whole infinite aggregate of things.

At the same time, and still without turning into an Absolute, matter is absolute — precisely as identical with objective reality. Unlike all its concrete manifestations, it is indestructible, eternal. The individual arises, acquired certain qualities, loses others, develops, perishes, and transforms into something else. But in all this infinite reincarnation, objective reality is preserved — matter, which is present in the most insignificant changes of things as well as in the cosmic catastrophes on the largest scale. Matter as an objective reality is eternal, infinite, ubiquitous.

Most of the representatives of classical and modern subjectivist and objectivist idealistic philosophy recognize the existence of matter as the 'sphere of sense objects', or as 'something immediately given', or as 'the other being of the spirit'. But they direct all their endeavors towards rejecting the objective reality of matter, its existence outside and independently of any consciousness and any spirit. This is precisely the rejection of matter as substance. In this sense Russell is quite right in noting that philosophers like Berkeley, Leibniz, and others, though they deny matter as opposed to mind, nevertheless in another sense, admit matter.[11] And as to Russell himself, in his view, "The collection of all physical objects is called matter."[12]

At first sight it seems, then, that matter has found recognition even in the most extreme idealistic teachings. But this recognition is not satisfactory since the existence thus admitted is explained in a distorted way, namely through the mediation of the spirit. Indeed, in this way matter has lost everything, because it has lost its very own, its objective reality.

As substance, i.e., as objective reality, matter is not determined by any other. It can logically be counterposed only to consciousness, which is not a substance or an essence because it does not possess the mark of objective reality; it is only a subjective reflection of reality.

Consciousness is man's attitude towards his environment; it is made possible only as a result of the capacity of the nervous system, and specifically the brain matter to elaborate external stimuli into more or less adequate reflections of external things. And when matter is defined as objective reality, existing outside and independently of consciousness, its prior significance is thereby secured.

The identification of the substantive character of nature with its objective reality reveals the specific philosophical significance of the problem of substance. It is because of its universal meaning that the problem of substance cannot be solved with the methods of the special sciences. If substance is identified with some of the manifestations of matter, which are known to modern physics, we shall obtain not a link of the theory of substance with the achievements of science, but a return to the old, metaphysical transformation of substance into a bearer of the concrete diversity of reality. The atom, and the proton, the neutron, and the photon, as well as all other well-known and still unknown elementary particles, can be 'logically' transformed into complexes of human perceptions or into some manifestation of social experience — unless they are determined as objectively real. It is this universal feature of the existing things that convinces us of their material nature, however uncommon it may be when compared with our initial notions of material things.

Substance is a philosophical concept, the most general content of which lies at the basis of every knowledge. All efforts of science to attain and master reality are devoid of sense, if it is impossible to prove the reality of the objects which the laws and theories of science speak of. Besides philosophy there is no other science having for its object objective reality as such.

Since the time of Plato, philosophy has revealed the dependence of knowledge upon the general essence of reality. Scientific laws (natural and social), even when their domain of application is limited to only a certain kind of phenomena and to certain conditions, have a universal meaning for that very kind of phenomena and those very conditions. (All bodies expand under the action of heat; all bodies fall towards the earth's center; the value of all goods is determined by the quantity of abstract labor necessary for their production; all things have causes . . .) If it were not for its universality, a law would not be a law. No scientific forecasting and no planning would then be possible. Knowledge would never be in a position to rid us of the shackles of empiricism and to pass on from description to explanation of facts. Science would be nothing but blind faith in bare facts.

The unity of the individual, which is a unity of the particular and the general, is the foundation for the understanding of the cognitive significance of processes and all laws of human thinking. Abstraction would be senseless, were not the abstracted and thereby general also contained objectively in the individual and the particular. The process of concept formation through which the essence of the object is reached would be impossible, were the relation between the general and particular not of substantial significance, i.e., were it not able to reveal the nature of things as they are in themselves.

NOTES

[1] Spinoza's *Short Treatise on God, Man, and His Well-Being*, translated and edited ... by A. Wolf (London: A. and C. Black, 1910), pp. 26–27. This is a translation of *Korte verhandeling van God / de mensch en des zelfs welstand* (c. 1661) in Spinoza's *Opera* edited by C. Gebhardt (Heidelberg: Winters Universitätsbuchhandlung, 1925), Vol. I, p. 23.
[2] *The Ethics and Selected Letters*, translated by S. Shirley (Indianapolis: Hackett, 1982).
[3] *Ethics*, Part I, Proposition 1, p. 32.
[4] *Ethics* [Passage not identified – Ed.]
[5] *Ethics*, Part I, Definition 5, p. 31.
[6] *Ethics*, Part I, Proposition 25, Corollary, p. 49.
[7] *Short Treatise* ... p. 146.
[8] *Ethics* [Passage not identified – Ed.]
[9] *Ethics*, Part II, Proposition 7, Scholium, p. 67.
[10] *Ethica* in *Opera*, edited by C. Gebhardt (Heidelberg: Winters Universitätsbuchhandlung, 1925), Vol. II: Pars II, Proposition VII, p. 89. ("The order and connection of ideas is the same as the order and connection of things." *Ethics* ... translated by S. Shirley (Indianapolis: Hackett, 1982), Part II, Proposition 7, p. 66).
[11] Bertrand Russell, *The Problems of Philosophy* (London: Williams and Norgate, 1912), p. 22.
[12] *Ibid.*, p. 18.

Dr. Elena Panova
Dept. of Philosophy
Sofia University
Sofia 1000
Bulgaria

ARMAND SIEGEL

TRACKING DOWN THE MISPLACED CONCRETON IN THE NEUROSCIENCES

My aim in this essay is to show how far it is possible to defend the quality of 'humanness' against the scientific world view, under stringent (and, I would say, appropriate) conditions.[1] The conditions are: first, not to resort to anything in the nature of vitalism or teleology — by modern standards, a not particularly stringent condition; second, which is, I think, quite stringent, not to resort to anything in the nature of holism or organicism, nor appeal to hypothetical schemes of thought or laws of nature of a non-empirical kind, like dialectics. In effect, I propose to defend this quality, which we may think of for purposes of this discussion as tied in with such felt, relatively noncognitive entities as 'the spark of life' or free will, sense of self or sense of one's uniqueness, without challenging the scientific reducibility of life to the laws of physics, and without postulating any laws of physics not in the textbooks today. The defense I speak of is not new. It was first put forth some fifty years ago by Alfred North Whitehead, in the early chapters of his book *Science and the Modern World* (Whitehead [1925]). I shall take up only Whitehead's very simple proposals of his early chapters, without the speculative material further on. I shall argue, first, that this stark message of Whitehead's deserves new attention because it is all that one can reasonably expect; and second, by means of various observations including a proposition that I call a 'quasi-theorem', which is a continuation of Whitehead's thought, that it is really enough, if properly and not very complicatedly developed, and therefore deserves the further attention of philosophers.

On this issue I am an absolutist: I refuse any recourse to such devices as lodging the essence of humanness, or of life, in an infinite limit unattainable in finite time, or the similar device of construing them as infinite-dimensional, with reduction being a projection onto finite-dimensional spaces (Smythies [1969]; Frankl [1969]). To me, these are essentially reductionist devices. If life, or human identity, are not to be reducible, they must be so absolutely, here and now, and there must be no flaw in the title. It is not that I am entirely out of sympathy with these devices; I am myself a part-time reductionist like any other scientist, and I do consider them important buffers against the everyday excesses of reductionism. But I want to see how far we can get by refusing the use of crutches, by facing the minimal nature of our defenses. I want to find the measure of the irreducible kernel of human existence.

First, I would like to express my sense of the urgency of the problem, by enumerating and characterizing some of the spectacular recent successes of the life sciences and particularly the neurosciences.

(1) The generalization, and the theoretical and practical manipulation of the immune system.

(2) Genetics and heredity: genes and their replication, the process of determination of tissue and embryonic development, and the central role of identified substances such as RNA and DNA. Genetic engineering, already past the threshold of reality.

(3) Hodgkin and Huxley's physical–chemical explanation of the process of nerve conduction in the neuron membrane.

(4) Biochemical correlates in mental or neurological disorders: L-dopa in Parkinsonism, lithium in manic-depressive psychosis, neurotransmitters.

(5) Neurosurgery: the localization in the brain of such functions as language, spatial perception and affect.

(6) Biofeedback: a paradigm based on the parallelism of the body to a cybernetic machine.

(7) Piaget's discoveries in cognitive development in children.

(8) The statistical study establishing a genetic component in the etiology of schizophrenia.

To this list I add one area which has not yet acquired the status of work achieved, but whose potentialities have struck me particularly because of their profound reductive import and because of my first-hand knowledge of them through my own research: the phenomena of petit-mal epilepsy. In seizures of petit-mal type, practically the entire bulk of the brain goes into enormous electrical oscillations of incredible regularity (relative to its normal electrical activity, as recorded by electroencephalography), for all the world like a fairly simply constituted electrically active bowl of jelly. As the EEG trace is seen on the oscilloscope: the mainly random normal wavering of the moving point of light, then suddenly and without warning a burst of rhythmic pulsations consisting of a characteristic rounded dome followed by a sharp spike, repeated and repeated some dozen to sixty or seventy times, no two pattern units exactly alike to be sure, but all variations of a recognizably common theme. During this seizure the patient is motionless, staring fixedly as if into space, and unconscious. What is important about this unconsciousness is that it is tied in with the seizure and, so to speak in a coequal way: It is not an unconsciousness due to some invasion of the sphere of consciousness by the seizure as a disturbance external to itself; it appears that consciousness and the seizure activity may be manifestations of the same or of closely allied mechanisms. Thus the syndrome of petit-mal epilepsy forces us, at least provisionally, to think of the highest, most 'human' of all mental functions as inextricable from one of the most mechanical forms of brain behavior.

The point about the discoveries I have enumerated is that they are reductive, and they suggest, if indeed they do not demonstrate, a relentlessly accelerating momentum of reductive explanation that will invade and encompass every function that we consider characteristic of human life. There are examples among these discoveries of every kind and nearly every level of reduction, and the list, as a whole, constitutes a stunning demonstration of reduction, as a whole. We find physico-chemical reduction, indeed reduction to molecular terms, in the discoveries in genetics, the biochemical correlates and the lithium treatment; reduction to organismic, 'modular' entities, such as the immune system; logico-mathematical reduction of variables, in the case of nerve conduction; positivist reduction in the biofeedback paradigm, which is tied in with Skinnerian behaviorism; reduction to spatial terms, in the neurosurgical localization results; reduction to explicitly non-linear explanatory systems, in the Piagetian cognitive models. I lump these all together: reduction is reduction, and (at the primary level, at least) I do not consider some kinds more harmless than others. Let me explain.

In the first place, the method of scientific theory is to proceed by the abstraction of common qualities from individuals, followed by the expression of these results in terms of abstract elements such as number, space and time, or forces or tendencies, and finally the manipulation of these by abstract logical and/or mathematical rules. All of the discoveries I have enumerated consist of relations induced, i.e. abstracted, from the unique individual occurrences that gave rise to them. I consider it reasonable to refer to this process as reductive; it reduces a set of unique experiences to certain elements they have in common. This reduction, the rendition of that which is unique and concrete in terms of abstractions, is the essence of the scientific world view. Hence, for example, the degree of complicatedness of a scientific model is hardly a measure of its non-reductive character, at least in an absolute sense.

Pursuing the same train of thought, the most drastic kind of reduction is the most general kind, *viz.*, the basic Galilean–Cartesian–Lockean reduction of secondary to primary qualities, the essential paradigm of physical science. Here is the real locus of the problem. This was realized by the Vitalists, who sought to interpose a non-reducible element into the chain of reasoning between the body as a collection of molecules and the living organism. It is also realized, at least implicitly, by materialist scientists, non-Vitalists, who object to a mechanistic atomic-molecular model for biological life. (For numerous expressions of this kind, see Matson's *The Broken Image* (Matson [1964]; also P. A. Weiss [1969] and L. von Bertalanffy [1969]). Their principles prevent them from interposing an irreducible element so they would interpose instead difficultly-reducible elements. But it is hard to see why their presumed more intricate explanations are needed. With a collection of approximately 10^{25} molecules, composed of some 30 atomic species, at one's disposal, even if (contrary to fact, of course) they all behaved deterministically, what reason could there possibly be to doubt such a model's capacity to explain the complexity of life at least on the macroscopic level, provided we restrict ourselves to the abstract, i.e., scientifically measurable, properties? Do the people who belabor us with protests against physical–chemical explanations have any conception of what such magnitudes as 10^{25} really mean? As one example of what they mean, let me merely cite the fact that, by electron microscope observation, it has been found that there are 1.4×10^9 synaptic junctions per mm^3 of the superficial cortex of rat cerebrum! (cited by Bullock [1972]). Try combining this with the fact that there are of the order of 10^{10} neurons in the human brain: who can say that such complex physical–chemical systems could not form organismic entities to one's heart's content? What is the utility of protests, which one can commonly read in the writings of eminent and respected life-scientists (see Koestler and Smythies [1969]), that the entropy concept needs to be improved in some fundamental way in order to be capable of underpinning the natural-selection theory? Do the protesters have any conception of the meaning of the vast quantities of free energy – i.e., information in the abstract sense – pouring in on us from the sun every second, only a minute proportion of which can explain (abstractly again, of course) the organization, i.e., the negentropic properties, of evolved living organisms? (I infer this from the fact that a supply of free energy, which is obviously minute to a much higher order, is sufficient to supply the negentropic needs of ongoing human civilization.)

Organismic biology is thus not logically necessary to explain life in the abstract. And, since it interposes only a complicatedness of mechanism – often belied by the facts, and in any case not really greater than that of the physico-chemical explanation – it is not in any absolute sense a shield against Galilean–Cartesian–Lockean reduction.

What property of life, or of humanness, is then absolutely beyond the reach of scientific reduction? Obviously, only that which is not abstract, i.e., that which is concrete. The alternative to an emphasis on concreteness can only be, if one refuses to embrace mysticism, a more or less complete surrender to reductionism. And what I perceive, on the part of practically all scientists and scientifically literate laymen, is that the overwhelming success of scientific methods has in fact brought about, even in the most humane-minded, a complete, implicit acceptance of the ultimate reducibility of human life; an acceptance tempered only by the belief, or hope, that the practical achievement of complete reduction will never be finished in finite time, by virtue of the previously mentioned suppositions of infinite dimensionality and/or infinite detailedness of the universe that is to be reduced. However, my feeling is that these logical devices are like Trojan horses by which the complete conviction of imminent, if not already achieved, reduction is invading the human psyche. The power inside this Trojan horse, of course, lies in the accelerating pace of scientific discovery, and in the apparently never-failing power of logico-mathematical reduction to penetrate the most complicated systems once the 'right' framework is found: witness the following statement by a young biophysicist, William A. Seitz, that recently came to my attention more or less by chance:

The ... conformational state [of biopolymers such as DNA] is determined by the laws of statistical physics. A mystery, therefore, in biology is the seemingly mechanistic rather than statistical behavior of many systems, and the problem is one of developing statistical justifications ... for this behavior.

Nature is so eager to be reduced that she refuses to be bothered with the elaborate logico-mathematical courtship of statistical mechanics, displaying herself straightaway in the framework of a far less sophisticated reductive theory. This is why I wish to explore the philosophical tenability of a defense against reduction based wholly on concreteness.

Here is Whitehead's summary of the reductive method:

The primary qualities are the essential qualities of substances whose spatio-temporal relationships constitute nature ... The occurrences of nature are in some way apprehended by minds, which are associated with living bodies. Primarily, the mental apprehension is aroused by the occurrences in certain parts of the correlated body, the occurrences in the brain, for instance. But the mind in apprehending also experiences sensations which, properly speaking, are qualities of the mind alone. These sensations are projected by the mind so as to clothe appropriate bodies in external nature. Thus the bodies are perceived as with qualities which in reality do not belong to them, qualities which in fact are purely the offspring of the mind.

And further on:

> The advantage of confining attention to a definite group of abstractions, is that you confine your thoughts to clear-cut definite things, with clear-cut definite relations ... If the abstractions ... do not abstract from everything that is important in experience, the scientific thought which confines itself to these abstractions will arrive at a variety of important truths relating to our experience of nature ... The disadvantage of exclusive attention to a group of abstractions, however well-founded, is that, by the nature of the case, you have abstracted from the remainder of things. In so far as the excluded things are important in your experience, your modes of thought are not fitted to deal with them. [Yet] you cannot think without abstractions.

A corollary of this critique is that those who are under the spell of the scientific world view are prone to a benign cognitive disorder which Whitehead calls the *fallacy of misplaced concreteness*. They misplace concreteness in that they attribute concreteness to the elements of the abstract schemes of scientific theories (and to the elements of mind and body in the Lockean scheme of primary and secondary qualities as well). They literally project the mental constructs of scientific theory into the concrete events they behold: When they look at a material object they imagine it peopled by neutrons, protons, and electrons all dancing to the tune of Schrödinger's equation. And we all do this, of course.

To give the world of the concrete a convincing meaning in the age of science we must answer the two following questions: does it have a general existence? And, if so, what is its importance? Taking up the first question, to avoid triviality we must emphasize the word 'general'. That is, we must go beyond the *personal* experience of unique events. Thus I should like to prove the general existence of irreducible facts, i.e., facts beyond the power of abstraction to account for. I can do this, at least to a limited extent, and in a way suggesting the nature of the boundary between the concrete and the abstract. I shall show, reversing (as it were) the normal procedure, how in some cases concreteness emerges from the abstract.

The construction goes as follows: As everyone knows, the spatial universe can, among other transformations, be reflected as if in a mirror. Thus there exists another universe conceptually possible, in which my reflected twin might be giving this talk while turning the pages with his left hand. Mathematically, the only difference between us is a reflexive one; there is no way of identifying a left-handed or a right-handed person as such, there is no such thing as left-handed or right-handed, there are only two kinds of handedness which are reflections of one another. Only appeal to a concrete operation (see Figure 1) can rescue us from this hopeless roundabout: I can draw a

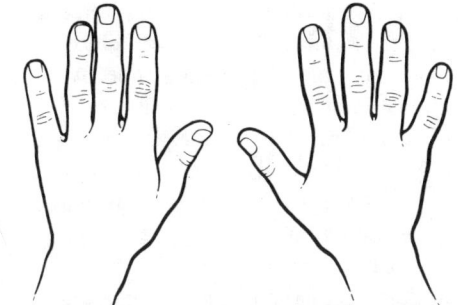

ABSTRACTLY: TWO OPPOSITE HANDS
(MUTUALLY REFLECTED)

ABSTRACTLY: TWO OPPOSITE SCREWS
(MUTUALLY REFLECTED)

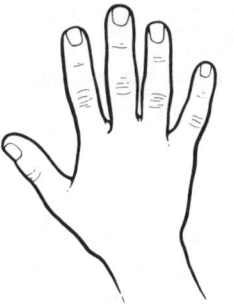   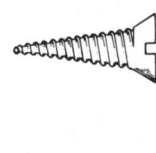

CONCRETELY: CONCRETELY:
A RIGHT HAND   A RIGHTHANDED
                     SCREW

Fig. 1.

picture of a right-handed person, even present one live. This, by ancient convention, defines right-handedness; the definition of left-handedness then follows.

A great many other abstractions used in physics require for their realization a concrete sample; for example, the units of physical quantities — length, time, mass, etc. — or location in space and time. All these cases involve the abstract systems known in mathematics as groups.

Now, to be sure, the concreteness here demonstrated is of a very modest sort, and even has a certain partial element of abstractness: right-handedness is a common property of all right-handed persons, and is therefore an abstraction. But the fact remains that it cannot be defined without resort to the concrete: I must display a drawing, or some kind of sample; no symbolic representation will do it; for example, if I were communicating by radio with beings on another planet, I could convey to them what a left-handed person is only if I knew where this planet was, so that I could speak to my *vis-à-vis* in terms of the configuration of the stars or nebulae visible to him.

Let us now admit the worst that can be said about the above demonstration, which I call a quasi-theorem: evaluated from the abstract point of view, it shows only the thinnest kind of uniqueness in the events of my universe or yours: There are two sets of irreducibly different events (in the case of reflection), those occurring to us and those occurring to our reflected twins; if my consciousness is reducible to space–time atomic–molecular events, it hardly matters whether it is left-handed or right-handed.

But this was to be expected. What has been proved is only the extent of the concrete, the irreducible, *from the point of view of the abstract universe*. Why have I started here? For the sake of drawing what advantage I can from temporarily (in Whitehead's words) "confining attention to abstractions ... [thereby] confin[ing] ... thoughts to clear-cut definite things, with clear-cut definite relations." It is because the demonstration is situated at the limit of the rigorously provable that I call it a "quasi-theorem." Obviously, the claims of the concrete universe, to be of importance, must extend beyond what has been proved here. It is reasonable to suppose that there are entities similar to mathematical groups, but more amorphous and only incompletely abstract in their nature, which account for a much richer area of logical messiness, or concreteness in the world than can abstract groups under my theorem. But the handling of arguments of this kind is a matter of much philosophical delicacy. It is hard to see how the argument can be convincing without becoming abstract and therefore betraying its cause; it cannot proceed by theorems. Therefore I shall have to limit myself to an assertion, not rigorously supported at this time, that the world of the irreducible contains a totality of elements not less important than the world of the abstract.

One useful property of my demonstration is that it says something about the location of the meeting-place between the abstract and the concrete. In this respect it is related to two other general situations in which the concrete and the abstract confront each other: (1) The carrying out of an experiment, regarded as an operation in which the possible outcomes of the experiment predicted from theory are compared to the realized, concrete outcome. (2) The alignment of theory with experiment, i.e., the establishment of correspondences between the elements of the theory and the various possible observable quantities.

Despite these difficulties that abstraction encounters at its borders, I have not put the problem of the alignment of theory with experiment at the forefront of my anti-reductionist formulation: all such difficulties will be solved, in the sense that ingenuity will be unremittingly applied by abstractionists and progress will be made indefinitely. Moreover the nature of these problems depends on current theory and experiment; a problem today can become an un-problem tomorrow. Finally, to rest the case for anti-reductionism on the theory—experiment alignment problem is to commit the fallacy of misplaced concreteness in spades: It assumes that the very essence of life and humanness, which is concrete, is immanent in the findings of science, which is abstract.

I can now comment on the supposed anti-reductionist conceptions of science as an infinite process converging on the ultimate, true picture of nature, and as a process that at any moment gives us a finite-dimensional projection of a true nature lying beyond, which is infinite-dimensional. I don't object to either of these intuitive pictures as reductionist *hilfsmittel* — scientifically they can be very helpful. But it must, of course, be remembered — to be abstractionist about it for the moment — that an infinite process in mathematics does not necessarily converge at all, i.e., it can fluctuate indefinitely, never approaching a definite value; and even if it does converge, it may not converge on what one thinks it converges on. If we assume convergence, and convergence on a true picture of nature, this is a misplacing of concreteness if nature is supposed to include an explanation of life and all our human qualities. The same holds for the projectionist picture, since it also postulates an infinite process; but here the misplacing of concreteness is even more apparent, as the properties of life are logically postulated to exist in a spatial manifold, an image of a visual type.

In consistency with this position, I must disagree with Abner Shimony when, in his paper 'Comments on Two Epistemological Theses of Thomas Kuhn' (Shimony [1976]), he takes issue with Kuhn's assertion that (in Shimony's words) "the progress of science ought not to be construed as the approach to a fixed goal which is the truth about nature." It follows quite clearly from all that I have said that science cannot ever, in any sense, provide 'the truth about nature' because it cannot formulate any conclusions about the part of nature that is irreducible. In particular, I would refuse to accept Shimony's "trapped-in-a-box" characterization of the refusal to regard science as converging toward the truth about nature (regarded in her totality, including her irreducible part); it is not we that are trapped in a box, it is science.

With respect to the second of our questions about the world of the concrete, namely, what is its importance?, I turn back to Whitehead, to his finding that

the ultimate appeal is to naïve experience and that is why I lay such stress on the evidence of poetry.

In effect, poetry (and other arts, to the extent that they function poetically) is the last refuge of the concrete in modern intellectual life.

One finds Whitehead's intuition in this respect to be backed by the critical positions of Cleanth Brooks and Robert Penn Warren (1950) in their famous textbook, *Understanding Poetry* and by Archibald MacLeish (1956) in an important essay, 'Why Do We Teach Poetry?' Brooks and Warren, and MacLeish, against prevailing trends, dwell on the function of poetry of expressing the concrete — as against such things as 'pure realization', or 'the beautiful expression of some high truth'; and, very explicitly, as against science and its method of abstraction. Woven into MacLeish's essay is the thesis that an ethical principle emerges from poetry, with its appreciation of the concrete character of individual events, whereas abstract habits of thought — related to the scientific world view although not characteristic of scientists as individuals — tend to undermine the ethical basis of behavior.

Let me now present a few corollaries of the above. Abstraction is fashionable (and MacLeish realizes this). But the cultivation of abstraction in art represents a surrender to the scientific world view. And if the celebration of the concrete world by art has an ethical function, any abandonment of this function is a serious ethical loss.

In this same connection, I think it is all-important to deny J. Bronowski's thesis (Bronowski [1965]) of the close kinship of science and art. There is only one limited aspect of this thesis that is valid: the scientific theory, or the classic scientific experiment, is indeed a concrete object in the same sense as a poem, a painting or a symphony, and it may also possess a certain esthetic value. But the notion of close kinship can only be sustained by ignoring the overriding difference between the content of science, which is an abstract scheme, and the content of poetry and painting, which is concrete reality. (Instrumental music here seems to occupy a kind of intermediate ground.)

Returning to the ethical aspects of MacLeish's essay, I find it gratifying that a critique of the bounds of the reductionism of science can find itself in such harmony with a fundamental esthetic and a fundamental ethical statement. The unity of philosophy is in this way once more made manifest (I am aware of a close similarity here to existential philosophy). Science contains no values — notwithstanding the sleight-of-hand with which Bronowski brings them in by confusing the *concrete* mode of behavior of scientists with the abstract scheme that is the subject matter. Ethics can only be brought into being by an appreciation of the irreducible reality that is your existence and mine. But going beyond this, I maintain that just as scientism implies the nullity of ethical values, this is equally true of any form of anti-reductionism which accepts the scientific method as capable of comprehending all of reality, and seeks humanness only in some complicated feature of the abstract explanation of life.

Having covered the generalities of my subject, I am now ready to apply my conclusions to the neurosciences and psychology, as well as a recently transplanted physicist may be able. It is clear, I think, from what has gone before, where the 'misplaced concreton' of my title lies in these sciences: it lies in the mechanistic self-image we harbor, resulting from our belief in the reducibility of life, of our personality, our consciousness, our free will, to abstract processes. In this respect, it does not matter how integrated, how non-linear, how organismic (in the purely biological, abstract sense) these processes are; the randomness of quantum mechanics and the integratedness of cognitive behavior as elucidated by Piaget, however valid they are scientifically, constitute misplaced elements of abstractness when projected into a human subject who is then held to be equivalent to these projections.

In this respect, it seems as if I am saying that 'concrete reality' is the famous 'something more' that has to be added to the 'nothing but' of scientific reductionism. But, from my point of view, to 'add something' is an illusory procedure; such an attitude again embodies the fallacy of misplaced concreteness. We cannot *add anything* to an abstract scheme that will make it equal to reality. What we have to do is to *restore* something that has been removed.

Let me now go over the roster of great issues that we all know to be at stake in the struggle over reductionism.

First we may discuss consciousness. *One's* consciousness is a datum of 'naïve experience', and, as such, is irreducible. Consciousness *as a generality* must also have a legitimate existence as a phenomenon common to nearly all human beings, since it is such a datum. It is therefore a proper subject for abstract scientific study. While it will be very difficult to study, because of its nature and its location at the boundary between the concrete and the abstract, it will not be this purely relative difficulty that will salvage our sense of human identity and uniqueness; rather it will be the status of consciousness as an existence prior to study or abstraction. We need not be concerned that the notion that, for example, 'consciousness is the result of mental functions elaborated by a process of biological evolution from lower forms of life', need damage anyone's sense of his worth. While this may be true of consciousness in general, it is not true of *one's* consciousness, which has an *a priori* existence, and can be identified with consciousness in general only by misplacing its concreteness.

Another naïve datum is the perception of free will or of the lack of it. Whether free will 'exists' or not, the denial of free will on grounds of atomic—molecular determination of behavior is a misplacement of concreteness. This conclusion is crucially necessary, of course, if a true ethics is to exist. Without it, the government of human behavior becomes a simple matter of the application of power to the guidance of deterministic processes, as advocated by B. F. Skinner.

Ethics requires more than the negation of the denial of free will, of course. It requires that free will exist. But free will is obviously not a naïve datum; like predetermination, its opposite, it is an abstraction. Thus a choice — an existential act — is needed to assert its existence.

The eminent neurophysiologist J. C. Eccles, in *The Physiology and Physics of the Free Will Problem* (1974) summarizes the impressive recent advances that have been made in the neurophysiology of higher brain function. Eccles clearly regards this work as leading to important challenges to the validity of the naïve interpretations of consciousness and free will. He appears to regard himself as anti-reductionist: "I state emphatically that to deny free will is neither a rational nor a logical act." Nonetheless, he postulates mechanisms by which mental functions and physiological functions act together:

> The neurophysiological hypothesis is that the causal action of the pure ego modifies the spatio-temporal activity in the modules of the liaison zone of the dominant hemisphere. It will be noted that this module detector hypothesis assumes that the pure ego has itself some spatio-temporal patterned character . . .

Eccles quotes, as closely related to his own conclusions, a passage by R. W. Sperry:

... the conscious phenomena of subjective experience do interact on the brain processes exerting an active causal influence ... The present interpretation would tend to restore mind to its old prestigious position over matter, in the sense that the mental phenomena are seen to transcend the phenomena of physiology and biochemistry.

This work is an undeniably valid scientific enterprise, and essential for the abstract understanding of consciousness and will (free or otherwise), i.e., for their elucidation as phenomena. What is disturbing about this essay, and similar ones I have read elsewhere, is the apparent absence in the author's thoughts of any awareness that their explanations are reductive. In saying this, I am resolving an ambiguity in Eccles' reasoning or wording, in which 'pure ego' is left dangling, not a product of physiology and biochemistry, not arising elsewhere either, but somehow 'transcendent'. If we think of it as incorporated in the space-time scheme, to this extent it is reduced (if it is not, we could dismiss the theory as so hybridized as to be devoid of internal consistency).

A straightforwardly reductive theory will unavoidably tend to generate in one's mind a mechanistic self-image, but any resulting misplacement of concreteness that results is apparent and readily corrected. However, an abstract theory that claims to generate anti-reductive bodies residing in itself must be rejected — this is a mechanistic self-image claiming immunity to prosecution.

Eccles's analysis tries to perform an impossible straddling of two worlds: 'pure ego', which *should* be *reduced* in any self-respecting scientific analysis, is presented as a gift from science to the 'humanness' camp. But as a common phenomenon it is not usable; naïve experience needs only *each* person's irreducible sense of self, for which the intercession of science is unnecessary and unavailing. If abstraction cannot even identify a left or a right hand, how can it account for my sense of self?

Future abstract analyses will take back this gift, and do their proper work of reduction on it. They will treat pure ego, as they should, as "A finer breath of spirits dancing in the tubes awhile, and then forever lost in vacant air."

Finally let me discuss psychological theories. The clear separation of abstract and concrete functions frees me to use phenomenological or logico-mathematical methods without any sense of guilt. This, of course, is what the best therapists do anyway, realizing the limitations of theory and the imperatives of self-hood. On behaviorism, I find myself, curiously, in partial agreement with Skinner: *my* freedom and *my* dignity, and *yours*, are not scientifically definable. We can only be thankful that Skinner wants no part of them: If he did, we would be confronted with one more mechanistic self-image claiming immunity to prosecution.

As for Skinner's claim that freedom and dignity do not exist at all, this scientific assertion is easily seen to be wrong in view of the existence and importance of a concrete irreducible universe.

Concerning the debate over neo-behaviorism that has generated so much heat in recent years, the view I have presented regards this as essentially an intra-scientific matter: the two points of view will compete by whatever process it is that determines the success of scientific theories.

Now, I am conscious that neo-behaviorism is not merely a mechanistic theory, but an *extremely* mechanistic theory. The correspondingly extreme mechanistic self-image it tends to generate gives rise to a legitimate need for a psychological theory more adapted to human qualities. It is my impression that Freudian psychology, including its continuation by such successors as Anna Freud and Erik H. Erikson, is such a theory, and one which goes about as far in the 'human' direction as an abstract theory can legitimately go at the present time. This statement is meant to be, as far as may be possible, independent of any judgment on the scientific validity of this theory.

I realize that by saying this I risk being crushed between the lower millstone of the behaviorists, who find Freudian theory 'unscientific', and the upper millstone of those who decry the 'reductionism' of the Freudian theory. With regard to the former, any comment I could make would not fall within the subject of this paper. Concerning the latter, I have already said nearly all there is to say, and I will therefore only summarize briefly. Let me put this summary in terms of a commentary on certain typical manifestations of the anti-reductionist critique of Freudianism contained in the book *Ego and Instinct* by Daniel Yankelovich and William Barrett (1970). These authors bitterly reproach Freud for his deterministic approach:

A mental phenomenon is meaningful only if it occurs as a result of conditions, is clearly predictable, and could always result from those conditions and objects. Why should this be so? Why should it be 'meaningless' for such an event to occur outside such a chain? ... Physicists a century ago would have found it 'meaningless' to speak of physical events other than within a strict scheme of Newtonian determinism. Today, such non-deterministic events are the physicist's daily bread.

But physics is an inappropriate analogy here. Physicists did not welcome indeterminacy as such. Like Freud, and like good reductionists, they pushed the concept of causality as far as they could before abandoning it; and even then they abandoned it only when the principle of indeterminacy was put in an *abstract* form that successfully *reduced* crucial bodies of data that had previously resisted all theoretical explanation. As I have emphasized, determinacy is not the heart of mechanism, nor is indeterminacy the heart of humanness. The physicist's reaction to indeterminacy, like that of the engineer and the psychologist, is to 'plot it up'.

And when these authors express a desire to include the 'human spirit' in the formalism of psychoanalytic theory, I react as before, with a preference that this quality not be compromised by inclusion in an abstract scientific scheme. Does Skinner tell me my sense of consciousness, free will, dignity, are epiphenomena? Free-wheeling psychiatrists tell me they are quite real, and fit subjects of therapy. I find both equally dangerous.

Freud spoke of psychoanalytic theory as the third shattering event, for human sensibility (narcissism), of the scientific era, after the Newtonian and Darwinian revolutions. The reduction to materialistic explanation of the heavens and of man's physical form was followed by that of his psyche. At this point it is time for me to appeal again to my observation that a scientific theory, although an abstract form, is itself a concrete object. And not only is Freud's theory such a concrete object, but our perception of it is part of the tragic sense of life. Let the 'therapy' for this be left to art, or, if you prefer, to religion. F. R. Leavis (1963), in his reply to C. P. Snow's plea for the status of science as a culture, tells us that the function of art is to work at the question: "What for — what ultimately for? What do men live by?" The 'naïve experience' of this questioning eludes scientific explanation, but science is tied in with the asking of it.

NOTE

[1] The use of the awkward word 'humanness' is forced on us by the fact that all other seemingly usable terms, e.g., 'humanism', have been more or less adopted by philosophies which accept the scientific world view in a way which seems to me uncritical.

REFERENCES

von Bertalanffy, L. 1969. In: *Beyond Reductionism*, ed. by A. Koestler and J. R. Smythies. New York: Macmillan.
Bronowski, J. 1956. *Science and Human Values*. New York: Julian Messner, Inc.
Brooks, Cleanth, and Warren, Robert Penn. 1950. *Understanding Poetry*. New York: Henry Holt and Co.
Bullock, T. H. 1972. In: *Brain Research Institute Tenth Anniversary Symposia*, ed. by Margaret Babb, *et al*. Los Angeles: Brain Information Service, University of California.
Eccles, J. C. 1974. In: *Progress in the Neuro-Sciences and Related Fields*, ed. by S. L. Mintz and S. M. Widmayer. New York: Plenum Press.

Frankl, Victor. 1969. In: *Beyond Reductionism*, ed. by A. Koestler and J. R. Smythies. New York: Macmillan.
Koestler, Arthur, and Smythies, J. R. (Eds.) 1969. *Beyond Reductionism*. New York: Macmillan.
Leavis, F. R. 1963. *Two Cultures – The Significance of C. P. Snow*. New York: Pantheon Books.
MacLeish, Archibald. 1956. 'Why Do We Teach Poetry?' *Atlantic Monthly* 197 48–53.
Matson, Floyd W. 1964. *The Broken Image*. New York: George Braziller.
Shimony, Abner. 1976. 'Comments on Two Epistemological Theses of Thomas Kuhn'. In: *Essays in Memory of Imre Lakatos. Boston Studies in the Philosophy of Science*, Vol. 39, ed. by Cohen, Robert S., Feyerabend, P. and Wartofsky, M. W. Dordrecht and Boston: D. Reidel Publishing Company.
Smythies, J. R. 1969. In: *Beyond Reductionism*, ed. by A. Koestler and J. R. Smythies. New York: Macmillan.
Weiss, P. A. 1969. In: *Beyond Reductionism*, ed. by A. Koestler and J. R. Smythies. New York: Macmillan.
Whitehead, A. N. 1925. *Science and the Modern World*. New York: Macmillan.
Yankelovich, Daniel and Barrett, William. 1970. *Ego and Instinct*. New York: Random House.

Prof. Armand Siegel
Dept. of Physics
Boston University
Boston, Ma. 02215
U.S.A.

ELISABETH STRÖKER

DOES POPPER'S CONVENTIONALISM CONTRADICT HIS CRITICAL RATIONALISM? OBJECTIONS AGAINST POPPER IN GERMAN PHILOSOPHY AND SOME METACRITICAL REMARKS

Popper's philosophy has evoked much criticism in contemporary German philosophy that is directed in particular against the conventionalist features of his methodology of science. The critics mean that Popper, in these features, breaks his own principles of critical rationality. They accuse him of an 'uncritical dogmatism' and a 'blind decisionism', which is held to be all the more dangerous as it remains hidden under the mask of the claim that philosophical theories — just as scientific theories — are always exposed to the risk of failure. In reality, however, Popper's so-called critical rationalism is supposed to be simply a pseudorationalism, at most a 'halved rationalism' with a 'prematurely finished reflection' on its own presuppositions, which could justify only its own errors and mistakes.

This kind of critique gained far-reaching influence, especially because it was brought forth by a social philosophy in Germany which calls itself *Critical Theory*.[1] It descends from the tradition of the Left-Hegelians and Marx, and distances itself from the traditional understanding of philosophical theory in that it investigates the existent not only as an outcome of philosophical or scientific ideas and tries to understand it as existent, but insofar as it tries to understand it, also from, or within, its social and political context.

This critique has as one of its tasks the elucidation of the hidden presuppositions and the inexplicit social implications of the analytical philosophy of science, especially Popper's. In this respect the propositions of Critical Theory are crystallized in the thesis that scientific theories could be nothing but products and instruments of social practice. The center of the critique is the thesis that particularly the natural sciences were exclusively rooted in a technological interest, that, in other words, the interest which guides scientific knowledge is only an interest in technical availability.[2] The usual self-understanding of science, that science serves knowledge, is supposed to be a self-deception, by which it veils its true interests. The target of this sort of critique is particularly Popper's philosophy which, according to this critique, not only perpetuates this self-deception, but even legitimates it.

To prepare for a discussion of this critique, I would first like to differentiate the general version of *Popper's conventionalism* a bit more exactly and to distinguish three different components in Popper's conventionalist concept. Accordingly, this paper will have three parts:

Part I: Popper's conventionalism of the empirical basis of science ('b-conventionalism').

Part II: Methodological rules as conventions in Popper's philosophy ('r-conventionalism').

Part III: Popper's general 'decision' for critical rationalism ('d-conventionalism').

Evidently, the terms 'convention' and 'decision' have quite different meaning here, so that for each of these parts we are to investigate individually —

(1) what, respectively, 'convention' and 'conventionalism' exactly mean;
(2) how far Popper himself has described and analyzed the problem situation adequately;
(3) whether, and to what extent, the critique against Popper holds water.

I. THE PROBLEM OF POPPER'S B-CONVENTIONALISM

(1) Popper's philosophy, in its point of departure, is decisively determined by a critical discussion with the neopositivists of the Vienna Circle. Against their hopeless attempt to base empirical science inductively on protocol sentences, Popper maintains the insight of the primacy of theory, since observations are only to be 'made' — verbally — in a theoretical framework of inquiry. Accordingly, observations play a role for Popper in science only insofar as they are test-instances for theories. It is not how observations arise that matters here, but how statements about something observable can be justified; and it is this question of justification that Popper deals with as the 'problem of the empirical basis'.

Popper did this extensively only in his *Logic of Scientific Discovery*.[3] But the problem of the scientific basis remains relevant not only in his early falsificationism, with its originally supposed immediate confrontation of a theory with observation data, but also in his later widening to fallibilistic rationalism, since observation also represents a first-rate test instance in Popper's later comparative criteria for the valuation of theories.

Here it is remarkable that, in his treatment of the basic statements, Popper raises not only the *methodological* question of the systematic function of observations for theories in science, but also the *epistemological* question as to the grounds of the validity of observation statements.[4] As far as I can see, this is the only place where Popper makes this important distinction. But this is understandable, too, from his denial of neopositivistic philosophy. As is well known, Carnap and Neurath had first suggested that observation statements could be 'based upon' or 'founded by' perceptions, and that they neither require, nor are capable of, further foundation. Popper makes two objections against this:

(a) He regards this as an inadmissible psychologism. Besides the fact that the relation between perception and the corresponding statements about something perceived seems to be dark and unclear, Popper argues that something non-verbal may not be brought to bear on the foundation of something verbal, for — thus the argument — statements can be logically justified only by statements.

Yet this objection is not sound for several reasons. (I will just mention them here; later on I will turn a bit more thoroughly to a second and more important Popperian objection.) First, the question of relation between perception and perception statements is not a psychological, but an epistemological question. This remains concealed, to be sure, by the inexact formulation of the problem Popper takes over from the Vienna Circle. But as a matter of fact Popper is not concerned with the relation between perception statements and perceptions *qua* psychological data, but with the relation between a perception statement and the perceived fact. Furthermore, Popper achieves nothing against the psychologism of the Vienna Circle by his argument that statements can *logically* be founded only by statements. For this argument is trivial; even the neopositivists would have agreed to it, and one may assume that nobody ever claimed that the relation between perception statements and perceptions as mental processes is a logical relation.

(b) But Popper has a more important counter-argument against the neopositivists: every perception statement, he states, contains 'universals' and thus, to that extent, transcends perception. Yet universals are not reducible to classes of mental processes. Thus, Popper suggests, singular observation statements are also *hypotheses* which have 'the character of a theory'.[5] As hypotheses they are not finally verifiable and require further examination. Since these examinations cannot be continued *ad infinitum*, they must be given up at any given time. Clearly basic statements can only be 'recognized' and thus turn out to be the result of conventions or even 'decisions'.[6] That they are unavoidable in the foundation of science, Popper defends with a pragmatic argument, which is apparently forced by the dilemma of a threatening infinite regress: since one cannot check until the end of the world, but is nevertheless in need of test results for theories, in order to be able to do scientific research at all, there is nothing to do but to recognize certain basic statements, and this is equivalent to breaking off the procedures of testing.

But since they are basic statements which, in a final analysis, confirm a scientific system or refute it, science on the whole is based on conventions.

It is especially this b-conventionalism in Popper's philosophy which led to the objection against him that he had brought an element of arbitrariness into scientific methods, which nonetheless contradicted critical rationality as he has postulated it.

First we have to realize that it is of no help against this critique when Popper emphasizes that basic statements could never be 'founded' by those conventions, but just agreed upon, and that such agreements would be always preliminary and, in principle, revocable. For the fact that the basic statements cannot always be completely revoked, if there is to be science, and also that the accepted basic statements factually are not permanently revoked, shows sufficiently clearly, to the critics' opinion, that Popper — by his concession of b-conventions — has dispensed with any rendering of an account of the 'true' reasons for its acceptance.

(2) To examine this criticism we have first to discuss the question of whether Popper's exposition of the basis-problem is adequate at all. (Of course it is not necessary that the eventual negation of this question implies the incompatibility of b-conventionalism and critical rationalism.) It turns out, indeed, that Popper's b-conventionalism suffers from certain obscurities, and my thesis is that they essentially result from the fact that Popper, though explicitly pretending to differentiate between the methodological and the epistemological question, nevertheless confuses the treatment of both questions unjustifiably. I confine myself to three critical points which I want to bring against the way Popper handles the problem of basic statements.

(a) Popper emphasizes that basic conventions are in principle not to be justified, for the reason that universals occur in them. Therefore even singular observation statements are supposed to be 'hypotheses'. Indubitably it is correct that perceptual evidence is in principle in need of correction, since it is only subjective certainty of the existence of the stated fact. This will not be doubted here. Nevertheless, the way Popper grasps this need for criticism raises my suspicion that there is something wrong with the structure of his argument.

First, we have to keep in mind that the possible doubt of the truth of singular statements — the content of which is claimed to be 'evident' — is, as a matter of fact, ineffective insofar as we do not actually mistrust such observations, as long as no counterevidence appears. It is simply a fact that singular statements, on the basis of perceptual evidence, *are* accepted, and they are, by such evidence, not only 'motivated', as Popper occasionally states, but even justified. For instance, in Popper's famous example 'Here is a glass of water'; the reason 'I see it' is accepted as completely sufficient, and there cannot be another meaningful reason. To talk of a 'convention' in this context — only because this reason does not have the structure of a logical deduction and since both statements quoted above are not connected logically — would be absurd. When, nevertheless, Popper talks of 'convention' and even of 'decision' in this connection, he not only ignores the situational context in which such a sentence is spoken — and, perhaps, even attacked — but also overlooks the epistemological fact that there are other possibilities for legitimation of statements than logical derivations.

Furthermore Popper's reference to the 'transcendence of verbal representation' is powerless, too. There is, of course, something like 'transcendence', but it is important to remark that it occurs not only in the statements about the perceived matter, but already in the perception itself; perception is not only a mental process *qua* a singular datum of consciousness, but also and always perception *of something*. And its object or fact, though being something singular, is nevertheless apperceived as a 'case' or 'instance' of something general, for example, of a class of objects. Hence not only are the *terms* (in the statement) 'water' and 'glass' universals, but also the perceiving of the *objects* water and glass is apperceiving of something general, insofar as I perceive this 'as' water and that 'as' glass. Since Popper falsely assumes that there are, on the one hand, merely perceptions *qua* singular psychological events, and, on the other hand, merely statements with general terms, and since he completely overlooks the perceived object, he fails to take into consideration the intentional structure of perception. Tearing perception and perception statement from one another, Popper is of course unable to find a connection of legitimation between them. So he has to substitute conventions for the missing logical connection of language and perception.[7] Yet this is nothing but simply a rejection of a problematic which Popper was unable to solve. He started from inadequate presuppositions and thus failed to do justice to epistemological and phenomenological facts of perception.

(b) This rejection led Popper to false consequences. By this wrong description of perception he came to the opinion that singular statements would be hypotheses in the same sense as general hypothetical statements. However, that here there exists a fundamental distinction will become clear from the following.

The hypothetical character of general statements (like scientific laws) arises from the truth-functional relations between a general statement and the predictions which can be derived from it, at least by means of certain random conditions. The verification of a general hypothesis would be equivalent to the verification of all predictions derivable from it, for in their predictive content – which is, to be sure, infinite – the content of the general hypothesis would be exhausted. That is to say, the hypothetical character of general statements consists in their non-verifiability and *therefore* is equivalent to the possibility of their infinite testability, i.e., to the possibility of the verification of future observation statements which contradict the general hypothesis.

Yet according to Popper, even singular statements maintain more than what could be covered in a singular perceptual act. He holds that it is only for this reason that predictions could also be 'derived' from observation statements which must be tested by novel observations *in infinitum*.

However, the testability of observation statements is of a totally different kind; it does not depend on logical relations. The truth of the predictions of a singular observation statement is neither logically sufficient nor necessary for the truth of the singular statement to be tested. This is due to the fact that singular observation statements are not truth-functions of other observation statements. If this were the case – as it would have to be according to Popper's argument concerning the hypothetical character of those statements, analogous to that of general statements – then, furthermore, non-falsifiability of the theories would follow from the non-verifiability of the observation statements. But then the decisive argument in favor of Popper's fallibilistic methodology could no longer carry weight. For its claim is based precisely on the presupposition that singular observation statements, unlike general hypotheses, are verifiable. By his claim of the infinite testability of observation statements Popper deprives his own theory of its foundation.[8]

But singular statements are not indefinitely testable like general statements. Needless to say they are, in principle, criticizable by other observation statements. Yet such criticism has another structure of argumentation than the examination of law statements. For this reason singular observation statements should not be characterized as 'hypothetical' but perhaps, to avoid misleading connotations, simply as 'assertions'.[9] Thus it would be said that a singular observation statement only claims that something is the case. For it does not say any more, unlike general statements. An observation statement claims only that its truth-condition are fulfilled, i.e., that the proposed fact exists; yet it does not say how this fact is related to other facts. Only theories do that. In particular an observation statement does not say that something would be the case in the future; that, too, occurs first with laws and theories. Hence Popper is also mistaken if he reduces the acceptance of singular observation statements to conventions or even to decisions, for the reason that, since universals occur in them, they have hypothetical validity like general statements.

Yet my thesis is not only that Popper's arguments are epistemologically insufficient, but also that he has confused the epistemological problematic of basic statements and their methodological questions. This leads to my third critical remark.

(c) The proposed confusion becomes evident in the fact that in Popper's argumentation it is not clear what should, in fact, actually be based on conventions. As his example with the glass of water shows, it seemed up to now as if the conventions in question should be about the fulfillment of the truth-conditions of observation statements, and that can only mean: about the asserted observational fact. Yet the absurdity of such a 'conventionalism' would result from the fact that Popper does not make a clear distinction between the epistemological question of the *content* of observation statements as such, on the one hand, and the question of the *methodological relevance* of such statements, on the other. Certainly in science the two questions are not completely independent of each other, and the term 'content' itself becomes ambiguous here.

One meaning could refer to what is immediately or directly perceived, like the glass of water. This meaning of 'observation content' would correspond in science, for instance, to the moving of a pointer on an instrument, a signal of light, a track in a Wilson cloud-chamber, etc.; in short, everything perceivable during an experiment. But 'content' of a *scientific* observation can also mean – and means above all – that which is interpreted *'as'* that and that. For example, does this movement of a pointer refer to a flow of electricity, or does it measure something else completely? Must it be interpreted perhaps even as a meaningless contingency?

Such scientific interpretation is always within the framework of theories (and also of theories of the instruments), and only insofar as theories are concerned can the content of observations be a matter of convention. Hence one has to distinguish: it cannot be a matter of convention whether and what is to be immediately perceived in an experiment; this is just as little a matter of convention as the truth of the statement about the glass of water. A question of convention is only how the perceived in the experiment is to be interpreted and, furthermore, whether the interpretation is relevant for the examination of a certain theory or not. Basic statements of science must not be identified with observation statements; rather they are observation statements only in the context of a test situation for theories. This is, of course, also well known to Popper, but unfortunately he did not make enough use of this distinction in his argumentation.

(3) In test situations observation statements must be, indeed, recognized and stipulated by conventions. Here is the legitimate place of Popper's b-conventionalism. For whether the fact concerned is at all methodically relevant and how it is to be interpreted is only to be decided by a theory.[10]

Whoever accuses Popper of having finished philosophical reflection on the basis problem too early and of having replaced it by blind decisions, is correct only if he focuses on Popper's 'conventionalism' as an escape from the validity of singular statements. But then he is not aware of the fact that this escape is not a conventionalism at all, but simply epistemological nonsense.

On the other hand, however, one has to state that Popper's methodological b-conventionalism not only harmonizes with the postulates of his critical rationalism, but is even an integral part of it. Two reasons may be pointed out in favor of this proposition. Firstly, Popper's b-conventionalism becomes eminently critical owing to the fact that it is not referred to theories (like traditional conventionalism), but rather to the empirical basis of science which is to be chosen in such a way that theories should be exposed to the most rigorous tests possible. To this extent Popper's b-conventionalism exactly fulfils the fallibilistic program of his rationalism. Secondly, Popper's b-conventionalism disposes of the naive empiricist's mistake that basic statements are supposed to be available, so to speak, once and for all, naturally given by the organization either of our sense organs or of nature as such, or of both together.

As a matter of fact, observation never forces us to accept a basic statement of science; that must indeed be stipulated in the context of theories and the goals which are to be pursued with them. Hence, Popper with his b-conventionalism satisfies the further requirement of critical rationalism *to make manifest* the character of decision in the choice of basic statements, so that the moment of decision in all scientific observation can be laid open and thus becomes available to critical discussion.

Up to now I have not yet discussed the objection of critical theory to Popper that nevertheless the realm of such decisions is not determined by a technical interest. I would like to postpone the discussion of this thesis until the end, for this criticism will occur in modified forms also in the other parts of the conventionalist problematic in Popper's philosophy. Consequently, I do not claim for the moment that the fundamental thesis of Critical Theory against Popper has shown itself powerless. What I propose is that Popper's methodological b-conventionalism does not contradict the principles of his critical rationalism, but rather is an important part of it, and furthermore that the arguments which Critical Theory brings to bear against this b-conventionalism are inadequate, since they do not consider Popper's own misleading exposition of the basis problem.

II. METHODOLOGICAL RULES AS CONVENTIONS IN POPPER'S PHILOSOPHY (R-CONVENTIONALISM)

(1) It is a trivial statement that scientific research needs methodological rules. Less trivial, however, is the inquiry into their methodological status: is it concerned with rules which have been worked out during scientific development in the actual procedures of scientists, and which have proved to be useful; or are rules considered as norms which are supposed to determine how the scientific procedures according to those rules are to be judged, so that it can be decided how science ought to proceed or how it must not proceed?

This clear-cut alternative usually does not arise for the scientists themselves. The failure to distinguish the descriptive and normative character of their procedural rules is closely related to the fact that usually these rules are not explicitly formulated in normal scientific work. This does not occur before the frictionless process of scientific research — for whatever reasons — is no longer guaranteed. But then, the explication of the rules can have completely different meaning. If one explicates a rule because somebody has broken it and therefore meets with rejection or disapproval, then the rule functions as a *norm* one becomes conscious of. Usually it is the elementary prohibitions of scientific discussion which become conscious in this way, e.g., inadmissibility of logical contradictions, or of holding on to something refuted, rejection of vague terms, inconsistent speech, etc.

It could be different in another case in which material difficulties suddenly make the confirmation of past procedures questionable. Then the formulation has primarily the function of clarifying the rules practised thus far. The explication of the rules has this *informative* function especially when specific rules for certain fields of research are concerned. For instance, falsification or corroboration of hypotheses imply a series of singular rules which are usually explicated only if their use becomes controversial. This informative explication of the rules which have been used also serves, secondarily, the understanding of how one *should* further proceed or what one *should* not do in respective crisis situations.

There is also the question of the justification of these sorts of rules which remains in the background for the scientists themselves. But philosophy of science, which reflects on such rules, has to distinguish clearly: does it want only to discover the factual rules of science and analyse them? In this case the point of reference remains the factual work of scientists, and philosophy of science then becomes a descriptive methodology. Or, does philosophy of science intend to standardize scientific methods? This clearly does not mean that it dictates methodological imperatives to the scientists, which they would not accept. Rather it can only mean that methodology develops ideas as to how the existing sciences could be corrected and improved with respect to their goals. Even in this case, however, the point of reference remains the science as it is actually practiced, yet in such a way that now science is regarded as an incomplete realization of an *ideal* of science.

Popper's methodology comprehends itself as a normative methodology. All norms Popper deals with are centered around the principle of the greatest possible risk for scientific theories. The stimulus of criticism against Popper in this point was again Popper's version of the methodological rules *qua* 'conventions'. It should be observed that with this version Popper not only intended to explicate the rules which are actually handled *in* science, but also wanted to bring to bear his fallibilistic methodology as a canon of norms *for* the scientific procedures.[11]

But, hence, the question of the legitimation of these conventions must arise: why, after all, should science be obliged to proceed fallibilistically, and how far could a fallibilistic (and, according to Popper, only a fallibilistic) methodology satisfy the postulate of critical rationalism?

The critics again miss Popper's self-critical reflection on the stipulation of his rules. They reinforce their first objection against Popper's b-conventionalism in the direction of holding not only that Popper, in the case just mentioned, has dispensed with critical reflection too early, but also that an account of his r-conventionalism is totally missing and thus a product of arbitrariness for which critical rationality cannot be claimed at all. The substantial claim of Critical Theory is again that Popper's conventionalist methodology does not merely continue the screening of the fundamental technical interest and its social context, but rather tries to force it normatively.

(2) It must certainly be admitted that Popper has made it too easy here for his opponents, and this, as I see it, with respect to three points.

(a) The conventions intended here by Popper can hardly escape the suspicion that they are indeed to some extent arbitrary. Popper introduces this term 'convention' in express opposition to Dingler's conventionalism, but since Popper concedes that that kind of conventionalism can be consistently carried out, he obviously felt himself required to offer his methodological rules as mere resolutions, made in order to postulate a different concept of science.[12] The impression that these rules are arbitrary is further strengthened by the fact that Popper often speaks of the 'game' of science.[13] So his r-conventions appear at first as certain game-rules which could, in principle, have been stipulated differently.

(b) Certainly, Popper soon attempted to correct this initial arbitrariness of his rules by indicating concrete situations of scientific inquiry in which his rules are followed. In this way, however, his r-conventionalism ends up in an irritating ambiguity between descriptive and normative claims. His proposition that scientific method should be described by his rules 'in the sense here understood' is precisely the expression of this ambiguity.[14] For evidently Popper understands this method in such a way that his rules are actually to be discovered in factual scientific work; namely, exactly where decisive scientific successes have been obtained. We may set aside here how many historical anachronisms play a role in Popper's reports. But if he could hold the opinion that his conventions do reflect precisely the progressive procedures of science, so that they have to be allowed the status of norms exclusively as to those procedures in which science stagnates or even regresses, then it could easily appear that a justification of his conventions, with respect to their effectiveness for the growth of science, would be dispensable.

But it seems to me that it is still an open question whether in the end Popper was correct in speaking of his methodological rules simply as 'conventions', without telling us more about this issue. This leads me to my third critical point, and it is at the same time a metacritical remark against Popper's critics.

(c) It is most certainly not the case that Popper has totally ignored the requirement of legitimizing his methodological norms. In the course of the development of his philosophy it becomes increasingly clear – even if it is not explicitly said – that his norms are the methodological consequence of his 'negative epistemology' according to which absolute knowledge of truth is in principle impossible. But nevertheless it is the idea of objective truth which guides us in our scientific goals to come closer and closer to the truth. Since Popper does not dispense with the idea of objective truth, and since he regards science as dedicated to knowledge – and though he is nevertheless aware that even science is only an all-too-human affair and that error and fallibility belong inevitably to human reason – it appears to be nothing but consistent when he orientates scientific research on his principle of maximizing criticism.

Surely I do not want to say that normative statements are deducible from epistemological propositions. Neither do I want to state that there are no objections possible against Popper's epistemology. But it is not its requirement for correction which matters here but its role for the legitimation of methodological norms. And so far it is quite rational to think, exactly in Popper's sense, that in the frame of such an epistemology a fallibilistic methodology is urged. In other words, if one accepts Popper's epistemology, then its norms appear as *reasonable norms* precisely with regard to a reason which, as a finite reason, cannot further extend its limits than by consciousness of the only preliminary status of what is achieved, since it can best be activated by critical examination of that which has been gained. Whether one can also do justice to the limitation of human knowledge activity by some methodology other than a fallibilist one remains open. But, in any case, if there are possibly rational rules for research other than fallibilistic ones, it clearly does not make Popper's rules irrational, as his critics propose. Popper's rules turn out to be rational even in a twofold sense: the basic principle of maximal risk for scientific theories takes into account the necessity of critical methodological discussion in science, and furthermore it can be formally and rationally maintained as a consequence of Popper's epistemology.

(3) It is for these reasons that Popper's methodological norms should not be offered simply as 'conventions'; rather, they are revealed to be his methodological versions of his epistemological presuppositions. Clearly that does not mean that, on this epistemological basis, Popper's conventions are fully justified, and the same question arises as to his epistemology itself, which by the way has so far not been thoroughly investigated.

Whoever wants to discriminate against these rules as blind resolutions makes it too easy for himself. For he would first have to investigate Popper's epistemology in order to reach a judgement of it, instead of basing his objection merely on the term 'convention'. Yet I suppose that Critical Theory did not dispense simply accidentally with a detailed analysis of Popper's philosophy. For in that case its adherents would have also had to inquire into Popper's claim of scientific knowledge, instead of simply rejecting this claim. Up to now they only propose their own thesis and vigorously deny claims of pure knowledge in science without supporting this denial.

Nevertheless there remains a sticky aspect as to the question of Popper's r-conventionalism. Is Popper not, nevertheless, correct to speak of conventions here, despite the clear motivation of his rules by certain presuppositions of his epistemology? This question brings me to the third part.

III. POPPER'S 'DECISION' FOR CRITICAL RATIONALISM

(1) It is especially in Popper's later works that it becomes increasingly manifest that he did not want to have critical rationalism understood only as a certain methodology of science but that he has ascribed a more general significance to it. More often Popper uses the term *critical or rational attitude* — an attitude, by the way, which shall be obligatory not only for science but also for all domains of life.

I cannot investigate here whether this generalization for areas outside of science can be meaningfully demanded. Yet, however far one wants to draw the lines, the question remains how far Popper is able to justify such wide-reaching claims of stipulating norms as critical rationalism. It turns out that this question leads Popper into a dilemma.

According to his own principles, an absolute foundation of critical rationalism cannot be possible; such a foundation would mean its immunization and thus its self-refutation. But if Popper remains in agreement with his own principles and denies the possibility of an unshakable basis for his own conception, then he must also allow the risk of failure for his critical rationalism, and even in principle admit that it can be brought to failure by 'noncritical', that is, in his sense, by dogmatic, conceptions.

Evidently it is the falsificational kernel in Popper's critical rationalism which entails this consequence: critical rationality would, then, prove itself as such only if it would be ready to give itself up even against dogmatic positions and, thus, would turn against itself. It seems that this dilemma was not completely hidden from Popper and, apparently, it is not by chance that he does not characterize his rationalism as a 'theory' of rationality, as he remarkably does with the older concepts of rationalism from which he distances himself. For such a theory, too, would be in need of critical discussion, whereas Popper calls his rationalism a *general attitude* and requires nothing for it but a *decision*.[15] Thus Popper escapes from the first dilemma, but lands himself into a second one, and even into a paradox. For is it not required that this decision for critical rationality is itself a *rational* decision, should it not be arbitrary? But then there arises the further problem of rationality now being claimed for something that is already preliminary to the application of rational arguments, *viz.* rationality must already be presupposed in order to explain what rationality is supposed to be.

Popper was only able to escape from this difficulty by declaring the decision for critical rationalism to be *irrational*. This happened, to be sure, explicitly only in *The Open Society* and Popper's followers warn against the overestimation of this phrase.[16] Nevertheless ethics is not a science for Popper; and that fits with a proposition Popper quite recently gave in an interview with a German newspaper, saying that even critical rationalism has its limits — for instance, in ethics.[17] By this proposition Popper wanted to claim that ethical decisions are not subject to the rules of the fallibilist methodology, and that it is impossible to argue in favor of any ethical decision in just the same manner we do in favor of scientific statements. But the decision for critical rationalism is an *ethical decision* and, thus, falls out of the possibilities for critical discussion Popper postulates for scientific theories. That is to say that a decision for critical rationalism, which in itself is not a rational decision, implies a concession that critical rationality cannot itself be rationally legitimated and therefore cannot be justified at all. This paradox undoubtedly reveals a weak point in Popper's philosophy, and it is exactly here that the arguments of Critical Theory against Popper gain their support. In this case the main objection of Popper's opponents cannot be so easily dismissed; and Critical Theory is certainly not mistaken in its assumption that a positivistic trace in Popper's philosophy is to be found here.

(2) Nevertheless to my mind an adequate and more fruitful critique of Popper ought to be different from just accusing him of a blind decisionism; the difficulties in Popper's concept of rationality lie deeper than a superficial objection against his 'irrationalism' is able to grasp. It is important to see that this decisionism is the consequence of an intrinsic inconsistency in Popper's philosophy.

As a matter of fact, Popper does make a sharp distinction between arbitrary resolutions and reasonable decisions, owing to the fact that the latter are subject to possible critical discussions, as Popper himself concedes. That he nevertheless admits them to be 'irrational' results from the circumstance that his own concept of rationalism, as he has formulated it *explicitly*, indeed does not permit the decision for his critical rationalism to be a rational decision. But one has to differentiate between an explicit and an implicit concept of rationality in Popper's philosophy — the latter obviously belonging to Popper's background knowledge which has never been thematized by him. Yet both concepts seem to me inconsistent, particularly as to the question of decisions. The inexplicit concept turns out to be much broader and breaks out of the frame of Popper's expressly represented concept of rationality, yet without his having become aware of it.

The aforementioned difference between mere resolutions and reasonable decisions does not fit into Popper's explicitly defended rationalistic program, because it remains confined to fallibilistic arguments; and, furthermore, because the 'critical' part of his critical rationalism only admits of systematic attempts at refutation, whereas, on the other hand, for Popper's inexplicit conception this evidently does not apply. It might not be by chance that Popper, when conceding the possibility of critical discussion for standards, values, and decisions, never refers to fallibilistic arguments, but – admitting differences in the logical status of theoretical propositions and normative proposals – very cautiously, and even vaguely, accepts even 'listening to the arguments of the others' as a 'critical attitude' in this field.[18]

Here indeed it is to be asked whether the concept of critical rationality should not be broadened in such a way that – while the fallibilistic kernel remains intact – nevertheless other strategies of argumentation are also admitted to satisfy the requirement of critical rationality. For it is not evident why, for instance, positive attempts to prove and to ground statements should not also be ascribed the predicate 'rational' under certain limited conditions, instead of stamping them immediately with the label 'dogmatic' as Popper does. The principle of critical rationality would be broken only if one's own opinion should be seen to be applicable by stubborn will. In such a broadened concept of rationality, however, decisions and ethical positions would also be rationally discussable, as most of Popper's followers – remarkably against Popper – demand and regard as possible.

These remarks are, to be sure, only highly imperfect hints, referring to an ambiguity in Popper's philosophy which could deserve more philosophical attention. But at the moment I only want to emphasize that Popper's opponents have made it too simple for themselves when they just reject Popper's apparent blind decisionism.

But, on the other hand, the critics' objection to the prematurely finished reflection, with regard to Popper's own explicitly given statements, is undoubtedly valid; this ought to be conceded even by Popper's defenders. I think it is a merit of Critical Theory to have emphasized this fundamental lack in Popper's critical rationalism. It is here that future discussion of critical rationalism still has a lot to do if it wants it to remain alive, and if it does not want either to destroy it simply by negative criticism of its weaknesses or – as many of Popper's followers unfortunately do – merely to dogmatize it; a danger, by the way, Popper himself has clearly seen.[19]

The task is to take Popper literally and to replace his theory with a better program of rationality, which does not simply falsify the older Popperian one but rather modifies and corrects it, so that it demonstrates Popper's concept of critical rationality within the limits of its achievements, and makes these achievements, as well as its errors and mistakes, understandable.

(3) However, to return again to Critical Theory, one has to remark that it does not claim to make such corrections of Popper's critical rationalism. Popper's missing self-critical reflection is not compensated for in Critical Theory in the sense that Critical Theory tries to reveal the aforementioned intrinsic inconsistency in Popper's philosophy in order to remove it and to improve his rationalism. Rather, Critical Theory expounds a fundamental technological interest in which science is rooted and that, allegedly, remains concealed by Popper's 'pseudorationalism'. Concerning this problem, it extends far beyond mere criticism of Popper's philosophy and is, in its principal significance, in need of fundamental discussion, not only within the framework of the rationalist methodology of science. I would like to make only a few concluding remarks on this issue.

In order to do justice to both sides of the controversy one has again to make a distinction, which Popper's defenders do not take into consideration. They just get excited at the instrumental interpretation of science in the neo-Marxists' camp and merely insist on the scientific claim of pure knowledge; but apparently they believe that, just by doing this, they have already done everything necessary for the rejection of the fundamental thesis of Critical Theory.

But we cannot set this thesis aside so easily. One has to see that Popper's critical rationalism certainly has defects and shortcomings which Critical Theory has correctly pointed out. Yet the objection concerning the veiling of the technological interest in Popper's philosophy is a different matter. These are two different issues which must be kept in mind: first, what is wrong, inadequate, in need of correction and complementation in a theory, and, second, what is brought forth or made impossible by such defects. The latter claim goes beyond the former, and here Critical Theory is challenged to prove its instrumentalist interpretation of science. That has not occurred up to now. But this is not so easy to do, and not in the manner in which Popperians would like it to be done.

There has been too little attention paid to the following point: that the thesis of the technological interest, which supposedly guides scientific research, does not have the status of an empirical proposition which could be examined with respect to certain scientific problems and results from the history of science. The thesis in question is often misinterpreted and inadmissibly vulgarized — even, to be sure, by adherents of Critical Theory.

But Habermas, who has formulated the thesis, did not intend it to describe *how* science factually proceeds, but to explain *why* it proceeds just in that way it does — which is by no means self-evident.

The process of inquiry in the natural sciences is organized in the *transcendental framework* of instrumental action, so that nature *necessarily* becomes the object of knowledge from the point of possible technical control.[20]

In short, this thesis is supposed to answer the transcendental question of the 'conditions of the possibility' of scientific knowledge in a Kantian sense.

This is a new and exciting problem which deserves profound investigation, all the more so because transcendental propositions, too, have to be shown to be acceptable. Its transcendental status specifically requires proof that they are capable of accounting for factual science, i.e., that the transcendental regress to the 'conditions of the possibility' of natural science leads back to the technical interest *with necessity*, as has been maintained.

Yet such a proof of Critical Theory is missing. The plausibility this thesis may have for many people seems to lie in the misunderstood vulgarization. Usually the thesis is taken as an empirical statement, and one believes it to be confirmed by the close interlacing of science and technology, which, by the way, has not yet been disentangled sufficiently in philosophy. On the other hand this thesis harmonizes with, and lends credibility to, a specific interest of the Marxists.

But, above all, it seems to me that this thesis has been motivated by a characteristic confusion which occurs several times in the writings of Critical Theory: it unreflectedly identifies instrumental action of science with technical practice. So, for example, Popper's claim that the selection of basic statements is a part of the application of the theory and that they guide our action is often taken as Popper's unintended concession to the critics' own thesis.[21] But then they overlook the fact that Popper's selection of basic statements is not intended to be 'practice' with regard to the technical availability of scientific theories but, on the contrary, experimental practice is regarded here as an instrument for testing a theory which is to be examined as to its truth content. What Critical Theory confuses here is the distinction between an instrumental activity aimed at *examining theories and their claimed knowledge of reality* and a technical activity by which confirmed theories and the results to be gained by their application are *transformed into technical products*. We have to be mindful that in both cases 'action' is not only differently motivated, but also differently structured. The instrumental practice of science is — and, to be sure, only in the limits in which it has turned out to be successful — already a necessary precondition for the realization of technical practice.

Whoever wants to hold the transcendental thesis of the technical interest of science, would also have to do justice to the fact *that a technical interest is itself already a guided* interest — namely guided *by scientific knowledge*. That such knowledge can only be won by experimental activity forces us, admittedly, to re-examine the concept of knowledge in natural science which traditional epistemology has too exclusively attached to the achievements of thought. But it does not force us to regard knowledge in natural science as something secondary and merely derivable from technical interests.

Also in this respect Critical Theory is a serious challenge to contemporary philosophy of science, especially Popper's. The Popperians should accept this challenge — according to their own principles — so that a critical and rational discussion between the two camps would become possible. We have been waiting for this until now.

NOTES

This paper is based upon lectures I gave in April 1976 at the Department of Philosophy at the University of Pittsburgh and at the Center for Philosophy and History of Science at Boston University. I am very grateful to Prof. Adolf Grünbaum and Prof. Robert S. Cohen for their invitation and want to thank all the participants for their interest and fruitful remarks.

[1] The term 'Critical Theory' was coined by Max Horkheimer, in *Kritische Theorie. Eine Dokumentation*, 2 volumes (Fischer, Frankfurt 1968); [English translation by Matthew O'Connell, *et al., Critical Theory* (Herder and Herder, New York, 1972)] and in *Kritik der instrumentellen Vernunft* (Fischer, Frankfurt 1967); [English translation, *Critique of Instrumental Reason* (Seabury, New York, 1974)]. Furthermore, the critical discussion of Popper's philosophy was motivated by Jürgen Habermas in his 'Gegen einen positivistisch halbierten Rationalismus', *Kölner Zeitschrift für Soziologie und Sozialpsychologie* **16** (1964) 636–359; *Technik und Wissenschaft als Ideologie* (Frankfurt, 1968); and in his *Erkenntnis und Interesse* (Frankfurt, 1968); [English translation by J. J. Shapiro, *Knowledge and Human Interests* (Beacon Press, Boston, 1971)].

As to the historical development of Critical Theory see Martin Jay, *The Dialectical Imagination: A History of the Frankfurt School and the Institute of Social Research 1923–1950* (Little, Brown, Boston, 1973); and 'The Frankfurt School and the Genesis of Critical Theory', in Dick Howard, Carl Klare (eds.), *The Unknown Dimension: European Marxism Since Lenin* (Basic Books, New York, 1972), pp. 225–249.

[2] Jürgen Habermas, 'Erkenntnis und Interesse', in: *Technik und Wissenschaft als Ideologie*, pp. 146–168, Engl. translation in *Knowledge and Human Interests*, pp. 301–317.

[3] Karl R. Popper, *The Logic of Scientific Discovery* (Harper & Row, New York and Evanston, 1968), p. 43, p. 93 ff. (Henceforth referred to as *LSD*).

[4] *LSD*, p. 43.

[5] *LSD*, p. 95.

[6] *LSD*, p. 104; see also p. 106: "Basic statements are accepted as the result of a decision or agreement and to that extent they are conventions."

7 This has already been claimed by Albrecht Wellmer, *Methodologie als Erkenntnistheorie* (Frankfurt, 1967), cf. pp. 139 ff. I agree with Wellmer in several points of his criticism of Popper's treatment of the basic statements; but his objection on p. 172 in his above-mentioned work that Popper, by his b-conventionalism, conveys a blind decisionism to which he gives an illusion of rationality, seems to me not sufficiently founded. (See also my arguments in the following text below.)

8 At least in *LSD* Popper seems to hold that theoretical systems are *definitely* falsifiable. Note his assertion of the possibility of crucial experiments on p. 78. Though they are only permitted as falsifying experiments, they nevertheless require basic statements which are not just agreed upon but have their *fundamentum in re*. See also Popper's further criticism of traditional conventionalism, particularly his opposition against the denial of a sharp divisibility of theories between falsifiable and non-falsifiable ones, pp. 79–81. Cf. also his *Conjectures and Refutations* (Harper & Row, New York, 1968), p. 33 ff., p. 36.

9 Popper himself mentions that basic statements are "statements *asserting* that an observable event is occurring in a certain individual region of space and time". *LSD*, p. 103 (my emphasis).

There is another source of confusion that Popper sometimes 'neopositivistically' identifies basic statements with singular observation statements about *occurrences*, yet, on the other hand, correctly takes basic statements as general observation statements about an observable *event*, especially so in his epistemological arguments (cf. p. 84 *et al.*). See my criticism in my paper 'Falsifizierbarkeit als Kennzeichen naturwissenschaftlicher Theorien' in: *Kant-Studien* 59 (1968) 498–516.

10 Cf. also Lothar Schäfer, *Erfahrung und Konvention. Zum Theoriebegriff der empirischen Wissenschaften* (Stuttgart-Bad Cannstatt, 1974), p. 61. Schäfer emphasizes that Popper's conventionalism is not a result of arbitrariness but the price one has to pay for scientific progress. It might not be by chance that Schäfer comes to a positive evaluation of Popper's conventionalism – contrary to Wellmer, *op. cit.*, which Schäfer refers to – since he has exclusively in mind the methodological question of relevance, whereas Wellmer oriented his critique against Popper only around the epistemological question of the basic statements.

11 *LSD*, p. 48 ff..

12 *LSD*, p. 53 ff., p. 81 ff..

13 *LSD*, p. 37.

14 *LSD*, p. 53 "thus I shall try to establish the rules, or if you will the norms, by which the scientist is guided when he is engaged in research or in discovery, *in the sense here understood.*" (My emphasis). The original German version is still more ambiguous; "... the way we imagine it", instead of "... in the sense here understood."

15 *LSD*, p. 16 ff.; *Conjectures and Refutations*, p. 26 ff., 49; *The Open Society and Its Enemies* (Princeton University Press, Princeton, 1971), p. 228 ff.

16 *The Open Society and its Enemies*, p. 231 ff.; cf. also *LSD*, p. 37.

This phrase caused criticism even among Popper's followers. It was especially Bartley who focussed on the possibility of critical arguments also in the domain of decisions, norms and values. Popper showed himself impressed by Bartley's elaborations and subsequently made corrections in Chapter 24 of *The Open Society*. It is all the more remarkable that Popper nevertheless stuck to his former view, stating again (in *The Open Society*, p. 231): "We are ... free to choose a critical form of rationalism, one which

frankly admits its origin in an irrational decision (and which, to that extent, admits a certain priority of irrationalism.)"
[17] "Even the critical attitude has its limits. They lie, for example, in the domain of ethics". *Deutsche Zeitung, Christ und Welt*, 2 (January, 1976). (My translation.)
[18] *The Open Society and Its Enemies*, p. 225.
[19] In: *Deutsche Zeitung*, Popper states that he does not feel happy about the term 'critical rationalism'. Such a concept appears to me to have the implicit danger of a new dogmatism. If I were to call myself a critical rationalist, this could perhaps lead to a new dogma. The main thing is to avoid what is dogmatic, the incessant critical attitude, *even vis-à-vis the critical attitude itself.* (Italics and translation are mine.)
[20] Jürgen Habermas, *Knowledge and Human Interests*, p. 286.
[21] *LSD*, p. 106, 108. Cf. Jürgen Habermas, 'Analytische Wissenschaftstheorie und Dialektik' in: *Der Positivismusstreit in der deutschen Soziologie* (Neuwied-Berlin, 1969); [Eng. tr., *The Positivist Dispute in German Sociology*, by Theodor Adorno *et al.*, Glyn Adey and David Frisby trans. (Heinemann, London, 1976).], p. 179. Cf. also Albrecht Wellmer, *op. cit.*, p. 150 ff.

Prof. Dr. Elisabeth Ströker
Philosophisches Seminar der Universität zu Köln
Köln
West Germany

A. SZABÓ

HOW TO EXPLORE THE HISTORY OF ANCIENT MATHEMATICS?

I started investigations in the last two decades, which shed — I hope — some new light upon several problems of the *history*, as well as of the *philosophy, of mathematics*. Now I shall try to characterize here not so much the results of this research, as some of the problems dealt with and the methods I followed in these investigations.

Let me begin with a well-known classification, which led me to one of my most important problems. The different branches of scientific inquiry may be divided into two major groups: the *empirical* and the *non-empirical sciences*. The former seek to explore, to describe, to explain, and to predict the occurrences in the world we live in. Their statements are checked against the facts of our experience, and they are acceptable only if they are supported properly by *empirical evidence*. The empirical sciences are generally divided into the natural and the social sciences. (The behavioral sciences are, as I think, mainly social sciences, even if they do not exclude some very important influences on the part of the natural sciences. But this is not my problem now.) In any case, the dependence on *empirical evidence* distinguishes the empirical sciences from the non-empirical disciplines: logic and pure mathematics. For the propositions of logic and the mathematical theorems are proven *without essential reference to empirical findings*. The same division can, of course, be formulated also by saying that the method of all empirical sciences is *induction*, whereas both non-empirical disciplines, logic and mathematics, are *deductive*. Now, so far as mathematics is concerned, the last statement can be completed by a simple observation, particularly stressed in recent times by the mathematician Pólya, who reminded us that

> mathematics has two faces; it is the rigorous science of Euclid but it is also something else. Mathematics presented in the Euclidean way appears as a *systematic, deductive science*; but mathematics in the making appears as an *empirical*, experimental, inductive science.

Indeed, most mathematical propositions were initially conjectures, mere guesswork on the basis of some empirical experience, and they became genuine theorems only later on, when mathematicians eventually succeeded in proving them, i.e., deducing them from some principles of mathematics.

The question which now raises itself is a double one: a philosophical and a historical one. (1) Philosophers have to answer the question: why is it that we appreciate the *deductive certainty*, that is, the proverbial certainty of proved mathematical propositions, more than the so-called *inductive uncertainty of the empirical sciences*? Speaking about this *uncertainty* I mean:

No matter how extensively an empirical hypothesis has been tested, and no matter how well it has been borne out by the test findings, it may yet fail in cases that have not been examined. Never does any empirical evidence suffice to verify a hypothesis, to establish its truth with *deductive certainty*; it can only lend the hypothesis more or less strong inductive support (Hempel).

No doubt, it is owing to our appreciation of 'deductive certainty' that mathematicians always strive to transform their conjectures, the mere guesswork, into genuine proved theorems. But I wonder whether the high esteem, in which we hold the so-called 'deductive certainty' is not exaggerated. So far as I know, there is not yet any satisfactory answer to the question, why mathematicians have to prove their statements, even in the case when these are obvious without any proof. As Pólya wrote:

We may say, with a little exaggeration, that humanity learned this idea — namely the idea of mathematical proof — from one man and one book: from Euclid and his *Elements*.

And so our second, the historical question: how did it come about that mathematics was transformed into a systematic, deductive science? For it was originally, in the ancient, pre-Hellenic cultures of Egypt and Babylon, only a highly developed practical, *empirical knowledge*. Indeed, it is generally accepted today that the Babylonians were able to find quite good approximations to the solutions of relatively complicated mathematical problems about a thousand years *before* the beginnings of Greek mathematics. Nevertheless, there is absolutely no evidence to suggest that the Babylonians (not to mention the Egyptians) ever tried to deduce mathematical theorems rigorously from first principles. It remains an open question, whether such fundamental scientific concepts as 'theorem', 'proof', 'deduction', 'definition', 'postulate', 'axiom', etc., were known to Egyptian and Babylonian mathematicians at all. It is not even certain that the Babylonians knew how to formulate general theorems. Mathematics prior to the ancient Greek civilization was nothing more than a useful, even sometimes an ingenious collection of practical rules, prescriptions as to how some particular mathematical tasks are to be carried out. The fundamental change in the history of mathematics, the transformation of the earlier practical—empirical knowledge, took place in the course of the development of ancient Greek culture. The question is therefore: what was the reason that the Greeks did not rest satisfied with empirical knowledge? Why did they replace the 'collection of prescriptions' by a systematically constructed deductive science? What prompted them suddenly to put more trust in what they could prove by theory — demonstrate or refute — than in what pure practice showed to be correct or incorrect? Obviously, deductive mathematics is born when knowledge acquired by *experience* alone is no longer accepted; theoretical considerations are required even for what experience seems invariably to corroborate.

I think that the most striking feature of Greek mathematics which distinguished it at first glance from its Oriental counterpart is the presence of genuine *proofs*. Greek science is concerned not only with stating propositions, but with providing proofs for them as well. The role played by proofs is no less important in Greek, than it is in contemporary mathematics; in fact it is the same. Furthermore, Euclid's proofs are for the most part models of their kind and have set the standards of mathematical rigor for generations. Notwithstanding the fact that modern mathematicians sometimes find them wanting in one or another respect, the proofs in the *Elements* are by and large exemplary.

Now I can summarize my reconstruction of the historical development of mathematical *proof* in the following. The technical term for '*proof*' and '*to prove*' in Greek mathematics is the verb δείκνυμι. It is used by Euclid at the end of every demonstration. He closes the discussion of his theorems in every case with the words ὅπερ ἔδει δεῖξαι, '*quod erat demonstrandum*' = (being) 'what it was required to *prove*'.

The fact, that the Greek verb δείκνυμι originally means '*to point out*', '*to show*', '*to make visible*' suggests the idea that the earliest 'proofs' in mathematics may have involved some kind of 'pointing out' or '*making visible*' of the facts. We can even support this idea by an example taken from Plato.

In the dialogue *Meno*, Socrates asks a simple, uneducated slave to tell him how the area of a square whose sides are two feet long can be doubled. So as not to be misunderstood, Socrates immediately draws the square whose sides are each supposed to represent a length of two feet. (In other words: he makes visible the square which is to be doubled.) Since Socrates explicitly mentioned that the sides of the original square are *two feet long*, the first idea of the young man is that the area of the square might perhaps be doubled by doubling the length of its sides. Socrates, however, draws a second diagram to *show* that a square whose sides are twice as long as those of the original one has an area four times its size. The slave inevitably realizes that his first attempt at an answer to the question was wrong. He now thinks: a square whose area is twice of the original will clearly have longer sides as well; so the sides of the square being sought will have to be *longer than two feet*. They cannot, however, be four feet long, because a square with four feet long sides has an area which is four times that of the original. Thus the required length must be *less than four feet*. Now three lies between two and four, and so pehaps a square which has sides *three feet in length* will have twice the area of the original square. Once again Socrates responds with a diagram *showing* that a square whose sides are three units long, has an area of nine square units, and hence is too large, it cannot be double the original one. Finally, in his fourth diagram he shows that the square on the diagonal has exactly twice the original area.

I think that this passage from Plato provides an excellent illustration of the way in which statements were verified at an early stage in the development of Greek mathematics, i.e., it tells us how propositions were 'proved' in an archaic sense of the word.

"*A square constructed on the diagonal of another has twice the other's area*" — this was the proposition which Socrates put in the form of a question to the uneducated slave. Diagrams 2 and 3 (see Figure 1) refute the first two answers and make their defects visible. Diagram 4 on the other hand not only shows the correct answer (i.e., formulates the proposition in question), but is at the same time a visible 'proof' of its correctness.

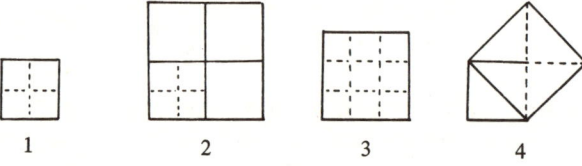

Fig. 1.

It would of course be wrong to claim that there was ever a time when mathematical *proofs* were nothing more than *making the facts visible*. We must not forget: a *proof* is obtained by reflecting upon what has been seen. It is reflection which transforms what we see into visible, empirical evidence. I should like to emphasize only that the *core* of the most ancient mathematical proofs was provided by making the facts visible.

This is, however, by no means the typical form of Euclidean proofs. Euclid, whose work represents for us the most ancient, and, at the same time, the classical form of Greek mathematics, is not concerned with making anything visible. He is more interested in convincing the reader of the truth of his theorems by a train of abstract arguments in which he strives to avoid every visual element. I illustrate this idea — the intended avoiding of visual elements — by the Euclidean proof of two very simple theorems about *even* and *odd numbers*. One of these propositions says:

If as many even numbers as we please be added together, the whole is even.

And the other one:

If as many odd numbers as we please be added together and their multitude be even, the whole will be even.

Both of these propositions must, of course, strike anyone who has had even the slightest acquaintance with arithmetic as too obvious to require proofs. This makes the manner in which they are proved all the more interesting. In both cases the first step is to state a typical instance of the proposition in question. So, for example, in the first case:

Let as many even numbers as we please, AB, BC, CD, DE be added together; I say that the whole AE is even.

It is clear that Euclid thought of the said numbers as line segments, since he denoted each of them by a pair of letters. Similarly, the fact that he called their sum AE, indicates that he must have imagined it as being obtained by laying these segments down one after another in such a way:

A   B   C   D   E

The proof now continues:

For since each of the numbers AB, BC, CD, DE is even, it has a half part; so that the whole AE also has a half part. But *an even number is that which is divisible into two equal parts*. Therefore AE is even. Q. e. d.

It should be apparent that this proof has nothing to do with that kind of 'making visible' which I discussed earlier. It can hardly be claimed that the line segments **AB**, **BC**, etc., are accurate, visible representations of 'even numbers', which *show* us that their sum, the segment AE, can only represent an *'even'* number'. For it is nonsense to suggest that a segment of any kind can be the visible representation of an *'even'* number'. After all, in the proof of the second proposition the same segments (denoted by the same letters) are supposed to represent *odd* numbers. There is in fact no way to distinguish between odd and even numbers by using line segments in the representation. For such segments can always be cut in half. By definition, however, odd numbers cannot be halved, although even numbers can.

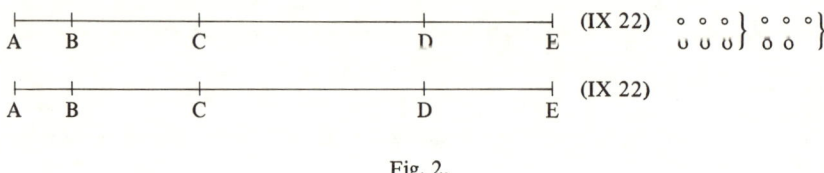

Fig. 2.

So it is true that Euclid routinely observes the convention of illustrating numbers by line segments, but his use of the verb 'δεῖξαι' ('to show', 'to point out') is merely figurative. He talks of *pointing* things *out*, but his arguments are essentially abstract, their steps cannot be seen.

It is even easier to recognize how far Euclid falls short of making the truth of these simple propositions visible (i.e., of 'proving' them, in the most rudimentary sense of word), if his method of demonstrating them (with the help of line segments) is compared with the original one. Then we know that the theorems about odd and even numbers were originally illustrated with the help of *pebbles*, in such a way that even numbers contained as many black as white pebbles, while odd numbers were obtained from even ones by adding or subtracting a pebble of either color. This is a rather primitive technique of proof, but it does make the essential content of the theorem visible. We need only glance at the rows of pebbles to convince ourselves whether the proposition is true. For the pebbles enable us to see whether or not a number is even.

It seems that Euclid abandoned visual demonstrations, because he wanted his proof to be valid *for all possible cases*. This is also why he substituted line segments for the pebbles. For let us not forget: it is not possible to illustrate mixed numbers with the help of pebbles ($5\frac{2}{3}$, $\frac{3}{5}$, etc.). Instead, the line segments are not intended to stand for mixed numbers, but for *arbitrary* even or odd numbers. It was just for the sake of greater *generality* that Euclid also turned to logical arguments. This is why his proof proceeds by recalling that an even number is, by definition, one which can be divided into two equal parts. From this it obviously follows that the sum of even numbers is even, etc., etc.

Now, the idea that Euclid was unable to illustrate even and odd numbers in an accurate manner because he was striving for generality, brings to mind an interesting passage from Plato. I am referring to the one in which Socrates explains that, as far as arithmetic is concerned, numbers cannot have visible or tangible bodies; they are merely ideal elements which cannot be grasped except by means of pure thought.

So we see that Greek mathematics, as exemplified by Euclid's proofs with their avoidance and suppression of the visual, also endeavored to treat its subject matter as something which belonged exclusively to the domain of pure thought. This trend in the development of science was responsible for the most remarkable Euclidean proofs. My third example is just such a genuine Euclidean proof.

There is a proposition in the *Elements* (VII 31) which establishes an interesting property of composite numbers. Before examining it more closely however, let us recall three definitions which are indispensable for its understanding. These are:

VII 2: "A *number* is a finite multitude composed of units."
VII 11: "A *prime number* is that which is divided by a unit alone."
VII 13: "A *composite number* (i.e., one which *is not* prime) is that which is divided by some number."

Now the theorem, the proof of which I should like to examine here more closely, states that "*Any composite number is divided by some prime number.*" The proof of it runs as follows:

Proof. Let a be some composite number. We have to show that some prime number divides a. Since a is composite, there must be at least one number, say b, which divides it — by the definition of composite number itself: 'A *composite* number is that, which is divided by some number'. Now b is either prime or composite, for both definitions of '*prime* number' and '*composite* number' exclude any other (third) possibility. In the former case (b is prime) the theorem is proved, since b divides a, and b is prime. In the latter case, on the other hand, there must be some number, say c, which divides b (again by the definition of '*composite* number'). Now c obviously divides a as well; so if c is prime the theorem is proved. If on the other hand c is composite, there must be some other number which divides it, etc., etc. The proof emphasizes that this process must eventually terminate; otherwise we would obtain an *infinite* decreasing sequence of composite divisors of a. (According to the Greeks a number was always greater than any of its divisors.) The existence of such an *infinite* sequence of divisors for any number *a* would however contradict the definition of *number*: 'A *number* is a *finite* multitude composed of units'. (The *unit* is the smallest divisor of every number.) Hence *a* must be divided by some prime number and the theorem is proved.

a, b, c, ...   (VII 31)   δείκνυμι (δεῖξαι)
                                                      dico, digitus
                                                      zeigen

Fig. 3.

Nothing is '*shown*' or '*made visible*' in this proof. It is true that Euclid uses line segments to illustrate a typical instance of his proposition; he represents an arbitrary composite number (*a*) by one such segment, and its divisor (*b*) by another, shorter one — but this indicates nothing more than his observance of a traditional and outmoded convention. His chain of arguments cannot be seen. If we want to understand it, there is no point in *looking* at the numbers (as represented by line segments); we need instead to keep in mind the *definitions* of the different kinds of number, since the proof depends solely upon these.

Besides the argument not being visible, there are at least *two* other interesting features of the Euclidean proof just examined, the importance of which I would particularly stress here. The first one is: we had to demonstrate the *existence* of some prime number, divisor of an arbitrary composite number, *a*, and we got the decisive proof of it by having *refuted* the 'non-existence' of it. That is, we have made clear that its *non-existence* is not possible. In other words: our proof was a so-called 'indirect proof', a very often used procedure in mathematics, namely one which establishes the truth of an assertion by showing the falsity of the opposite assumption.

The other feature of the former demonstration to be stressed is the following. To get an indirect proof of our statement: "the arbitrary composite number *a* is divided by some prime number," we formulated first the opposite of this statement, in order to be able to refute it, because the refutation of the latter is the proof of the former one. Therefore we said: "number *a* has infinitely many divisors, which are themselves all composite numbers." It is now of interest to realize what makes the incorrectness of the latter statement evident: the fact that it is an obvious *self-contradiction*. For, saying "*a* is a number" means: "it is a *finite* set composed of units"; and it is contradictory to the statement: "it has *infinitely many* divisors." Having accepted the former of both contradictory assertions as *true*, we cannot but reject the other one as *false*.

Now I think that it was precisely the application of indirect proofs that was the most important historical step which prepared the transformation of earlier *experimental* and *inductive* mathematics into a *deductive system*. Before speaking, however, about the origins of this ingenious method, I should like to discuss here, at least in broad outlines, the most famous case of its application in Euclid. (I think many of us will remember that same Euclidean proof, for we all had learned it once in school.)

A well-known theorem of arithmetic (IX 20) states: "*the series of prime numbers is infinite.*" The indirect proof of this proposition runs as follows. Let us formulate the opposite of the statement we want to prove, and we will show that it is false, and therefore the theorem originally formulated must be true.

If the series of primes *were not* infinite, we could write down the complete series of all primes

$$2, 3, 5, \ldots, P$$

where P is allegedly the 'last and greatest prime number'. And now let us see what is wrong with this statement? We can construct the number Q, $Q = (2 \cdot 3 \cdot 5 \cdot \ldots, P) + 1$ (multiplying all primes with each other and adding the unit to the product). Number Q is greater than P and therefore, allegedly, Q cannot be a prime, it must be a composite. (The greatest prime is, supposedly, P.) Consequently, Q must be divisible by a prime. Every composite number is divisible by some prime number. However, all primes at our disposal are, hypothetically, the numbers in the series. Q, however, divided by any of these numbers, leaves the remaining *one*; and therefore the prime number in question, which divides Q, is not any of the allegedly complete series of primes. So we have here a contradiction; namely, we stated at the beginning: "Our series of prime numbers *is complete*." And we must now admit: "The same series *is not* complete."

This contradiction reveals the incorrectness of the statement, namely that we could write down the complete series of prime numbers. No, that statement cannot be true, and so the truth of the opposite statement is evident: "The series of prime numbers *is* infinite."

$$\left. \begin{array}{l} 2, 3, 5, \ldots P \\ = (2 \cdot 3 \cdot 5 \cdot \ldots \cdot P) + 1 \end{array} \right\} \text{(IX 20)} \qquad \text{ὅπερ ἔδει δεῖξαι} \quad \text{ὅπερ ἐστὶν ἀδύνατον ἐν ἀριμοῖς}$$

Fig. 4.

Now, my historical conjecture is that the method of indirect demonstration was not created by the mathematicians, nor were they the first to use it. They took it over, so to speak, ready-made, from the philosophers of Elea. As you are likely to know, the philosophers of Elea used this method to prove a lot of such paradoxical statements as were diametrically opposed to every common-sense experience. Zeno, for instance, proved indirectly the impossibility of motion, although sober experience showed that motion, of course, is real.

I cannot now enlarge here upon the philosophy of Elea. I summarize only some of its most important features that determined the framework of deductive mathematics as well.

First of all, these philosophers contested the reliability of the senses. They said: truth cannot depend on sense-experiences. No matter what kind of testimony is borne out by our senses and by everyday experience, reality and true statements about it can be established only by means of pure thought, because senses and experience are misleading. This is also why Euclid tried to avoid visual arguments in his mathematical proofs. Numbers cannot have visible and tangible bodies, and the truth about them cannot be grasped except by means of thought.

A second feature is this: Parmenides, and in general the philosophers of Elea, considered the absence of contradiction to be the unique criterion of a true statement. I need not emphasize here how, in mathematics, too, the consistency of every closed system must be guaranteed just by the absence of any contradiction.

The decisive role of the contradiction in every investigation of truth also explains why the philosophers of Elea attributed such great importance to the indirect proof in their system. They never proved any of their statements; instead they refuted the opposite statements, pointing out that the latter ones implied some contradiction. For example, Parmenides distinguished three ways of investigation: (1) 'being exists', (2) 'being does not exist', and (3) 'being exists and does not exist'. Then he discarded the second and third way as obvious self-contradictions, and this was the convincing proof of correctness of the first way: 'being exists'.

Now what prompted mathematicians to adopt this peculiar form of demonstration from the philosophers? I think it became possible in this way to prove a surprising fact of mathematics, which otherwise, without this form of thinking, could never have been settled. The point is, namely: early Greek mathematicians observed the difficulty (or better, the *impossibility*) of expressing side and diagonal of the same square by numbers. This follows from the fact that both linear magnitudes, side and diagonal, are incommensurable.

But let us not forget why this concept (incommensurability) — which originates rather in the theory and not in empirical practice — had been introduced into mathematics. The proof says: if side and diagonal of the square were commensurable, the same number had to be *even and odd*. In order to avoid this absurd self-contradiction mathematicians introduced the notion '*incommensurable*'. So we see that the application of indirect proof led to the creation of a new scientific concept. But this was also the origin of the foundation of mathematics upon definitions and axioms.

As I conjecture: it was under the influence of the philosophy of Elea that Greek mathematicians transformed their science into a deductive system. In order to get this into a consistent form, i.e., into one which does not contain any contradiction, they started with some principles which they accepted in advance without any proof. Then the propositions (the theorems) had to harmonize with the unproved principles, fit them without contradiction.

I here illustrate my conjecture with the following examples.

Earlier we discussed the arithmetical proposition: "any composite number *a* is divided by some prime number." We read in its proof: if the prime number sought is not found, that means: an infinite series of numbers divides the number *a*, each of which is less than the other: *which is impossible in numbers*. Now, reading these words in the original, the question which suggests itself is: Why does Euclid emphasize that an infinite process of the sort he has described would be impossible '*in numbers*'? His words seem to imply that there is some other domain in which something similar could happen. This consideration brings to mind Zeno of Elea. For it was he who argued that a body in motion had first to cover half the distance to its destination; before doing so, however, it would need to cover half of this half, and so on to infinity. Zeno was trying to make the point that a body in motion travels a path made up of *infinitely many distances*, each one shorter than the last. The argument can, however, be interpreted as also asserting that any straight line, **AB**, can be split up into infinitely many segments, each one shorter than the last; but this is just another way of saying that **AB** has an *infinite decreasing sequence of divisors*. It would appear therefore, that the proof of our arithmetical proposition bears traces of Zeno's influence. For its author went out of his way to emphasize that the proof was only valid for numbers, and his motive for doing so was probably that he wanted to exclude the case considered by Zeno.

I mentioned at the beginning the division of the different branches of scientific inquiry into two major groups: *empirical* and *non-empirical sciences*. No doubt, all empirical sciences derive ultimately from sense experiences.

However, our senses are unreliable. Besides, the method of empirical sciences is, of course, induction, which involves a great deal of uncertainty. Therefore, what are our empirical sciences if not, as some philosophers put it today, research programs, bold conjectures and honest efforts first to corroborate, and then to refute the same conjectures. On the other hand, of what does the proverbial certainty of deductive mathematics consist? No doubt, the mathematical propositions are established by proof, which shows that the propositions in question follow logically from the principles, so that, if the principles are true, the theorems are certain to be true as well. The certainty of the theorems is thus relative to the principles. There are, however, famous cases in which the decision about the truth of some mathematical principles is more or less *arbitrary*. (This was well known even in earlier antiquity, at least *in times prior to Aristotle*.) I am not referring only to the opposition of Euclidean and non-Euclidean geometry (to the famous postulate of parallels, which is not to be decided *a priori*). There is also another less known Euclidean axiom: "*the whole is greater than the part*." This axiom seems to take up a position not only against Zeno's paradoxical argument (otherwise irrefutable): "*the half of the time is equal to its double*," but also, so to speak, against that way of thinking which is characteristic for modern set-theory: "an infinite subset, a real part of another infinite set, is equivalent to the whole set."

Now, I think that the history of ancient mathematics explored in such a way may essentially contribute to our most actual, present-day problems in the philosophy of science as well.

REFERENCES

Hempel, C. G. 1966. *Philosophy of Natural Science*. New York: Prentice-Hall.
Hempel, C. G. 1973. 'Science Unlimited?' *Annals of the Japan Association for Philosophy of Science* 3 31–46.
Pólya, G. 1945. *How To Solve It*? Princeton: Princeton University Press.
Lakatos, I. 1970. 'Falsification and the Methodology of Scientific Research Programmes.' In *Criticism and the Growth of Knowledge*, edited by I. Lakatos and A. Musgrave, pp. 91–95. Cambridge: Cambridge University Press.
Szabó, Arpad. 1969. *Anfänge der Griechischen Mathematik*. Munich and Vienna: Oldenbourg.

Prof. Dr. Arpad K. Szabó
Pasaréti ut 60/a H-1026
Budapest
Hungary

E. V. WALTER

NATURE ON TRIAL: THE CASE OF THE ROOSTER THAT LAID AN EGG

In 1474 a chicken passing for a rooster laid an egg, and was prosecuted by law in the city of Basel. Now, we are inclined to dismiss the event as fowl play, but in those days *lusus naturae* was no joke. The animal was sentenced in a solemn judicial proceeding and condemned to be burned alive "for the heinous and unnatural crime of laying an egg." The execution took place "with as great solemnity as would have been observed in consigning a heretic to the flames, and was witnessed by an immense crowd of townsmen and peasants" (Evans [1906], p. 162; Needham [1969a], p. 574). The same kind of prosecution took place in Switzerland again as late as 1730.

In the case of the Rooster of Basel, the executioner found three more eggs in him, according to a chronicle of the city. A recent historian, E. P. Evans, reporting the case, refused to believe that part of the chronicle, declaring it absurd, and regarding the event "not [as] a freak of nature, but [as] the freak of an excited imagination tainted with superstition" (Evans 1906, p. 162). Evans knew that eggs presumed to come out of roosters caused panic because people used to believe that cocks' eggs were used in witchcraft. Moreover, the same egg, according to an ancient folk belief, might produce a dreaded monster known as the basilisk or cockatrice — a malignant, winged reptile with the head of a cock and the tail of a serpent which destroyed men and things by its breath and its glance (Robin [1932], pp. 84–95). Regarding the superstition as absurd, Evans considered it equally absurd to expect eggs from roosters. Like most of us, he assumed that any egg under suspicion could ultimately be traced to a hen.

Yet, neither the history of poultry nor the intimate record of the barnyard will support the dogmatism of Evans. In the winter of 1921–22 the *National Poultry Journal* of London was full of news about a chicken being exhibited in a poultry show at Westminster which attracted a great deal of interest. A Buff Orphington with the voice and manners of a hen but with the plumage of a rooster was reported to have laid eggs. The truth was obscured because a practical joker kept putting all kinds of eggs of different colors from different breeds into the pen. But a student at the Eyresford Training Centre, overcome by the zeal to know the truth, performed a crucial experiment, which has almost been lost to history. In the January 6, 1922 issue of the

National Poultry Journal, Mr. H. M. B. Spurr reported his observations. On December 22, a fellow trainee had noticed that the bird in question had laid an egg. On the 24th, Mr. Spurr, who was on trapnesting duty, found the bird in No. 4 nest box, No. 6 pen, at 11:45 a.m.. He immediately took possession of the poultry house key, locked the house, and continued his tour of duty. Returning at 12:25, he released the bird, and removed a small but typical Buff Orphington egg. In his own words, as reported in the *National Poultry Journal* of Great Britain, "Seeing is believing, Sir, and although previously doubtful I am now assured ... that this 'cock of the south' does not lie — it lays" (Cole [1927], pp. 98–99).

On this side of the Atlantic, the case of the Rooster of Madison was reported in the *Journal of Heredity* in 1927. The bird had the plumage of a brown leghorn cock, but was laying eggs, and was acquired as a curiosity by the Poultry Husbandry Department of the Wisconsin Experiment Station in 1922, the same year that the Buff Orphington was attracting attention in England. To a casual observer, the Wisconsin bird had the full plumage of a rooster, but anyone intimate with poultry would have recognized that the head and the body of the bird were rather effeminate. In order to settle the question of egg production conclusively, the bird was put into a padlocked wire cage, with a screen that would not allow eggs to be sneaked into it, in a closed room. An unlaid egg in the oviduct had already been detected, and the following morning, a normal brown leghorn egg was found in the cage. Following that, the alleged rooster laid eggs about every other day (Cole [1927], p. 99).

Plate I is a picture of the Rooster of Madison, taken on February 17, 1922, together with three of its eggs, arranged from left to right in the order they were laid on February 13, 15, and 17, respectively. Cole concluded that if Evans, writing in 1906, could have seen this picture, he would not have been so ready to doubt the chronicle telling that the executioner cut open the Rooster of Basel in 1474 and "found three more eggs in him." The Rooster of Madison continued to lay regularly over a long period of time (Cole [1927], p. 99).

An explanation may be found in the dynamics of sex expression in poultry. Under certain conditions, a fowl may take on secondary sex characteristics that contradict its reproductive anatomy. Farmers know that sometimes hens exhibit male plumage and other masculine attributes, and studies have shown that this condition is associated invariably with tumors or some other diseased condition of the ovary. If the ovary is removed, a hen grows male feathers, and the changes that accompany progressive ovarian tumors resemble the sequelae of ovariotomy. Some hens may experience a temporary disturbance of the ovarian function and grow male plumage but, as the ovary returns to normal, they continue to lay eggs. Meanwhile, the bird's dress will be out of harmony with its physiology, and it must wait until the next molt before it looks like a hen once more.

The alleged Rooster of Madison in the spring of 1922 did not fool a certain white leghorn male. She was mated with that authentic rooster, behaving, in spite of her appearance, like a normal female, and lived happily ever after. The eggs hatched out of that union produced chicks who grew into ordinary barnyard citizens and lived uneventful lives. She recovered the normal plumage of a female in the fall molt of 1922, returned to the existence of an ordinary hen, and disappeared from history. Plate II shows her, photographed on November 11 of that year, nine months after the previous photograph (Cole [1927], p. 100).

L. J. Cole, a geneticist at the University of Wisconsin, suggested that such a return to normal, while probably less common than a progressive course of ovarian disease, happened often enough to explain the references to 'cocks eggs' in ancient and medieval times. Contrary to Evans's assumption that the Rooster of Basel was framed, and that the eggs were really produced by some other bird, Cole believed that the accused bird had actually laid those eggs. He concluded that "its guilt lay in looking like a cock when it was in reality a hen" (Cole [1927], p. 105).

Let us examine comparatively the social impact of alleged roosters who have thrust upon them the reputation of laying eggs. The Rooster of Madison provokes little more than a chuckle – perhaps appearing as a curiosity in the pages of a newspaper, if the events were to happen today, or possibly inspiring a notice in Ripley's *Believe It Or Not*. A mechanistic explanation drawn from biology calms any disturbance we might feel. In contrast, the Rooster of Basel in the 15th century had gathered an enormous crowd and had generated a wave of fear and excitement. But in traditional China, Joseph Needham tells us, when an apparent rooster laid an egg, the chicken would go unharmed, but the provincial governor or even the emperor might be in serious trouble. He could be impeached and removed from office ([1969a], p. 575), for such a rare and frightening event would be regarded as a reprimand from Heaven. Although one finds numerous accounts of sex reversals in man and animals in Chinese literature, and even enlightened discussions of the phenomena, these occurrences remained prodigies. Seers and diviners pondered their implications for the future and for the affairs of state (Needham [1970], p. 310). Nature and society were expected to

Plates I and II. The alleged Rooster of Madison.

remain in a condition of organic harmony, and if the harmony were disturbed by the appearance of biological anomalies, it was often assumed that the emperor or some other great official was at fault. Needham argues that animal trials were unthinkable in China because the Chinese were never so presumptuous as to pretend to know what God had in mind for delinquent roosters. Besides, the notion of the law of nature as a command that should be enforced was alien to the Chinese.

In Western civilization, Needham reminds us, the laws of nature (in a scientific sense) and natural law (in a juristic sense) shared a common root ([1969a], p. 518) which had coercive implications. Things and animals and people were commanded to behave according to the rules given by the transcendent legislator, subject to divine sanctions. The Chinese sense of natural order, in contrast, depended on an idea of inevitable cooperation. In Needham's words:

The harmonious cooperation of all beings arose, not from the orders of a superior authority external to themselves, but from the fact that they were all parts in a hierarchy of wholes forming a cosmic pattern, and what they obeyed were the internal dictates of their own natures ([1969a], p. 582).

Law of nature, in the Western sense, Needham suggests, may have reached the limits of usefulness. Western science is abandoning mechanical causation for organic causality in a

great movement of our time towards a rectification of the mechanical Newtonian universe by a better understanding of the meaning of natural organisation [1969a], p. 291).

Modern science is being obliged "to incorporate into its own structure" an organic view of the world that is typically Chinese ([1969a], p. 286). Yet, Needham suspects that the old Western view of natural law may have been an essential phase in the rise of modern science. He wonders if

the recognition of ... statistical regularities and their mathematical expression could have been reached by any other road than that which Western science actually travelled.

He concludes with an intriguing question:

Was perhaps the state of mind in which an egg-laying cock could be prosecuted at law necessary in a culture which should later have the property of producing a Kepler? ([1969a], p. 582; [1969b], p. 330).

I intend to take the question seriously and explore the road that led from the Rooster of Basel to Kepler. It is my way of responding to what Benjamin

Nelson calls 'Needham's Challenge'. This program of inquiry seeks to explain, by comparative method, and by differentiating and clarifying specific factors, why modern science developed originally in Western Europe and nowhere else. My own inquiry takes a different turn and searches for the occasions that led up to the spiritual revolution that produced not only modern science, but also what Weber called "the disenchantment of the world." On the way, I have turned up what I believe are some modest clues to the great question posed by Needham's Challenge as well. Needham tells us, "historically the question remains whether natural science could ever have reached its present state of development without passing through a 'theological' stage" ([1969b], p. 330). In this paper I have enlarged that observation to understand that the word 'theological' includes the term 'demonological' as well. I shall go so far as to propose that demonology is a link that fastens, a hyphen that binds the idea of law to the idea of nature, at least until the end of the 17th century.

Because Needham's attention remains with the rational side of natural law, he prefers to neglect the demonic side. But in the minds of the people who executed them, egg-laying cocks were no ordinary lawbreakers. They inspired sacred dread, for they were possessed by evil spirits, their eggs might be used in witchcraft, or they might hatch preternatural monsters. Their moral and religious relationship to the demonic world obliged Christians to try those chickens and execute them. The Chinese also believed in demons and bad spirits, but their system of belief differed in crucial ways. Unlike them, Western Christians were urged by a moral imperative to scrutinize nature and to put unusual phenomena on trial.

Until the modern worldview made its familiar impact, all the great civilizations experienced nature through a system of perceptions and ideas that is usually called 'animism'. In this mode of experience, nature was full of spirits and natural objects endowed with a living principle that also vitalized the human soul. In the most familiar varieties of animism, nature was a society of souls, often including minerals, plants, animals, and humans. The forces of nature, therefore, were understood as personal forces, and spirits held personally responsible for causing natural phenomena and their good or evil consequences. It does not matter if Durkheim is right and totemism was older than animism. Nor does it matter if Marett is right and animism was preceded by 'pre-animism'. For our purposes, it is enough to recognize that animism was widespread and, one is tempted to say, universal. However, it is also important to recognize that animistic systems differed from one another in important ways, and that these differences had consequences.

Nineteenth-century writers liked to associate animism with primitivity, but that connection loses its meaning if we recognize the varieties of Greek, Roman, Buddhist, Chinese, Christian, Egyptian, Hebrew, Indian, Japanese, Muslim, and Persian animism. The Hastings *Encyclopedia of Religion and Ethics* describes twenty demonological systems, setting 'civilized' cheek by jowl with 'primitive' forms. Previously, comparative inquiries have tended to stress the common features of animistic thinking. Now I am suggesting that we look at them differentially, and I shall argue that the peculiar features that distinguish Occidental Christian animism from other systems may have an important bearing on Needham's question about the emergence of modern science in the West.

The history of science is the history of human relationships with nature. Werner Heisenberg writes that "science is but a link in the infinite chain of man's argument with nature," and science "cannot simply speak of nature 'in itself'" ([1958], p. 15, italics removed). He insists:

When we speak of the picture of nature in the exact science of our age, we do not mean a picture of nature so much as a *picture of our relationships with nature*. . . . Science, we find, is now focused on the network of relationships between man and nature, on the framework which makes us as living beings dependent parts of nature, and which we as human beings have simultaneously made the object of our thoughts and actions. Science no longer confronts nature as an objective observer, but sees itself as an actor in this interplay between man and nature ([1958], p. 29; italics in the original).

Animistic thinking understood the interplay as a set of moral relationships. As the American Indian writer, Vine DeLoria, observes, when the white man wants to stop polluting the river, he does not stop thinking of the river as a mechanism. In contrast, the traditional Indian asks about his responsibility to the river as a living being. The modern European assumes that moral ideas are not relevant to the workings of inanimate things. He has drawn a boundary and placed the river on the other side of it. The history of the boundary settlement underlying such European assumptions is the subject of this paper. It is the history of the great transition from animism to mechanism.

In every human society, people believe that some unobservable order, personal or impersonal, includes causes and reasons that transcend and explain the phenomena of experience. Nathan Sivin has observed that even though the idea of the Unseen Order does not fit the standard categories of intellectual history, it is one of the greatest of man's imaginative conceptions. As William James observed:

Such is the human ontological imagination, and such is the convincingness of what brings it to birth. Unpicturable beings are realized, and realized with an intensity almost like that of an hallucination. . . . They are as convincing to those who have them as any direct sensible experiences can be, and they are, as a rule, much more convincing than results established by mere logic ever are ([1902], p. 71–72).

The Unseen Order in traditional China was understood by impersonal, abstract concepts, such as *yang* and *yin* and the five elements. It was also filled with personal spirits: ghosts, demons, gods, and so forth. In the world of Chinese medicine, Sivin has shown us, the impersonal abstract concepts belonged to the "great tradition" of China's tiny educated elite. But for vitality, it depended on the "small tradition" of the common folk,

whose world was not only much more intellectually restricted but full of personal forces, spirits and ghosts, which brought and took away sickness and other visitations of fate . . .

Although the two realms enjoyed a symbiotic relationship, it was always possible, over centuries, to discern the border between "the spiritualistic world view of folk medicine and the abstract speculative cosmology of classical medicine . . ." (Nakayama and Sivin [1973], p. 206–207).

In the Chinese scheme, the boundaries between ghosts, demons, and gods were fluid, and a single spirit might become all three. The Occidental scheme maintained rigid boundaries, and ghosts, demons, and gods always had distinct identities and did not cross class lines. Another difference was the presence or absence of moral segregation. Max Weber pointed out that the Chinese demonology lacked a principle of radical evil, and that spirits would commit good or evil deeds depending on their circumstances. Both Confucianism and Taoism, the major forms of religious expression in China, "lacked even traces of a satanic force of evil against which the pious Chinese, whether orthodox or heterodox, might have struggled for his salvation" (Weber [1951], pp. 153, 206, 228). In contrast, the Occidental demonology segregated the invisible world into realms of good and evil spirits. Christian dualistic animism imposed certain obligations on religious communities. One of these obligations was to stand guard in the realm of nature, ready to place natural phenomena on trial, testing them for good and evil. From the middle of the 14th century until the 18th century, Europeans tried to control the demonic forces in nature by trials for witchcraft. As Lea wrote:

All destructive elemental disturbances — droughts or flood, tempests or hail-storms, famine or pestilence — were ascribed to witchcraft ([1907], p. 207).

For religious reasons, then, European Christians were obliged to carry out experiments. They could not leave nature alone. These experiments, we shall see, were carried out not in laboratories, but in the courts. Christian dualistic animism inspired a conspiratorial view of the universe, leaving men confronted with the terrors of a vast spiritual underworld bent on ruining them for eternity. The church assumed the responsibility of exorcising the Devil in all his manifestations. It was the province of the courts to cooperate in this spiritual police action against *maleficia*, or evil magical actions, against diabolic agencies, malicious spirits, watchful fiends, and crowds of demons. Both church and judiciary expressed a horror of collaboration with evil spirits, trying to limit the power of demons over mankind by catching their agents *in flagrante delictu*. As Langton observes,

The belief in demons and the belief in witches are but two aspects of the same belief; for the witch is a person through whom the demon chooses to manifest itself" ([1949], p. 29).

Studies in comparative demonology will reveal no other animistic system in which law and judicial proceedings play such an important part. Like the trials for witchcraft, animal trials illustrate the unique and peculiar legalism of Occidental demonology.

Lynn White has shown the importance of the moral and "emotional basis for the objective investigation of nature" in the late Middle Ages ([1947], p. 434). He has also observed that modern science, emerging in that period,

was more than the product of a technological impulse: it was one result of a deep-seated mutation in the general attitude towards nature, of the change from a symbiotic–subjective to a naturalistic–objective view of the physical environment ([1947], p. 435).

Although scientific thought developed apart from the courtroom, the changing moral and emotional relation to nature may be traced in the trials that tested the presence or absence of demons in the behavior of animals as well as humans.

The moral imagination of the West is juridical, and the courts have remained near the center of moral and spiritual life, and never remote from the vital currents of intellectual concern. For centuries, courtroom debates enlarged or defined the boundaries of scientific as well as theological issues. As Coulton observed, "just as legal theories crept into medieval demonology, so did demonology creep into the law-courts" ([1923], p. 66). The legalistic demonology of the West made the Christian form of dualistic animism different from any other kind.

Before disenchantment, the natural world was not differentiated from the world, and the world was experienced and understood through categories that were not only moral but also theological and demonological. God may have 'owned' the world, but the Devil 'possessed' it, or at least a good part of it. For the early Christians, the Devil was the prince of this world. However, they also believed, as the Vulgate tells us, that " ... *princeps huius mundi iam iudicatus est*" (John 16; 11): the prince of this world has already been judged. The case was closed, but, as Tertullian put it in his *Apologia*, written around the beginning of the 3rd century, even though the evil demons had been condemned, it gave them some comfort before their ultimate punishment to act out their malignant dispositions. According to Tertullian, "Their great business is the ruin of mankind" (*Apol.* 22, 26). Sermons, tales, and other pious writings shaped a collective experience of the world that personified danger and evil. Through this literature, people learned to ascribe an uncanny or unusual experience to the Devil. As a medieval historian put it, the Devil

inspires evil thoughts, instigates crimes, and causes any unhappy or immoral happening. It is just as much a matter of course as if one should say to-day, I have a cold, or John stole a ring, or James misbehaved with So-and-So" (Taylor [1949], I, p. 504).

The world was an arena in which God and the Devil made competing claims on human loyalty. Men and women who chose to ally themselves with the Devil were guilty of spiritual treason. This is why witches in the 15th and 16th centuries drew such heavy retribution. In giving aid and comfort to the Eternal Enemy, they had committed high treason or *lèse majesté* against God.

Christian speculation about the world developed a legalistic theory about the relation between God and Satan. The world, lost through sin, had become the Devil's property through right of possession. In order to remain just, God would not injure the Devil or remove him by force. The world must be ransomed by something more valuable than the world, and God exchanged his Son for the world. Christ's death, therefore, bought Humanity back from Satan. Roman thought tended to legal formulations, and this theory, which began in the East, took hold in the West.

The Western church, therefore, took kindly to that view of the Atonement which represented it as the result of a lawsuit between God and the Devil (Coulton [1923], p. 62).

Medieval writers described imaginary dialogues in a cosmic courtroom between God and the Devil. In some of them, God and the Devil divided the real estate of the world. Usually, the Devil claimed the largest expanses, but those turned out to be deserts and arid mountain tops (Coulton [1923], pp. 466–67). That helped to explain why deserts and wild places were haunted by demons. Sometimes, the Devil went to court to defend his demons, witches, and other associates. Pierre de Lancre, a distinguished magistrate of Bordeaux, who investigated sorcery allegations in the Basque region, wrote a book in 1611. It included an anecdote about a Witch's Sabbath. The Devil had missed several previous Sabbaths, and when he finally reappeared, the witches and warlocks greeted him eagerly and asked where he had been. He replied that he had been in court, pleading their cause against the Savior, and that he had won the case, meaning that they would not be burned (Coulton [1923], pp. 65–66). In some of the trials, God and the Devil dispute claims to the souls of mankind in a series of legal quibbles. In others, God is the judge, Satan the prosecuting attorney, and the Blessed Virgin the advocate for the defense. In France, as in some other European kingdoms, the queen actually held an important position in the judicial system, and she could be petitioned to intercede for defendants. Terrestrial queens may have inspired the judicial role imagined for the Queen of Heaven (Schapiro [1939], p. 379). In a juristic fantasy, *The Trial of Satan*, ascribed to the great writer on Roman Law, Bartolus of Sassoferrato, who revived jurisprudence at Perugia, Satan objected that the Virgin Mary must not be admitted to the bar as an advocate: first, because no woman was eligible to be a barrister, and second, because of her kinship to the Judge. Since the Judge was her son, the hearing would be biased. Mary responded to the objection by warning her son to ignore that shyster, who made quibbling allegations to cause confusion, and she urged him to get on with the case (Coulton [1923], p. 64).

Legal imagery pervaded medieval ideas about the relation to Satan. In verse and fable, "the devil is very careful to establish his title to the soul of man by a faultless legal document," in later centuries signed in blood (Carus [1969], p. 414). The Faust legend is the most familiar story of a pact with the Devil, but that was preceded by the story of Theophilus the Penitent (Russell [1972], p. 19). Faust was a heroic figure of grand proportions, but Theophilus was a little man, not a gifted scholar in search of universal knowledge, but merely a frustrated bureaucrat. His story could serve as a model for the subtitle of Hannah Arendt's book on Adolph Eichmann: 'the banality of evil'.

According to the story, which was one of the best-known romances in the Middle Ages, frequently represented not only in folklore, but also in sculpture and on painted glass, Theophilus was a church administrator in Cilicia around 538 A.D., during the reign of Justinian and a few years before the Persian invasion. He was known for his piety, his competence as an administrator, and his liberality to the poor (Fryer [1935]; Baring-Gould [1914], II, pp. 88–91). When the post of bishop fell vacant, he was urged by the people as well as the church officials to occupy the office, but he refused out of feelings of humility. Someone else was raised to the seat, and later, hearing false rumors against Theophilus, the new bishop removed the latter from his administrative post. Hurt and brooding, Theophilus made a pact with Satan through a necromancer to get his job back. The pact with the Devil was inscribed on parchment and signed in blood. Subsequently, the bishop restored Theophilus to his old position, and the people cheered. Then, conscience got the better of Theophilus and allowed him no rest. He resolved on a solemn fast, praying in church all night. During his long vigil, the Blessed Virgin appeared one night, listened to his pleas for mercy, and agreed to intercede for him. The next night she reappeared and assured him that Christ had forgiven his sins. He woke with a cry of joy and found on his breast the document that had deeded his soul to the Devil. Without the contract, the Devil held no power over him. The next Sunday, Theophilus confessed in public during the liturgy, and displayed the contract, recovered from the Evil One by the mercy of the Mother of God. The bishop gave him absolution, and presided over the public burning of the document. Theophilus received communion, left the church in a fever, and died three days later (Baring-Gould [1868b], pp. 363–70).

Although the story is probably a religious romance, the *Acta Sanctorum* include Theophilus as a saint, honoring him on February 4. The legend was first written in Greek by Eutychius, who claimed to have lived in the house of Theophilus and to write from personal experience of the events. It was translated into Latin by Paul the Deacon in the 8th century, dramatized in the 10th century by Hrosvitha, the illustrious nun of Gandersheim in Saxony, and inspired a number of morality plays, perhaps ultimately suggesting the Faust theme. The iconography of Theophilus is extensive, and the legend often appears in stone, including two representations in the cathedral of Notre-Dame de Paris. Plate III, a detail of the central sculpture of the north transept tympanum of Notre Dame, made around 1250, shows the story in four scenes. The first scene shows Theophilus kneeling, pledging fealty by placing his folded hands between the palms of the Devil. The last scene shows the Virgin wielding a cross to threaten Satan, who crouches before her and surrenders the contract. The earliest image of the Theophilus legend

Plate III. Legend of Theophilus, Notre-Dame de Paris.

appears in a sculptured relief on the tympanum of a portal in the domed church of Souillac, a Romanesque structure completed around 1130 in southern France, (Schapiro [1939], p. 380; Fryer [1935], p. 289; Deschamps [1972], p. 27).

The sculptured relief at Souillac appears over an arched portal. Two lower figures flanking the arch are Old Testament representations: Joseph on the left, Isaiah on the right. To the right of Isaiah stands a sculptured column, an intricate *trumeau* of straining, tangled figures. The face of the *trumeau* is covered by interlocking beasts, men, reptiles, and monsters devouring one another. The central scene over the arch represents three episodes in the life of Theophilus. The flanking figures of St. Benedict on the left and St. Peter on the right place the scene in the context of ecclesiastical administration (Schapiro [1939], p. 378).

Plate IV, a detail of the central field, shows two pairs of figures in the lower register: two representations of the Devil and Theophilus. The images of the Devil show an emaciated body with visible ribs, the hideous head of a monster, and, indeed, the spurs of a cock on the calves of his legs. The feet are different in each scene, and the Devil on the right has the claw of a predatory bird for one foot and a cloven hoof for the other. In the left

Plate IV. The Devil and Theophilus.

pair, the Devil and Theophilus are holding the document and drawing up their legal contract. In the right pair, the Devil is grasping the hands of Theophilus between his own, making him his liege man in a ceremonial gesture of feudal homage. In the upper register, the final scene, which transcends the others, the Queen of Heaven and an angel are descending to the sleeping–praying Theophilus, who lies adjacent to the church in which he spent his vigil of forty nights. The Holy Mother is returning the contract, which she had recovered from the Devil, to the penitent, assisted by an angel who has one hand on her shoulder and the other on Theophilus. In all three scenes, the most important images are dominated by legal symbols — the document of the pact and the feudal oath of fealty.

The legalistic imagery of the Middle Ages turned the forces of evil into a vast spiritual underworld, permitted within limits to act in nature on men. The divine purpose was to test the loyalty of men and to strengthen their moral fiber in the crucible of temptation. The essentials of the theory of nature implied in the cosmology and demonology were spelled out by St.

Augustine. Centuries later, the treatises and manuals on witchcraft — Lea lists about forty of them in his materials on the history of witchcraft — may be understood as footnotes to the work of Augustine. The *Malleus Maleficarum*, or 'Hammer of Witches', published around 1486 by the inquisitors, Sprenger and Krämer, provided the model for this literature. The *Malleus*, incidentally, was one of the first books to be printed in pocket editions. Judges and lawyers questioned the accused with their copies of the *Malleus* ready for reference under the table or up their sleeves. Lea writes that the *Malleus* acquired such great authority that it

> fastened on European jurisprudence for nearly three centuries the duty of combating the devil and saving mankind from his clutches ([1939], I, p. 305).

During the peak of the witchcraft trials, animals shared some of the burden of persecution (Evans [1906], p. 165). Pigs suffered the most, since it was thought they were especially vulnerable to demonic possession. The legion of devils that had entered the herd of Gadarene swine in the New Testament story, it was remembered, had said to Jesus, "Send us among the pigs and let us go into them" (Mark 5; 12). Pigs ran freely in the streets of medieval towns and often got into trouble. Besides, animals were often distinguished as 'sweet beasts' or 'stenchy beasts'. The hart and the hind, panting after the flowing brooks as the soul thirsts for the living God, as the Psalmist said, led the list of sweet beasts. The pig, of course, led the stenchy beasts. Goats and polecats provided other stenchy habitats enjoyed by unclean spirits.

Pigs were often judged for injuring and sometimes killing children. In 1386, a sow of Falaise that had attacked and killed a child was mutilated and then executed in the village square, dressed up as a human being. Plate V is a picture of that execution. The expense of the case included a pair of new gloves for the executioner, so that he might come out with clean hands. Even though pigs were rarely shown mercy, in one case youth was a reason for clemency, when in 1457 at Lavegny, a sow and her litter were charged with having murdered and partially devoured a child. The sow was condemned to death, but her piglets were released because of their tender age and because their mother had set them a bad example (Westermarck [1912], I, p. 257).

An execution without a proper trial could stir a great deal of indignation. In 1576 in Schweinfurt in Franconia, a sow that had mutilated a child was delivered into custody. Without legal authority, the executioner "hanged it publicly to the disgrace and detriment of the city." The hangman was forced to flee and never dared to return. The case gave rise to the proverbial phrase,

Plate V. Execution of a pig.

'*Schweinfurter Sauhenker*', meaning 'sow hanger from Schweinfurt', used to characterize a ruffian and vile sort of fellow. As Evans wrote,

It was not the mere killing of the sow, but the execution without a judicial decision, the insult and contempt of the magistracy and the judicatory by arrogating their functions, that excited the public wrath and official indignation ([1906], p. 147).

As Needham has shown, the frequency of animal trials followed a curve rising from three recorded instances in the 9th century to a peak of about sixty in the 16th century, and then dropping to nine cases in the 19th century. They fall into three types: (1) actions against domestic animals for attacking human beings (e.g., the execution of pigs for devouring infants); (2) actions against swarms, resulting in anathemas or excommunicatory rituals — a kind of spiritual pesticide; and (3) the condemnation of *lusus naturae*: e.g., the laying of eggs by putative roosters (Needham [1969a], p. 574; [1969b], p. 328).

The animal trials collected by Evans (1906) account for more than two hundred cases extending over a thousand years. The latest in his record, the case of a dog executed in Switzerland for homicide, took place in 1906, the very year the book went to press. Some celebrated cases were located in Switzerland and France, but the list names a large number of Occidental countries, including Belgium, Denmark, Germany, Italy, Portugal, Russia, Spain, Turkey, England, Scotland, Canada, and the United States (Hyde [1916], p. 709). A whole range of insects and animals were brought to the bar of justice,

including asses, beetles, bloodsuckers, bulls, caterpillars, cockchafers ... cows, dogs, dolphins, eels, field mice, flies, goats, grasshoppers, horses, locusts, mice, moles, rats, serpents, sheep, slugs, snails, swine, termites, turtledoves, weevils, wolves, worms, and nondescript vermin (Hyde [1916], p. 708).

The most common defendants were pigs, for reasons I have discussed. The condemned animals were dispatched in various ways, depending on the local forms of punishment. The Russians, for example, continued to use banishment in one or two cases, and at the end of the 17th century, the record shows a billy goat exiled to Siberia (Hyde [1916], p. 712).

Karl von Amira, a historian of law writing at the end of the 19th century, insisted on a technical distinction between secular animal punishments for crimes such as homicide, and ecclesiastical animal trials. The trials, he showed, led back to the demonology of the Middle Ages, (von Amira [1891], p. 548), and were associated with certain formal adjurations, particularly the *maledictio* and the anathema found in the ritual of excommunication, as well as the more familiar rite of exorcism. These procedures were directed not primarily at the animals on trial, but at the evil spirits believed to inhabit them. The ritual was intended to prevent further devastation of orchards, vineyards, and fields, and to halt the depletion of soil and water by the action of noxious vermin possessed by demons. The effectiveness of the imprecation or interdiction depended on the proper judicial ritual. In other words, these supernatural sanctions were not expected to work without due process of law. Evans showed,

> Before fulminating an excommunication the whole machinery of justice was put into motion in order to establish the guilt of the accused, who were then warned, admonished, and threatened . . . (Evans [1906], p. 4).

In the 10th century, the pious Archbishop of Trèves was saying mass in the church of St. Peter while an irreverent swallow dipped and soared over his head. If he enjoyed a halo, it offered no protection against this winged creature, for it defecated on the venerable head, and the holy man transcended his piety to roar an excommunication. From that moment, swallows kept scrupulously out of the building, leaving in peace the worshippers within, and if one of them intruded over the entry, it promptly fell dead upon the pavement. A case still better known is recorded for the 11th century. St. Bernard, preaching in the monastery at Foigny, which he had founded, was tormented by the flies buzzing around his head. He shouted at them, "I excommunicate you!" The flies fell on the floor in heaps so high that shovels were needed to get rid of them (Lea [1883], p. 428–429). The case of the Flies of Foigny became so well known that the only point left open for speculation was the question of how long it took for the flies to experience the impact of the excommunication. The flies had been executed without due process of law, but the chronicler explains that the situation was desperate and no other remedy was at hand.

In the early part of the 16th century, Bartholomew Chasseneux, the leading authority on ritual procedures against animals, became one of the most distinguished jurists in France. Starting out as an advocate in Bourgogne, he was elected in 1531 to the rank of counselor in the *Parlement de Paris*, and in the following year appointed to the *Parlement de Provence*, where he held the post of Premier Président, a position equivalent to the rank of Chief Justice. Chasseneux was the author of a wide-ranging work he called *A Catalogue of the Glories of the World*, and he was also known as a commentator of the customary law of Burgundy (Pignot [1880]). A collection of seventy-nine of his principal *consilia* appeared in 1531 (Pignot [1880], p. 211), and the first *Consilium* in this collection became his most celebrated work. It was a lengthy, definitive treatise explaining and justifying the procedures of excommunication against animals and insects. He provided a long list of cases, beginning with the cursing of the serpent in the Garden of Eden, in which anathemas and excommunications had worked against creatures that crawl and creep and fly. This treatise, the *Consilium Primum*, established his eminence as a theorist, but he won his laurels as a barrister from his work in a celebrated trial before the ecclesiastical court of Autun. In that trial, and in similar cases that followed, he made a brilliant reputation defending rats. As Evans put it,

the ingenuity and acumen with which Chasseneux conducted the defence, the legal learning which he brought to bear on the case, and the eloquence of his plea enlisted the public interest and established his fame as a criminal lawyer and forensic orator (Evans [1906], p. 21).

Lest you think there was nothing extraordinary in a criminal lawyer making a brilliant career defending rats, you must understand that they were four-legged rats.

The rats were being charged with devouring the barley crop in the countryside of Burgundy. The people, complaining that the infestation was intolerable, petitioned the bishop to excommunicate the varmints. The episcopal court, knowing Chasseneux's reputation as an expert on spiritual pesticides, appointed him as defense attorney to the rats. He prepared the case with great skill. It was believed that no excommunication or other adjuration against animals could be effective unless the beasts had been provided with a proper and scrupulous legal defense.

Chasseneux's first maneuver was to challenge the summons. He argued that the rats had a bad name and suffered the disability of having public opinion against them. They were improperly summoned, because they were dispersed all over the countryside, dwelling in numerous villages, and a single summons was insufficient to notify them all. The second citation, then, was read from the pulpits of every parish inhabited by the rats. This proclamation took more time, and at the end of the period assigned, the rats still did not appear. Chasseneux argued that since there were so many rats living in so many places, great preparations were necessary for a mass migration, and this required more time. When the rats still did not appear, he got an additional postponement, and excused the default of his clients on the grounds that their journey was difficult and made hazardous by the presence of their natural enemies, the cats. These mortal foes of the rats, Chasseneux contended, watched all their movements and lay in wait for them at every turn. He showed that a proper summons implied the right of safe conduct, and that if the way were full of peril and without protection, the defendants were justified in not obeying the writ. Finally, he demanded that the plaintiffs — the farmers — be required under bond to prevent their cats from frightening the rats. The plaintiffs demurred, but the case moved from one delay to another. The record does not tell us who won, but it is safe to infer that the rats eventually lost by default, and that ultimately an excommunication was fulminated against them (Evans [1906], pp. 18—19; Hyde [1916], pp. 706—707; Lea [1883], p. 430; Baring-Gould [1896], p. 65).

Throughout the Middle Ages, treatises were written to protest the absurdity of animal trials, most of them criticizing the folly of maledictions, anathemas, and excommunications against pests. Occasionally, a prelate would forbid fulminations against animals without special permission or specific license (Baring-Gould [1896], p. 67). Some Spanish theologians were prone to dismiss the trials as vain and superstitious (Lea [1883], p. 432), arguing that insects, being devoid of reason, cannot comprehend the meaning of prayers and curses launched at them, and since their depradation is caused by their natural appetites, and since they have no free will, they were not guilty of sin. Joseph Needham suggests that the medieval attitude wavered:

Sometimes the field-mice or locusts were considered to be breaking God's laws, and therefore subject to prosecution and conviction by man, while at other times the view prevailed that they had been sent to admonish men to repentence and amendment ([1969b], p. 330).

I believe that the medieval attitude was not a single wavering viewpoint, but a triad of contrary positions. One position considered the animals hungry creatures of God, with neither reason nor responsibility, simply following the inclinations given to them by nature. The second position considered them instruments of God, sent to punish a community for some sin committed by the inhabitants. The third position viewed them as the temporary vehicles of demons or as instruments of the Devil. The first two positions implied decent treatment: they were persuaded to stop their devastation and given another place to go. The third required some kind of exorcism, or some kind of powerful intervention. The argument for the prosecution proceeded from the third position. The defense argued from the first or the second, sometimes both.

It was an empirical question in each case to determine if the animals were acting simply as creatures, as special instruments of God, or as the instruments of evil spirits. Anathemas hurled at the animals were directed inferentially at the Devil or at the demons contained within them. Thomas Aquinas argued that it was either blasphemous or vain to curse beasts if they were agents of God or simply creatures behaving according to instinct. They were properly cursed only if they were agents of Satan and inspired by the powers of hell (Evans [1906], pp. 54–55).

Every animal trial tested natural phenomena to ascertain if they represented divine agency, diabolic agency, or nature working alone. Just as witchcraft cases placed phenomena on trial, so also animal trials sought confirmation of one of the three hypotheses. Moreover, the trials of delinquent animals as well as witches were forums in which lawyers and judges argued the precise location of the boundaries between natural and supernatural events.

In his book on the decline of witchcraft trials in France, Robert Mandrou (1968) shows the importance of those trials in establishing a line of demarcation between the natural and the supernatural, and also shows how the trials provided the occasion for lawyers, judges, priests, physicians, and scientists to collaborate in that "collective adventure" that a "spiritual revolution" represents (Mandrou [1968], p. 564). Out of that collaboration in the 17th century there emerged a new jurisprudence, a new theory of abnormal psychological states, and a new view of natural processes. The magistrates of the sovereign courts, Mandrou demonstrates, in their new integration of medical knowledge and theology, occupied the first rank in the progress of rationalism in 17th century France.

In the past three centuries, the most striking change, Lecky tells us, may be found in the common response to the idea of the miraculous. Now, when the spirit of rationalism predisposes men to attribute all kinds of phenomena to natural rather than to miraculous causes, the account of a miracle would draw "an absolute and even derisive incredulity which dispenses with all examination of the evidence." To ascribe unexplainable phenomena to supernatural agency "is beyond the range of reasonable discussion." In contrast, a few centuries before, miraculous accounts were not only credible but ordinary (Lecky [1868], I, pp. 17, 27). The vocabulary of disenchantment provided alternative expressions for experiences that had been identified by names for supernatural or preternatural agency.

The great astronomer, Johannes Kepler, helped to invent the scientific idiom of disenchantment by repudiating the old animistic ideas of planetary motion. Before Kepler, celestial motion was believed to be the product of souls or minds, usually represented as divine agencies (Jammer [1957], p. 51). Kepler used the term "force or energy" — *vis seu energia* — to explain the movement of the planets (Jammer [1957], p. 87). He did not invent the Latin word *vis*, but before him Pliny in his *Natural History*, which appeared about 77 A.D. and remained one of the most important scientific works of the ancient world, has used the term *vis* in a very general and ambiguous way, to mean all kinds of forces, including psychic and occult effects as well as physical force (Jammer [1957], p. 45). Kepler restricted the concept to mechanical force. He explicitly distinguished it from any kind of psychic, spiritual, or mental force.

In his *Epitome of Copernican Astronomy*, which was published in 1621, Kepler concluded that the motion of the planets was not the work of mind, as the ancients believed, but the work of the natural power of bodies (Kepler [1866], p. 372). The common practice of reducing celestial movement to the hidden forces of some soul, he wrote, was the sanctuary of all ignorance and the death of all philosophy. He preferred to think of the cause of planetary motion as impetus only — that is, as movement produced by "a uniform exertion of forces" (for 'forces' he used the Latin word *virium*, the genitive plural of *vis*), without the work of mind (Kepler [1866], p. 393). In the same year (Jammer [1957], p. 90, refers to a note that appeared in the second edition of *Mysterium cosmographicum*) he proposed that the word *vis*, which means 'force', should replace the word *anima*, which means 'soul'. That substitution, Dijksterhuis observes, implies nothing less than "a radical revision of thought" ([1961], p. 310). The action of 'souls' in nature was understood by principles of magic. The laws of mechanics are expressed in the language of mathematics.

Kepler's substitution removed the magic from motion. In Collingwood's words, Kepler's "momentous step" of replacing *anima* by *vis* implied that "the conception of vital energy producing qualitative changes should be replaced by that of a mechanical energy, itself quantitative, and producing quantitative changes." Before that replacement, "man's mastery over nature was conceived not as the mastery of mind over mechanism but as the mastery of one soul over another soul, which implied magic . . . " ([1945], pp. 102, 96).

Albert Einstein, in his Preface to Kepler's *Life and Letters*, suggests that the two opposing principles of animism and mechanism struggled within Kepler, and that he never succeeded in entirely extricating himself from animistic thinking (Baumgardt [1951]). Max Caspar, Kepler's biographer, agrees that Kepler, who founded the mechanistic explanation of the heavenly motions, remained suspended between an animistic and a mechanistic view of nature ([1959], p. 383).

But Kepler remained enmeshed by animism in another way as well.

How does the Rooster of Basel, the victim of medieval animism, lead to Kepler, Needham asks. The answer is ironic, for Kepler suffered a narrow escape from similar victimization. In 1621, the *Epitome of Copernican Astronomy* completed publication, substituting the concept of *vis* for the concept of *anima*. But Kepler himself wrote about 1621: "I spent the whole year on my mother's trial" (Kepler [1967], xix). His mother was being tried for witchcraft, and he had assumed the burden of preparing her legal defense. In a document of 128 pages, he did not deny a belief in witches — just as his British contemporary, Francis Bacon, did not deny it. Edward Rosen observes that "like many another great man in his time Kepler never expressed any disbelief in the existence of witches" ([1966], p. 449). Caspar agrees that "the belief in demoniac influences and effects" remained part of Kepler's thinking ([1959], p. 240). Nevertheless, in his brief for the defense and in his bill of exceptions, he carefully accounted for every act for which his mother was being charged by referring it to a natural process. He drew the line and saved her life.

Caspar writes that after Kepler, a later era

raised the completely mechanistic explanation of the models of nature to a principle and, with a remarkable shyness of everything which is called soul, required, in the name of science, the weeding out of every psychic power ([1959], p. 384).

This passage from animism to mechanism evolved through specific and remarkable historic occasions — juridical occasions as well as scientific. In Kepler's own words, which I take the liberty to translate from Latin:

To me, the occasions by which men arrive at the knowledge of celestial things seem no less astonishing than the very nature of celestial things (Koyré [1973], p. 119).

The courts, I have argued, in the trials that tested demonic influences, provided some astonishing occasions for the progressive disenchantment of nature.

When nature is full of souls, their actions and the consequences of their behavior may be understood through moral categories. Spirits acting in nature were held personally responsible for certain natural phenomena. Kepler's substitution of *vis* for *anima*, extended from the celestial to the terrestrial sphere, does much more than submit nature to the language of mathematics. It is also a declaration of the innocence of nature, or at least a proclamation that moral conceptual categories are irrelevant to the understanding of natural phenomena.

In the 17th century, animals still went on trial, but Racine wrote *The Litigants*, his only comedy, about a trial in which a dog is charged with stealing a capon; when it was shown in 1668 it made Louis XIV laugh. In 1672, Colbert forbade the sovereign courts of France to hear cases of witchcraft (Michelet [1939], p. xx).

In the physical sciences, Newton drew a boundary line and stood on it like Janus, with faces to both centuries. Keynes called him "the last of the magicians" (1947). As Keynes observed, Newton dropped the 17th century behind him and became the 18th century figure, which is the unmagical, traditional Newton — the sage of the Age of Reason.

At the end of his eight-volume *History of Magic and Experimental Science*, Lynn Thorndike breathed a sigh of relief and concluded, "animism had been replaced by mechanism." As he explained,

a dividing line had been drawn between science and superstition which was sharper and more satisfactory than any that had been previous attempted. ... The boundaries of natural and experimental science seemed to be more distinctly defined than they ever had been before. They had been so drawn as to lie outside theology as well as of magic, and to exclude miracles, demons and diabolical or spiritual action as well as other forms of the occult (Thorndike [1923–58], VIII, p. 604).

The courtroom exploration of demonological issues had helped settle those boundaries. It had inspired a collective effort of reinterpretation and disenchantment: a radical revision of thought. It helped change the contours of animistic thinking so that it did not remain what Gaston Bachelard called an "epistemological obstacle" to the scientific world view (1969). Instead, the legalistic demonology of Christian animism shaped a forensic matrix for the expression of scientific thought and for its extension beyond the boundaries of science. Within that matrix, lawyers and judges, who were in touch with the changing scientific currents of the 16th and 17th centuries, carried on debates about the boundaries of nature. Within that matrix we may trace the changing map of the universe.

REFERENCES

Amira, Karl von. 1891. *Thierstrafen und Thierprocesse* (Verlag der Wagner'schen Universitäts-Buchhandlung, Innsbruck).
Bachelard, Gaston. 1969. *La formation de l'esprit scientifique* 6th ed. (J. Vrin, Paris).

Baring-Gould, S.
  1868a. *Curious Myths of the Middle Ages*, First Series (Rivingtons, London).
  1868b. *Curious Myths of the Middle Ages*, Second Series (Rivingtons, London).
  1896. *Curiosities of Olden Times*, rev. ed. (Grant, Edinburgh).
  1914. *The Lives of the Saints*, 16 vols. 3rd ed. Vol. II. (Grant, Edinburgh).
Baumgardt, Carola. 1951. *Johannes Kepler: Life and Letters* (Philosophical Library, New York).
Boyd, William K. 1905. *The Ecclesiastical Edicts of the Theodosian Code* (Columbia University Press, New York).
Brown, Peter. 1970. 'Sorcery, Demons, and the Rise of Christianity from Late Antiquity into the Middle Ages'. In Douglas, 1970, pp. 17–45.
Carus, Paul. 1969 (1900). *The History of the Devil and the Idea of Evil* (Land's End Press, New York).
Caspar, Max, 1959. *Kepler*. Tr. by C. D. Hellman (Abelard-Schuman, London and New York).
Catholic Encyclopedia. 1907 (Gilmary Society, New York).
Cohn, Norman. 1970. 'The Myth of Satan and His Human Servants'. In Douglas, 1970, pp. 3–16.
Cole, L. J. 1927. 'The Lay of the Rooster'. *Journal of Heredity* 18 97–105.
Collingwood, R. G.
  1940. *An Essay on Metaphysics* (Oxford University Press, London).
  1945. *The Idea of Nature* (Oxford University Press, London).
Coulton, G. G. 1923. *Five Centuries of Religion*, Vol. I. (Cambridge University Press, Cambridge).
Curry, W. C. 1933. *The Demonic Metaphysics of Macbeth* (Univ. of North Carolina Press, Chapel Hill).
Deschamps, Paul. 1972. *French Sculpture of the Romanesque Period: Eleventh and Twelfth Centuries* (Hacker Art Books, New York).
Dijksterhuis, E. J. 1961. *The Mechanization of the World Picture*, tr. by C. Dikshoorn, (Oxford University Press, London).
Douglas, Mary. 1970. *Witchcraft Confessions and Accusations* (Tavistock, London).
Evans, E. P. 1906. *The Criminal Prosecution and Capital Punishment of Animals* (Dutton, New York).
Farnell, Lewis R. 1911. *Greece and Babylon* (T & T Clark, Edinburgh).
Fauconnet, Paul. 1920. *La responsibilité* (Librairie Félix, Paris).
Fryer, A. C. 1935. 'Theophilus, the Penitent, as Represented in Art'. *Archaeological Journal* 92 287–333.
Gerth, H. H., and C. Wright Mills. 1946. *From Max Weber: Essays in Sociology* (Oxford University Press, New York).
Ginsberg, M. 1939. 'The Concepts of Juridical and Scientific Law'. *Politica* 4 1.
Grillot de Givry, Emile A. 1929. *Le musée des sorciers, mages et alchimistes* (Librairie de France, Paris).
Heisenberg, Werner. 1958. *The Physicist's Conception of Nature*, tr. by A. J. Pomerans, (Harcourt, Brace, New York).
Hyde, W. W. 1916. 'The Prosecution and Punishment of Animals and Lifeless Things in the Middle Ages and Modern Times'. *University of Pennsylvania Law Review* 64 696–730.

James, William. 1902. *The Varieties of Religious Experience* (Modern Library, New York).
Jammer, Max. 1957. *Concepts of Force* (Harvard University Press, Cambridge, Mass.).
Kepler, Johannes.
    1866.   *Epitome Astronomiae Copernicae* in *Opera Omnia*. Ed. by Ch. Frisch. Vol. VI (Heyder & Zimmer, Frankfurt and Erlangen).
    1952.   *Epitome of Copernican Astronomy*, Book IV. tr. by C. G. Wallis. In *Great Books of the Western World*, ed. by R. M. Hutchins (Encyclopaedia Britannica, Chicago).
    1967 (1634). *Somnium* tr. by Edward Rosen (University of Wisconsin Press, Madison).
Keynes, J. M. 1947. 'Newton the Man,' in *Newton Tercentenary Celebrations* [under the auspices of the Royal Society of London] (Cambridge: Cambridge University Press), pp. 27–34.
Koestler, Arthur. 1959. *The Sleepwalkers* (Macmillan, New York).
Koyré, Alexander, 1973. *The Astronomical Revolution*, tr. by R. E. W. Maddison (Methuen, London).
Langton, Edward.
    1942.   *Good and Evil Spirits* (S. P. C. K., London).
    1949.   *Essentials of Demonology* (Epworth, London).
Lea, Henry Charles.
    1883.   *Studies in Church History* (Henry C. Lea's Son & Co., Philadelphia).
    1907.   *A History of the Inquisition of Spain*, Vol. IV (Macmillan, New York).
    1939.   *Materials Toward a History of Witchcraft*, 3 vols. (University of Pennsylvania Press, Philadelphia).
    1956 (1887). *A History of the Inquisition of the Middle Ages*, 3 vols. (Russell, New York).
Lear, John (ed.). 1965. *Kepler's Dream*, tr. by P. F. Kirkwood, (University of California Press, Berkeley).
Lecky, W. E. H. 1868. *History of the Rise and Influence of the Spirit of Rationalism in Europe*, 2 vols. (Appleton, New York).
Lenormant, François. 1877. *Chaldean Magic* (Bagster, London).
Mandrou, Robert. 1968. *Magistrats et sorciers en France au XVIIe siècle* (Librairie Plon, Paris).
Michelet, Jules. 1939. *Satanism and Witchcraft*, tr. by A. R. Allinson. (Lyle Stuart, Secanus, N.J.).
Monter, E. William. 1969. *European Witchcraft* (Wiley, New York).
Nakayama, S., and N. Sivin. 1973. *Chinese Science* (M.I.T. Press, Cambridge Mass.).
Needham, Joseph.
    1969a.   *Science and Civilisation in China*, Vol. II (Cambridge University Press, Cambridge).
    1969b.   *The Grand Titration* (Allen & Unwin, London).
    1970   *Clerks and Craftsmen in China and the West* (Cambridge University Press, Cambridge).
New Catholic Encyclopedia. 1967 (McGraw-Hill, New York).
Pignot, J-H. 1880. *Un jurisconsulte au seizième siècle: Barthélemy de Chasseneuz*, (Larose, Paris).
Robin, P. A. 1932. *Animal Lore in English Literature* (Murray, London).
Rosen, Edward. 1966. 'Kepler and Witchcraft Trials'. *The Historian* 28 447–50.

Russell, J. B. 1972. *Witchcraft in the Middle Ages* (Cornell University Press, Ithaca, N.Y.).
Schapiro, Meyer. 1939. 'The Sculptures of Souillac'. In *Medieval Studies in Memory of A. Kingsley Porter*, ed. by W. R. W. Koehler. Vol. II (Harvard University Press, Cambridge, Mass.), pp. 359–387. Reprinted in Shapiro's *Selected Papers*. Vol. I: *Romanesque Art* (Braziller, New York), pp. 102–130).
Skeat, Walter William. 1966 (1900). *Malay Magic* (Barnes & Noble, New York).
Smith, W. Robertson. 1972 (1874). *The Religion of the Semites*, 2nd ed. (Schocken, New York).
Spiro, Melford E. 1967. *Burmese Supernaturalism* (Prentice-Hall, Englewood Cliffs, N.J.).
Sprenger, Jakob, and Heinrich Krämer (Henricus Institoris). 1928 (c. 1486). *Malleus Maleficarum*, tr. by Montague Summers (Bloom, New York).
Taylor, Henry Osborn. 1949. *The Mediaeval Mind*, 2 vols. 4th ed. (Harvard University Press, Cambridge, Mass.).
Tertullian. 1869. *Writings*. Vol. I. tr. by A. Roberts and J. Donaldson (T & T Clark, Edinburgh).
Thorndike, Lynn.
    1905.   *The Place of Magic in the Intellectual History of Europe* (Columbia University Press, New York).
    1923–1958. *A History of Magic and Experimental Science*, 8 vols. (Columbia University Press, New York).
Valois, Noël. 1880. *Guillaume d'Auvergne: Éveque de Paris* (Picard, Paris).
Waddell, Helen. 1957 (1936). *The Desert Fathers* (University of Michigan Press, Ann Arbor).
Weber, Max. 1951. *The Religion of China*, tr. by H. H. Gerth (Free Press, New York).
Westermarck, E. A. 1912. *The Origin and Development of the Moral Ideas*, 2 vols. (Macmillan, London).
White, Lynn, Jr. 1947. 'Natural Science and Naturalistic Art in the Middle Ages'. *American Historical Review*, 52 421–435.
Withington, E. T. 1917. 'Dr. John Weyer and the Witch Mania' in *Studies in the History and Method of Science*, ed. by Charles Singer, Vol. I (Clarendon Press, Oxford).

SOURCES OF PHOTOGRAPHS

Plates I and II: Cole (1927).
Plate III: Fryer (1935).
Plate IV: Grillot de Givry (1929).
Plate V: Evans (1906).

Prof. E. V. Walter
Dept. of Sociology
Boston University
College of Liberal Arts
Boston, Ma. 02215
U.S.A.

N. SIVIN

REFLECTIONS ON 'NATURE ON TRIAL'

Vic Walter's paper on chicken justice forces me to begin by addressing what I would have to call a meta-question; namely, why does a historian of science cross the road? In this case the answer is to show that there are two sides. Historians need to sit down with philosophers of science to understand how the development of science can be in one sense a gradual asymptotic approach to the physical reality of the world out there and at the same time can be one damned thing after another. When philosophy helps us to realize that complicated ideas are really simple, historians reciprocate by taking simple and logical intellectual transitions in the past and making them complicated. This evening the marriage of philosophy and history is being mediated by sociology, and in a way that seems to me thoughtful and scrupulous toward both bride and groom.

It is tempting to wander with you into the thickets of Chinese ritual, but I think it would be best to consider some general issues that are not often treated as well as we have just heard these treated.

I agree with Vic and with others here that the Scientific Revolution problem provides a rich and relatively concrete occasion to make the perspectives of sociology, philosophy, and history converge. Let me remind you of the classical form of the problem in Joseph Needham's words:

Why did modern science, the mathematization of hypotheses about Nature, with all its implications for advanced technology, take its meteoric rise *only* in the West at the time of Galileo?[1]

Now, the problem of why the revolution that gave rise to modern science did not take place in China did not spring full-blown from Needham's brow. I would want to trace it back to the Protestant Ethic problem as Max Weber defined it. Weber was convinced that, as he put it, "the varied conditions which externally favored the origin of capitalism in China did not suffice to create it."[2] He had already, for the European case, made the strongest case he could for economic causation and found it insufficient. He was therefore prepared, for the Chinese case, to locate the crucial factors in the realm of values rather than in the realm of, perish the thought, economics.

This finding of Weber's did not arise from the play of a mind untainted by presuppositions and responding only to empirical correlations. Weber was acutely sensitive to the operation of values in institutions, and convinced from the start of the predominant influence of ideal factors. He understood and was influenced by the power of Marx's positivistic analysis of social change in terms of economic factors, which took seriously the relation of changing conditions of production to changing views of the world, of nature, and of human options.

For our purpose, what I would call the intensive phase of the prehistory of the Scientific Revolution problem begins back beyond Weber, with Marx. It was an attempt to find *the* crucial factor, the sufficient condition, or at least to point to the field of human endeavor in which the springs of fundamental social change are bound to lie. For Marx, at least the Marx who has inspired a century of ideology, the true causal factors had to be economic and social; intellectual and theoretical factors were reflections of those. Weber was reacting to the idea that values belonged to what we call the superstructure. He made values the sufficient conditions, with economic factors demoted to necessary conditions. Weber did not react against the idea that there was *a* crucial factor, or a coherent clump of sufficient conditions located in one realm rather than another. He was no better prepared than Marx to see both the necessary and sufficient causes of the modern world diffused rather equitably across the whole field of human action and thought.

Neither Marx nor Weber was prepared to see every layer of human thought and action as interdependent, with the dynamic of history driven by their interplay. Both insisted on a *primum mobile* located at the bottommost layer. That was because the location of the bottommost layer for both thinkers was known — known by an illumination that began the quest.

Returning now to Needham, just as Weber's point of view was a reaction to Marxist history, in a sense Needham's point of view is a reaction to Weberian history. I say that with the fact in mind that in the 1071 pages of bibliography and the 437 pages of index in *Science and Civilisation in China* thus far, the name of Max Weber does not occur once, no more than does that of Tom Kuhn — or Sir Karl Popper. But that of Robert Merton most assuredly does. Rather than count citations, I would prefer to say that Weber's influence in sociology is too universal to be missed by people who ask the sorts of questions Needham does.

My impression from conversations with Needham is of a distaste for what he sees as arid attempts to explain the evolution of science in terms of ideas bombinating in the void or values divorced from their origins in work. Once again, the upshot is a search for a locus of sufficient conditions on one side or another.

In 1944 their location was unequivocal:

So we come to the fundamental question, why did modern science not arise in China? The key probably lies in the four factors: geographical, hydrological, social and economic.[3]

By 1964, two decades of additional reflection had led to slightly stronger wording than 'probably':

I believe that the analysable differences in social and economic pattern between China and Western Europe will in the end illuminate, as far as anything can ever throw light on it, both the earlier predominance of Chinese science and technology and also the later rise of modern science in Europe alone.[3]

The argument, for instance as presented in the famous section 19k, 'Mathematics and Science in China and the West,' of *Science and Civilisation in China* — I think many of you have read Robert Cohen's meditation on it[4] — is, keeping in mind Needham's focus on science rather than capitalism, a kind of parody of Weber's approach:

Interest in Nature was not enough, controlled experimentation was not enough, empirical induction was not enough, eclipse-prediction and calendar-calculation were not enough — all of these the Chinese had. Apparently a mercantile culture alone was able to do what agrarian bureaucratic civilisation could not — bring to fusion point the formerly separated disciplines of mathematics and nature-knowledge.[4]

I will not draw this thread out further, except to note in passing that mathematics and nature were never separated as a matter of principle in China, as they were in Europe. My point is that we have the option of breaking this chain of reaction and counter-reaction, and giving up pseudo-questions about which sphere of action is more fundamental than what other in the causation of change. It is the interconnection of intellectual and socio-economic factors that needs exploring.

What I found most educational about Vic's paper was its concern for the interplay of ideas and institutions, exploring the special quality of that interplay rather than just using instances as a prop for one-sided generalizations. Needham's rich and across-the-board proof of the high level of science and technology in ancient China has led to cascades and avalanches of hypotheses from one scholar or another about factors that inhibited the evolution of modern science in China, or characteristics unique in the West that furthered a major scientific revolution. Just over a decade ago David E. Mungello revived the old argument that, although Ch'ing dynasty thinkers took the world as observable, nominalistic fact, just as Bacon did, unlike him they did not develop a scientific methodology. The question of whether modern scientists use Bacon's scientific method was not even considered.[5] Shortly afterward Ho Peng Yoke assured us that if Chinese scientists

> were fully satisfied with an explanation they could find from the system of the *Book of Changes* they would go no further to look for mathematical formulations and experimental verifications in their scientific studies. Looking at the system of the *Book of Changes* in this light, one may regard it as one of the inhibiting factors in the development of scientific ideas in China.[6]

The fallacy is a matter of confusing for a cause what is merely a *description* of an earlier state of a culture, or of a culture's way of doing something. In its mirror image, the absence of the subsequent state is confused with an 'inhibiting factor', a negative cause. But exactly what does 'inhibiting factor' mean in such contexts? Consider one of these often adduced to explain the failure of China to beat us to the Scientific Revolution despite an early head start, namely the predominance of a scholar—bureaucrat class immersed in books, faced toward the past, and oriented toward human institutions rather than toward Nature as the matrix of the well-lived life. But in Europe at the onset of the Scientific Revolution we are faced with the predominance of the schoolmen and dons, immersed in books, faced toward the past, and oriented toward human institutions rather than toward Nature. They did not prevent the great changes that swept over Europe, and it would take a more imaginative historian than myself to say whether those changes would have taken place sooner had they never existed.[7]

The confusion about 'inhibiting factors' is no less a confusion when it has to do with ideas or techniques. One might just as well call Euclidean geometry an inhibiting factor for the development of non-Euclidean geometry, since so long as people were satisfied with it they didn't move on to a new step. It is unfortunate to see the remarkably interesting technical language of the *Book of Changes*, so powerful in systematically relating broader ranges of human experience than modern science could hope to, written off as an obstacle before anyone has taken the trouble to comprehend it thoroughly.

If we fault the Chinese for not developing a Baconian scientific method, we may as well go all the way with the historian Wolfram Eberhard and point out with all gravity that in the last two centuries B.C. Chinese scientists

> were not interested in applied technical sciences, e.g., in developing theoretical tools which could be used to control the flight of a cannon shell or to dirct ships safely across the sea.[8]

So much for the first civilization to note the declination of the compass needle.

There seem to be two spectacularly weak links in the chains of reasoning that generate such fallacies. One is the assumption that (to give an example) the horse and buggy delayed the invention of the automobile, with the further implication that the automobile would have emerged sooner if the buggy had never been invented. The second is a curious misapplication of the Parmenidean notion that being cannot be born out of non-being. I mean by that the idea that if an important aspect of our Scientific Revolution cannot be found in another culture, we assume that tells us why the whole shebang could not have happened there. There was, after all, a time not so long ago in Europe, too, when people were not interested in controlling the flight of a cannon shell. Why then has no one taken that as proof that modern dynamics should not have emerged there? If Weber were right that no rational capitalism could emerge from a traditionalist social order, we too would still be in the Middle Ages, when that was a correct description of European society.

The resolution of this conundrum, as Weber knew, is that the central values of a society may have no power to hinder a revolution that takes shape — as they so inconveniently tend to do — at the margins.

Both of the confusions I have described — blaming the earlier state for delaying the later state, and using the absence of something modern at one point to explain the unattainability of modernity later — confuse continuity with stasis. They are bad history because they are bad philosophy.

How do you avoid fallacies like these? Perhaps by a different model of historical change, in which there is no such thing as *the* cause. I am partial at the moment to the organismic model which the old Chinese historians used so effectively, which encouraged them to look at interplay rather than argue either/or propositions. This model sees most change as readjustments in a system where every part is in a sense a cause of every situation, where increasing articulation gives rise to new events in ways too complex to be traced to a single trigger or even to a single class of trigger. I will have to leave it at that, but let me sum up by saying that historians badly need the cooperation of philosophers in avoiding reductionist conceptions of causality in history based on bad analogies with physics. Such conceptions the better Chinese historians would have rejected as another exotic mannerism of the mysterious West.

Now a couple of remarks on Vic's paper. First, I would have been grateful for explicit attention to the way that, as contexts differ, animistic words can lose their vital connotations, and terms that seem objective to us can be rich in them. I am sure Vic does not mean to imply that the impetus-theoreticians of the late Middle Ages or Copernicus had a magical conception of motion. One might argue that an objective definition of force was not needed by people who were doing mathematical astronomy before Kepler. The concept of circular motion as eternal functioned as what might loosely be called a principle of inertia; it made force laws unnecessary. It was only when Kepler reluctantly relinquished circularity that he felt obliged to account for the eternity of motion. He did so by means of a force which radiated from the sun, acting radially on the planets to an extent which varied according to whether what he called their 'friendly' or 'unfriendly' faces were pointed inward — a sort of oscillatory gravitation, influenced by contemporary theories of magnetism. Whether this radiating force was occult in character, whether Kepler's fascination with the harmonies of the spheres was mystical, are questions that I find too dependent upon definitions to be very interesting.

Finally, I feel that the most appropriate expression of gratitude for Vic's paper would be a grand cross-cultural comparison on zoological fulmination. Space does not allow it, even if I had tried to do nothing else. Let me substitute the historian's ultimate offering — a document. Here is the famous crocodile edict, which the great man of letters Han Yü wrote in 819 when he was exiled to Ch'ao-chou in South China and found the inhabitants being terrorized by a crocodile. For all the rhetorical elegance (best preserved in this rather free Edwardian rendering), it is an exorcism, and yet there is not a breath of evil implied. Its style is clearly bureaucratic rather than juridical, which is precisely the point I wish to make about the mandarin approach to deviant fauna:

THE CROCODILE OF CH'AO-CHOU*

On a certain date, I, Han Yü, Governor of Ch'ao-chou, gave orders that a goat and a pig should be thrown into the river as prey for the crocodile, together with the following notification: —

"In days of yore, when our ancient rulers first undertook the administration of the empire, they cleared away the jungle by fire, and drove forth with net and spear such denizens of the marsh as were obnoxious to the prosperity of the human race, away beyond the boundaries of the Four Seas. But as years went on, the light of Imperial virtue began to pale; the circle of the empire was narrowed; and lands once subject to the divine sway passed under barbarian rule. Hence, the region of Ch'ao-chou, distant many hundred miles from the capital, was then a fitting spot for thee, O crocodile, in which to bask, and breed, and rear thy young. But now again the times are changed. We live under the auspices of an enlightened prince, who seeks to bring within the

Imperial fold all, even to the uttermost limits of sea and sky. Moreover, this is soil once trodden by the feet of the Great Yü himself; soil for which I, an officer of the State, am bound to make due return, in order to support the established worship of Heaven and Earth, in order to the maintenance of the Imperial shrines and temples of the Gods of our land.

"O crocodile! thou and I cannot rest together here. The Son of Heaven has confided this district and this people to my charge; and thou, O goggle-eyed, by disturbing the peace of this river and devouring the people and their domestic animals, the bears, the boars, and deer of the neighbourhood, in order to batten thyself and reproduce thy kind, – thou art challenging me to a struggle of life and death. And I, though of weakly frame, am I to bow the knee and yield before a crocodile? No! I am the lawful guardian of this place, and I would scorn to decline thy challenge, even were it to cost me my life.

"Still, in virtue of my commission from the Son of Heaven, I am bound to give fair warning; and thou, O crocodile, if thou art wise, will pay due heed to my words. There before thee lies the broad ocean, the domain alike of the whale and the shrimp. Go thither, and live in peace. It is but the journey of a day.

"And now I bid thee begone, thou and thy foul brood, within the space of three days, from the presence of the servant of the Son of Heaven. If not within three days, then within five; if not within five, then within seven. But if not within seven, then it is that thou wilt not go, but art ready for the fight. Or, may be, that thou hast not wit to seize the purport of my words; though whether it be wilful disobedience or stupid misapprehension, the punishment in each case is death. I will arm some cunning archer with trusty bow and poisoned arrow, and try the issue with thee, until thou and all thy likes have perished. Repent not then, for it will be too late."†[9]

---

* This diatribe has reference to the alleged expulsion of a crocodile which had been devastating the water-courses round Ch'ao-chou, whither Han Wên-kung had been sent in disgrace. The writer's general character and high literary attainments forbid us, indeed, to believe that he believed himself.
† The crocodile went.

NOTES

[1] Needham, Joseph, *The Grand Titration. Science and Society in East and West* (George Allen and Unwin, Ltd., London, 1969), p. 16.

[2] Weber, Max, *The Religion of China. Confucianism and Taoism*, tr. Hans H. Gerth (The Free Press, Glencoe, Illinois, 1951), p. 248. I have examined more closely the problems discussed in the next few paragraphs in "Chinesische Wissenschaft. Ein Vergleich der Ansätze von Max Weber und Joseph Needham," pp. 342–362 in Wolfgang Schluchter (ed.), *Max Webers Studie über Konfuzianismus und Taoismus* (Suhrkamp, Frankfurt-am-Main, 1983).

[3] Needham, Joseph, *The Grand Titration*, pp. 150, 248.

[4] Needham, Joseph, *Science and Civilisation in China*, Vol. 3 (Cambridge University Press, Cambridge, England, 1959), p. 168; Robert S. Cohen, 'The Problem of 19(k)', *Journal of Chinese Philosophy* **1** (1973), pp. 103–117.

⁵ Mungello, David E., 'On the Significance of the Question "Did China Have Science"?', *Philosophy East and West* **22** (1972), pp. 467–478. For a critique of this essay, which was apparently written without any attempt to study scientific writing, see my letter to the Editor in *Philosophy East and West* **23** (1973) 413–416.

⁶ Ho, 'The System of the Book of Changes and Chinese Science', *Japanese Studies in the History of Science* **11** (1972), pp. 23–39. The idea had emerged also in *Science and Civilisation in China*, Vol. 2 (Cambridge University Press, Cambridge, England, 1956), p. 340, where Needham speaks of "the evil effects which [the *Book of Changes*] had on scientific thinking."

⁷ I have taken up this and related problems at greater length in 'Why the Scientific Revolution Did Not Take Place in China – or Didn't It?' *Chinese Science* **5** (1982) 45–66.

⁸ Eberhard, Wolfram, 'The Political Function of Astronomy and Astronomers in Han China', in *Chinese Thought and Institutions*, John K. Fairbank, ed. (University of Chicago Press, Chicago, 1957), pp. 33–70, 345–352.

⁹ *Gems of Chinese Literature*, Herbert Giles, tr. (Bernard Quaritch, London, 1884), pp. 129–131.

Prof. N. Sivin
Dept. of History and Sociology of Science
University of Pennsylvania
215 South 34th Street D6
Philadelphia, Pa. 19104
U.S.A.

DONALD D. WEISS

TOWARD THE VINDICATION OF FRIEDRICH ENGELS*

1. THE PROBLEM

It is by now a commonplace that Marxism is going through an important period of theoretical (and no doubt, for the same reasons, practical) development. It has been noted by many that the 'crude' and 'mechanical'[1] application of Marxian categories which prevailed as orthodoxy for about a half century — starting, roughly, with the formulation of the Erfurt Program, and lasting even beyond the Stalin years — is, in many quarters, fast giving way to a far more sophisticated and sensitive means of cultural analysis. Indeed, even this occurrence of the word 'cultural' is itself quite significant. For the most fundamental aspect of Marxism's current transformation may perhaps be best stated as follows: we are beginning to appreciate what it means to say that *Marxian analysis is cultural analysis*. It has become clear that Marx's interest in the economic metabolism of society was rooted in his conviction that this is the way in which the interchange of *the products of human creativity* is to be most adequately studied. The importance of this formulation is that once Marxism is explicitly conceived as a theory concerning the production and distribution of *culture*, it becomes quite impossible to treat the ideational aspect of human history as it was treated by the older 'orthodoxy': as if it were some merely epiphenomenal dimension riding piggyback upon an essentially *non*-ideational 'base' of society, the latter 'material foundation' being something over which human consciousness can have no truly decisive influence.[2] If there really is a science in 'scientific socialism', its insistence upon the materiality of human existence must not be taken to imply that the categories peculiar to human — i.e., *conscious* — materiality are to be accorded secondary status.

This transformation has at least two basic sources. The first is historical and political: it has come to be acknowledged by many that the epiphenomenalist treatment of the categories of consciousness was at least one significant factor lying behind the passivity — and that is to say, the non-revolutionary stance — which dominated the Second International and so contributed to the rise of fascism. The notion that history does its own work in its own sweet time — despite *our* impotent attempts to direct its course — is now more clearly

R. Cohen and M. Wartofsky (eds.), Methodology, Metaphysics and the History of Science, 331–358.
© 1984 *by D. Reidel Publishing Company.*

understood to be a vicious misconception of Marxian thought, and one which contributed to what is probably the greatest disaster to which human experience has (thus far) borne witness.

The second source is textual: I refer, of course, to the publication (only in 1932) of Marx's 1844 manuscripts and the *German Ideology*. These works cannot but make the epiphenomenalist interpretation of Marx appear sadly distorted; those who would still wish to defend the old 'orthodoxy' have been forced to scurry about in search of some means of explaining them away. Before their appearance, on the other hand, the opponents of historical epiphenomenalism had to ask such a text as the *Theses on Feuerbach* — cryptic in its extreme brevity — to bear virtually the entire weight of their analysis, insofar as their analysis was rooted in the Marx—Engels opus itself.

All of this is well known, and it is not my purpose here to defend this anti-reductionist turn which some of the more creative exponents of Marxism have taken. I am satisfied that this particular battle has by now been won. What I wish to do, rather, is to contribute to the clarification of what this turn of thought should be taken to *mean*. I want to argue that the anti-reductionist trend in Marxian theory, in itself laudable, has created some rather unfortunate — and potentially dangerous — confusions of its own; that, in particular, it has led, in many quarters, to a complete misapprehension of the significance to Marxian thought of the materialist philosophy of Friedrich Engels — and indeed, for just the same reasons, of the status of the very concept of *materialism*. The anti-reductionist turn has, in fact, led some to believe that Marxism is not literally a materialism at all, and that Engels' explicit materialism — with its great attention to natural phenomena, and its insistence upon the thoroughgoing materiality of *human* phenomena — is a crude reductionism, and hence a 'betrayal of Marx'.[3] In this paper, I wish not only to refute this conception, but also to show why it is so important that it be refuted.

Now it is notorious that Engels devoted a good portion of his theoretical energies, especially during the decade 1873—83, to the investigation of the 'dialectics of nature', — i.e., of what has subsequently come to be called 'dialectical materialism'.[4] Just as notorious is the fact that Karl Marx devoted quite little of his own time to the study of these very same questions. But while no student of Marxism could ever have failed to notice this difference between the two friends, it was only with the publication of the young Lukacs's *History and Class Consciousness* that the assumption that Marx's and Engels' opinions together constituted a theoretical unity began to be seriously challenged. Lukacs maintained that Engels' ascription of a 'dialectic' to nonconscious nature betrayed a failure, of which Marx was not guilty, to understand that it is to "the dialectical relation *between subject and object in the historical process*" that all dialectical reasoning, properly so called, has essential reference. 'Dialectics', that is to say, concerns the interaction between historically produced reality, on the one hand, and the consciousness of those who participate in history, and hence whose task it is to produce a new historical reality, on the other. Engels, says Lukacs, "does not even mention" this central application of the concept, "let alone give it the prominence it deserves." The lamentable result is that "thought remains contemplative and fails to become practical; while for the dialectical method the central problem is *to change reality*." [5] The Engelsian dialectic of nature is thus seen as partly responsible for the unfortunate epiphenomenalist tendency already discussed.

Lukacs's criticism, it should immediately be noted, is quite brief. More importantly, he does not at any point reject the idea of a *materialist* dialectic; to the contrary, the importance of materialism is stressed numerous times throughout his important work. And this point is quite important, in view of the fact that the more recent criticisms of Engels' conception of dialectic — and it is this more recent criticism that I wish to consider in this paper — is, though rooted in Lukacs' remarks, far more radical than the latter. The more recent critics of Engels make bold to proclaim what Lukacs himself would never have dared; namely, that not merely is the agency of consciousness left out of Engels' 'dialectic' but, more profoundly yet, the root of this unfortunate tendency on Engels' part is that he was a *materialist* at all. This materialist tendency was not present, the partisans of this view maintain, in the (properly interpreted) writings of Karl Marx. It was Engels, after all, who taught that consciousness is but the "highest product of matter." [6] And it was this conception that led him, as contrasted with Marx, into an essentially reductionist emphasis upon entirely nonconscious factors. With Marx — these interpreters insist — things are otherwise. That Marx sometimes called himself a materialist cannot be denied. But it is nonetheless insisted that 'materialism' meant something else for Marx than what it is typically taken to mean; and thus that it would be more accurate to adopt one of the other terms that Marx occasionally used in order to characterize his fundamental philosophical position — 'naturalism' or 'humanism'.[7] Armed with the 1844 manuscripts, these critics maintain that Marx reserves an irreducible role for conscious activity, *as contrasted with purely 'material factors'*, in human life and hence in history; that, according to Marx, the nature of (for example) human discontent cannot correctly be conceived in purely materialistic categories, since all truly *human* discontent is a response to the *alienation of creativity* from people's lives. Thus, not wishing to deny the importance of this *ideal* aspect, Marx called himself a 'materialist' only so as to indicate that Hegel and other (putatively thoroughgoing) 'idealists' went wrong by *excluding* material factors from their historical interpretations. It is for this reason that, as compared (for example) with the Hegelians, Marx seems fairly materialistic; but it would be more accurate to say that (authentic) *Marx*ism amounts to an ingenious *synthesis* of the best elements of materialism *and* idealism.

George Lichtheim has devoted two chapters to this 'Marx *versus* Engels' wedge-driving. A representative passage asserts that:

[t]here is a fatal flaw in [Engels'] attempted synthesis of speculative philosophy and positive science: if nature is conceived in materialist terms it does not lend itself to the dialectical method, and if the dialectic is read back into nature, materialism goes by the board. Because he knew this, or sensed it, Marx wisely left nature (other than human nature) alone. Engels ventured where Marx had feared to tread, and the outcome was dialectical materialism: an incubus which has not ceased to weigh heavily upon his followers...[8]

The other side of this exegetical coin is then made explicit: since Engels erringly imported materialism into the human-social domain, and since materialism is incompatible with an insistence upon the importance of consciousness in history, it is Engels who is partially responsible for the vicious epiphenomenalist tendency to consider history as being impelled by a necessity over which we (who participate in the process) have no control, a tendency which has plagued much of the Marxist movement by substituting sanguine expectations for what was really needed: decisive action to capture the heart and mind of the people.[9]

Shlomo Avineri writes in a similar vein, beginning with the observation that

... Engels says in *Dialectics of Nature* not only that matter historically preceded spirit, but also that it is the cause and source of the evolution of consciousness. It has become commonplace and fashionable to credit Marx with such a reductionist view which sees in spirit a mere by-product of matter.[10]

On this basis Avineri alleges a sharp distinction between Engels' thought and Marx's:

Marx's views cannot be squared with Engels' theories as described in *Anti-Dühring* or *Dialectics of Nature*. Lukacs and his disciples are perfectly right in maintaining that the dialectics of nature, in Engels' sense of the term, has very little in common with the way Marx understood materialism, and that the origins of Engels' views must be sought in a vulgarized version of Darwinism and biology...[11]

Such critics clearly go much further than Lukacs in their indictment of Engels' philosophical orientation. For Lukacs had confined himself to an *en passant* criticism of what he regarded to be an obfuscating interpretation of the notion of 'dialectic'. Engels' more recent critics, on the other hand, have enlarged the attack into an attempt to show that Engels' thought is fundamentally different from Marx's. Marx is, in this interpretation, made to sound much like a *dualist*, one who rejects materialism in the usual acceptation of that term, in virtue of the fact that 'ideal factors' are, to him, as important as 'material factors'. Engels, on the other hand, is a full-blooded, dyed-in-the-wool materialist, a monist who preaches that all the world is reducible to the very same brute stuff; his approach must therefore be considered basically at variance with Marx's.

Lichtheim and Avineri are joined, on this general line, by Dupré,[12] by Tucker,[13] and — in a work of no less than 458 pages devoted to the exploration of this single theme — by Z. A. Jordan.[14] But perhaps the most terse statement of this entire interpretative tendency is provided by F. L. Bender:

The lost humanism of [the early, unpublished] Marx was all the more easily forgotten because, under the influence of Engels, an entire generation of Social Democrats had imbibed the doctrines that (a) there is nothing in the universe but matter; (b) therefore man is material, i.e., a collection of atoms and molecules which are in motion solely in accordance with the laws of physics; (c) therefore man is not and could never become free; and (d) since capitalism will collapse anyway, we socialist politicians shall await that event, await the reins of government falling into our hands . . .[15]

It is no exaggeration to say that we are today confronted with a veritable bandwagon of writers who insist that the (putative) crudities of Engels' materialism must not be confused with — but are, in fact, inconsistent with — the authentic teachings of Karl Marx. The present paper is a contribution to the project of bringing this bandwagon to a halt. I will argue that this recent large-scale attack upon Engels errs in holding that the latter is a 'crude reductionist', and in thinking that Engels' thought is inconsistent with Marx's. All the more emphatically, I will argue that the full significance of Marx's own theory of history, which is aptly called the 'materialist conception of history', cannot be understood without the Engelsian variant of materialism. To jettison Engels is to abandon the philosophical groundwork of Marxism. (This counterattack will equip the reader to draw out critical implications with respect to Lukacs's milder criticism.) There will be three stages in this vindication of Engels. In the first part of the argument (to be presented in the next section), I will show that Engels' philosophy — far from being (as Lichtheim, Avineri *et al.* maintain) a *reductionist* view that by its nature excludes the factor of 'consciousness' — is (on the contrary) the only philosophical framework within which essentially anti-reductionist attitudes can be presented in a manner that does not violate the materialist presuppositions of modern science. In the second stage of the argument, which will occupy Section 3, I will present a sketch of the basic Marxian culturological perspective in such a way that it will be at least intuitively plain why Marx, and not only Engels, regarded the Marxian historical theory as quite literally materialist. Finally, in Section 4, I will try to indicate why an insistence upon the materialist character of Marxism is not merely of intellectual interest, but rather has significant *practical* import.

2. THE IMPORTANCE OF ENGELS' GENETIC MATERIALISM

We may begin by observing that Lichtheim is quite right in referring to Engels' position as a 'materialist evolutionism'. Other accurate appellations would be 'developmental materialism' and 'genetic materialism'; these terms may be fairly arbitrarily interchanged. But, as the reader will have observed, the term '*dialectical* materialism' has been excluded from the above listing. This is not because I think that this term is entirely inappropriate, but rather because its use would be likely to constitute a stumbling-block in the way of an unprejudiced interpretation of Engels' philosophy. The point is that Lukacs was right at least in this much: that according to the historically dominant usage of the term 'dialectic', a usage having a fine philosophical pedigree stretching from Plato through Hegel and thence to Marx, a process can *by definition* be termed 'dialectical' only if it involves the activity of the human critical faculty. Now if this meaning is presupposed, as it very often has been, then Engels' attempts to show that all of reality, including nonconscious nature, involves a 'dialectic', will, *ab initio*, seem quite absurd. The result will be a refusal to consider the substantive philosophical position that lies clothed in (what one may admit to be) a potentially misleading vocabulary. When, on the other hand, Engels' position is stated (as he himself sometimes did state it) independently of that vocabulary, it becomes plain that his assertion that there is a 'dialectics of nature' is the foundation of a philosophical *attack* upon the very reductionist tendency that his critics consider him guilty of maintaining.

Now, in thus calling Engels' position 'genetic materialism' I am of course committed to the retention of one fundamental term — 'materialism'. This term is still in quite general philosophical usage, and a proper understanding of why, and in what precise sense, Engels' views are indeed *materialistic*, is clearly necessary for an understanding of the potency of his general position. For in fact much of the failure to understand Engels may be attributed to a failure to understand what materialism really does involve. Thus, Avineri, Lichtheim *et al.* appear to assume that a thoroughgoing materialist must be a 'reductionist' with respect to human consciousness; that to be a materialist is to believe that a wholly adequate science could be constituted of the terms and laws appropriate to natural science. Now of course if one believes this, one may readily move to an attack upon Engels for the latter's abandonment of the *Marxian* premise that the categories of the human and social sciences are to some extent *sui generis*, i.e., irreducible to categories of a non-human and non-social order. But, in fact, *'materialism' does not by itself connote a commitment to any form of reductionism*; and, in fact, the explicit defense of an *anti-reductionist materialism* is at the very core of Engels' approach.

Now, it would be quite fruitless to engage in a purely terminological squabble over the word 'materialism'. Like many other philosophical words, this one has taken on various distinguishable colorations. It may, therefore, be granted that 'materialism' has often been narrowly used with particular reference to physicalist reductionism. It has, that is, been used to connote the following twofold denial: first, of the proposition that there are *non-physical substances*; and, second, of the claim that there are various *levels of integration* of physical substances, the terms and/or laws appropriate to the science of 'higher' (e.g., biological or psychological) levels being not wholly replaceable by the terms and/or laws of physical science. (This distinction between two non-reductionist orientations, one asserting the existence of non-physical *substances*, the other asserting that there are various *levels of physical organization*, the science of the 'higher' being irreducible to the science of the 'lower', will figure importantly in the discussion below.) But it is a mistake to think that this reductionist meaning is the only one which has been attached to 'materialism'. And excessive attention to the reductionist variant can lead, in fact, to a failure to understand a more basic 'core meaning' of this philosophical term in virtue of which various views, reductionist and non-reductionist alike, have — with equal legitimacy — been called 'materialist'. Since Engels was a materialist in this latter, broader, sense, we must now clarify just what this sense is.

For Engels, any viewpoint is materialist if and only if it denies that there are any non-physical *substances*; i.e., if and only if it asserts that all existents — including those peculiar categories of existents; entities that live, and entities that think — are made up of the very same sorts of constituents. No doubt the differences between those things that are alive and those that are not, and between those that engage in mental activities and those that do not, are striking indeed. But the materialist position – as we are now construing it – nonetheless maintains that the theoretical apparatus peculiarly appropriate to living or mental phenomena is only an appropriate response to a particular manner in which entirely 'natural' substances happen to be organized. Thus, when these substances are arranged in a particular way, we have what is properly called a stone; when arranged in another way, an oak tree; when arranged yet otherwise, a well-functioning – and that is also to say *thinking* – human being. Materialists deny, for example, that there is a mysterious something that defies in principle our hubristic attempts to produce life in the laboratory, or to construct a system that really does think, out of such natural substances. To put this point more precisely, materialism is a twofold thesis maintaining both (a) that there *are* entities which neither live nor think nor presuppose the existence of anything that does live or think; and (b) that everything that exists may be correctly viewed as a particular arrangement or 'construction' of entities of this (nonliving and nonthinking) sort, even if – as is of course the case – our current technology is unable to make such constructions.

This constructibility hypothesis – we may call it 'Frankenstein's Hypothesis' – is what, at all events, we will hereinafter consider to be the necessary and sufficient condition of a doctrine's being truly materialist. And it is plain that all reductionists – e.g., physicalists[16] – are materialists in this sense, since one could not maintain that all phenomena are in principle tractable by physical science, at the same time as one maintained that the correct characterization (for example) of thinking entities requires postulation of some *further substance* than the *physical* constituents of the human organism. But, given our understanding, the converse implication does not hold: one can indeed be a materialist without being a reductionist. One can maintain without incoherence – and with a great deal of plausibility – that despite the fact, stressed by all materialists, that all of reality is made up of nothing but 'natural substances', reductionism is nonetheless wrong in denying that differing constellations of such substances require, at least sometimes, differing basic terms and/or laws if their properties, states and activities are to be correctly characterized.

For the remainder of this section, I wish to argue in defense of the following three theses: *first*, that Engels is indeed a materialist in the sense of Frankenstein's Hypothesis; *second*, that Frankenstein's Hypothesis is a basic article of modern scientific belief, and thus that Engels' anti-materialist detractors must themselves be attacked if they have meant to deny this hypothesis; and *third*, that Engels is clearly an *anti-reductionist* whose thought is in no sense inimical to, but rather explicitly supports, the view that the sciences of the human are in a clear sense *sui generis*.

I take it that with respect to the first of these issues, unlike the second and third, there is no disagreement whatever. It is agreed on all hands that Engels is a materialist, that he believed that living and thinking entities are constructible out of nonliving and nonthinking entities; and that such a process actually has occurred in the natural history of our planet. A lengthy exposition of this general perspective is given in the Introduction to *Dialectics of Nature*. Thus Engels says:

With progressive cooling [of the earth's surface] the interplay of the physical forms of motion which become transformed into one another comes more and more to the forefront until finally a point is reached from when on chemical affinity begins to make itself felt . . .

If, finally, the temperature becomes so far equalized that over a considerable portion of the surface at least it does not exceed the limits within which protein is capable of life, then, if other chemical conditions are favorable, living protoplasm is formed. What these conditions are, we do not yet know, which is not to be wondered at since so far not even the chemical formula of protein has been established.[17]

He then traces in broad outline the evolution of organic forms, an evolution finally resulting in *man's* becoming "distinct from the monkey" by means of the specialization of the hand, "the mighty development of the brain," etc.[18] Avineri is thus quite right (even if not terribly illuminating)[19] when he says that Engels propounded the view "that matter historically preceded spirit," that "spirit [is] a mere biological by-product of matter" (although, as will become plain, the word 'mere' is inappropriate in this context). It is clear that Engels is a materialist.

Now to the second point: I wish to claim that this materialist perspective is a *sine qua non* of membership in the modern scientific community. Modern science is materialist science. This is not to say that materialism has always been the most plausible scientific view: it is at the very least arguable that Cartesian dualism, according to which a human being is composed of two distinct sorts of substance — body and soul — was quite as sophisticated, given the state of science in Descartes's day, as the materialism so forcefully urged by his contemporary, Hobbes. A decisive resolution of the controversy between these two perspectives was in fact impossible within the framework of 17th century science; it has become possible only by means of more recent accomplishments.

I am not asking the reader mainly to consider the implications of recent achievements in biochemistry and cybernetics. For it is possible for one still to have reason to doubt, despite these accomplishments, that we will actually succeed in synthesizing living things 'in a test-tube', or in constructing a machine that, beyond any question, really does think. It is perhaps more important to understand that the modern doctrine of the *evolution* of all entities, including living and thinking entities, by means of more and more complex integrations of natural substances is a doctrine which no serious scientist would today take it upon himself to doubt. Whether or not human beings can construct living and thinking entities, what is *not* seriously dubitable is the proposition that all forms of life — including that species of which Shakespeare, Newton and G. W. F. Hegel have been members — did evolve from more primitive forms of life; that the most primitive forms of life developed, in turn, from a series of combinations of initially non-living chemicals; and that this whole process occurred within the scope of some three or four billion years. It, of course, would not make any difference (for *our* purposes at any rate) if the correct figure turned out to be five billion years; the point in principle is clear: modern science is committed to the proposition that living and thinking entities are, in principle, synthesizable from nonliving entities — *precisely because modern science has reached the conclusion that such entities were, in the actual course of natural history, so synthesized*. Thus, even if our species does not itself succeed in creating living and thinking things, this will only mean that we are for some reason unable to get the knack of doing what nature itself has already done. (We have the advantage, over 'blind' nature, of being quite intelligent; but nature has the advantage of having had an immense amount of time.)

It is worth observing that Darwin's notebooks leave no room for doubt that he was entirely clear concerning the materialist character of his account of evolution. Indeed, one writer has plausibly argued that the fact, long known but little understood, that Darwin achieved his decisive theoretical insight — the elaboration of the doctrine of natural selection — as early as October 1838, but withheld publication until more than 21 years later, was due to Darwin's extreme awareness of the penalties previous materialists had paid.[20] It was only in the privacy of his own notebook that he drew the inevitable conclusion that even the products of the human mind — even our loftiest thoughts and affections — are nothing more nor less than the 'effect' of a particular form of material 'organization':

Love of the deity, effect of organization, oh you materialist! ... Why is a thought being a secretion of brain, more wonderful than gravity a property of matter? It is our arrogance, our admiration of ourselves.[21]

Even in *The Origin of Species* Darwin is unwilling to make the materialist character of his thought anything more than implicit. Explicit materialist argumentation does not appear until a dozen years later, when acknowledgement of the 'heretical' nature of his theory is no longer avoidable.[22]

But however 'heretical' evolutionary theory may have been, or may still be, its place in modern science is nonetheless quite secure. I will not, of course, place myself in the ridiculous position of defending this doctrine here. The biologists have done, and will continue to do, well enough without me. A philosopher's prerogatives, in such matters, are quite limited. We must fairly humbly accept his place in the intellectual division of labor, which includes acknowledging conclusive scientific findings for what they are. It will, therefore, be enough for us to conclude that those who would dissociate Marx from Engels over the issue of 'materialism' quite plainly do him the disservice of setting him at odds with the Darwinian achievement which broke upon the scene in the very midst of Marx's own career – an achievement whose implications banished, thenceforward, all *non*-materialist accounts of life, and of human life, from the halls of science. Of such 'allies' as these, Marx – who explicitly admired (and indeed perceived his own connection with) Darwin – has little need.[23]

Why, then, have Marx's 'friends' been so anxious to reject the suggestion that he was a materialist? This question brings us to the third point I wish to make in this section: that the critics of Engels have been unable to understand precisely what Engels explicitly asserted and attempted to justify: that there is a quite clear – and let me add: quite 'hard-nosed' – sense in which one can be a strict 'materialist' and yet be profoundly anti-reductionist in orientation. The failure to understand how this might be so has led to the mistaken inference *from* the (correct) observation that Marx was not a reductionist, to the (erroneous) conclusion that he 'therefore' could not have been, like Engels, a materialist.

One of the implications of reductionism is, as we have seen, that the terms and laws appropriate to the study of complex entities (including human beings) are the same as are appropriate to the study of the substances of which these more complex entities are constituted. But Engels' insistence upon the 'dialectical' character of material reality is, by contrast, nothing more nor less than the insistence that such reality is inherently *developmental*; that, as Engels (following Hegel) puts it, "quantitative" changes in things in time give rise to "qualitatively" new sorts of things.[24] Engels maintains that the material organization of the world is constantly changing, and that, with new forms of material organization, there arise, correspondingly, novel properties and regularities. Thus, from the fact that a science S_1 is able to deal with entities a and b when the latter exist in isolation from one another, it does not follow that S_1 will be adequate when a and b combine in a particular way to constitute the complex substance aRb; for there may be characteristics and modes of behavior nonexistent before the advent of aRb which are brought about in virtue of aRb's coming into existence. To put this point in terms quite familiar to the Anglo-American philosopher: when a and b combine so as to form aRb, there may be *emergent* properties and regularities, novelties necessitating a more complex science, S_2, if they are to be comprehended. Now, Engels was in fact a contemporary of two of the most prominent members of the British school of 'emergent evolutionists', J. S. Mill[25] and G. H. Lewes;[26] and, like them, he was largely motivated by the conviction that the empirical evidence was accumulating in support of the twofold claim (a) that complex ('higher') forms *evolved from* less complex ('lower') forms, (b) *without* it being the case that the science of the former is *reducible* to that of the latter. (Let me say in passing that though there are many terminological and some substantive differences between Engels and the British emergentists, these differences are pallid when compared to the essential affinity of the respective doctrines. To my extreme surprise, I have found nowhere any discussion of this affinity. This is doubly remarkable in the light of a fact which I simply cannot bring myself to hide in a note: that Engels and Lewes not merely knew of each other's existence, but (as I have discovered) lived a short distance away from each other in the Regent's Park area of London.[27] History has produced lesser ironies than this.)

That Engels embraced the first part of the preceding twofold claim is a fact that has already been established. But he is just as insistent on the second point; mechanism (the Enlightenment brand of physicalist reductionism) is rejected in terms which make his general anti-reductionist orientation absolutely clear:

What the animal was to Descartes, man was to the materialists of the eighteenth century – a machine. This exclusive application of the standards of mechanics to processes of a chemical and organic nature – in which processes the laws of mechanics are, indeed, also valid, but are pushed into the background by other, higher laws – constitutes the first specific but at that time inevitable limitation of classical French materialism.[28]

Moreover, the suggestion that Engels was a reductionist with regard to specifically *mental* phenomena is nothing more nor less than a manifest absurdity; calling "the craze to reduce [Engels actually does use the term *'reduzieren'*] [29] everything to mechanical motion" a position based upon a "misunderstanding," he proceeds:

> This is not to say that each of the higher forms of motion is not always necessarily connected with ... mechanical ... motion, just as the higher forms of motion simultaneously also produce other forms; chemical action is not possible without change of temperature and electric changes, organic life without mechanical, molecular, chemical, thermal, electric, changes, etc. *But the presence of these subsidiary forms does not exhaust the essence of the main form in each case. One day we shall certainly 'reduce' thought experimentally to molecular and chemical motions in the brain; but does that exhaust the essence of thought?* [30]

So ends the passage; the supposition that thought can be adequately treated by means of the categories of natural science alone is dismissed with a scornfully rhetorical question. And the 'scare-quotes' around the word 'reduce' are, *nota bene*, Engels' own. As if he himself does not make this point with noonday clarity, we may paraphrase him thus:

> That mental processes are molecular and chemical motions in the brain, is a point that must be insisted upon; it is unnecessary to complicate our ontology by the addition of some sort of soulish substance. If that is all you mean by 'reduce', then 'reduction' is possible. But if this word is used as it typically is, as implying the superfluousness of specifically mental categories, then 'reduction' is obviously *not* possible.

In short: Engels' critics, working from a quite understandable *anti-reductionist* perspective, reject his theory because it is a *materialism*. This not only involves a fundamental confusion, since these are two entirely distinct issues — even worse, it leads to a rejection of what is, as we have seen, a fundamental presupposition of modern science. The only way for these critics to avert such a disaster is for them to espouse a position which is *both* materialistic *and* anti-reductionist. But if they do this, they will find themselves driven into the arms of a doctrine which had long ago already been embraced by no other personage than — Friedrich Engels.

3. THE ESSENTIAL AFFINITY OF ENGELS AND MARX

The critics of Engels must, however, be subjected to counter-attack on an even deeper level. For it can be shown that the most fundamental aspects of the materialist conception of history, a doctrine with which Marx's name is associated even more strongly than is Engels', cannot be grasped except on the basis of the sort of anti-reductionist materialism propounded by the latter.

Let us begin — quite tentatively — by observing that, as early as 1844,[31] Marx himself was apparently writing in defense of an anti-reductionist materialism. In a well-known passage in *The Holy Family*, Marx compares the philosophy of Francis Bacon favorably with that of Hobbes. Bacon, says Marx, conceives "motion" as "not only ... *mechanical* and *mathematical*, but even more as *impulse, vital spirit, tension* ..." Thus:

> In Bacon, its first creator, materialism conceals within itself still in a naive way, the germs of an all-sided development ... [M]atter smiles upon the whole of man in poetic-sensuous splendor ...
>
> In its further development materialism becomes *one-sided*. Hobbes is the *systematizer* of *Baconian* materialism. Sensuous knowledge loses its bloom and becomes the abstract sensuousness of the *geometer*. *Physical* motion is sacrificed to *mechanical* or *mathematical* motion; *geometry* is proclaimed to be the chief science. Materialism becomes *misanthropic*.[32]

This was written only a few months before Marx was to counterpose what he called a 'new materialism' to the above 'misanthropic' materialism of Hobbes *et al.* As is notorious, in the *Theses on Feuerbach* Marx maintained (among other things) that materialism must learn that human beings have a creative and 'subjective' side; that *materialism must be so understood that it does not deny the irreducible reality of — and indeed the transformative powers of — human intelligence.*[33]

But such passages as these are not by themselves sufficient to prove a Marx—Engels affinity. Indeed, we have already seen that some of the same critics that find Engels a reductionist consider texts like the *Theses* to constitute proof that Marx was not, strictly speaking, a materialist at all, but rather a thinker who 'synthesizes' materialism and idealism. (And one may suppose that a similar interpretation would be made of the passage quoted from *The Holy Family*.) Of course, we might *now* respond to such an interpretation by pointing out that its source is almost certainly the fallacious assumption that we have already exploded — namely, that materialism and reductionism come to the same thing. It might be argued that the same fallacy that leads these critics to consider an outright materialist like Engels a *reductionist*, leads them to interpret an outright anti-reductionist like Marx as not being, strictly speaking, a *materialist*. And thus it might be concluded that, once this fallacy is exposed, Marx's claim that he represents a 'new' form of what is nonetheless *materialism*, should be taken precisely at face value.

But the Marx—Engels affinity can be established even more decisively. For it can be rather easily shown that the fundamental *social* orientation of what is called 'Marxism' — i.e., *historical* materialism — presupposes the very anti-reductionist materialism — i.e., *dialectical* materialism — which Engels attempted to elucidate. Marx's own assertion that it is the 'mode of production' that provides the 'base' of a given society, other societal factors being theoretically secondary, cannot otherwise be understood.[34] I will accordingly now sketch — more than a sketch will not be possible — some of the fundamentals of the historical materialist thesis, in such a way as to make clear its essential connection with the philosophy of Engels.

The essential point is that the Marxian emphasis upon the factors of *production* would be incomprehensible — or at least would appear to be entirely arbitrary — were it not part of a quite general theoretical position according to which the development of the human powers must be viewed in the context of the development of *material culture*. According to this conception (first fully worked out in 1845, in the first part of *The German Ideology*, by Marx and Engels *jointly*),[35] the *differentia specifica* of the human species is our ability — an ability which is augmented by each successive historical stage — to modify our talents, and hence our needs, through the *production of useful objects*, or artifacts. The development of the human powers, of the totality of human culture, must be considered a process whereby human beings become ever more able to assimilate purposefully the material environment, to transform 'brute nature' into rational-intentional forms. Human culture has advanced insofar as the members of our species have literally extended their sensory-motor wherewithal, by making what was at one time *mere* nature into instruments wherein new sensory-motor talents can now be exercised.

In the (historically) first instance, the human powers were delimited by the equipment that belongs to the human body: sense organs, fingers, etc. Such powers were not, of course, insignificant; but insofar as the members of our species were restricted to such a meager physical means of interacting with their environment, their talents clearly could not be significantly more impressive than those of the 'lower animals'. For, confined to his natural body, a person has minimal *control over* his world, and hence also cannot *understand* it beyond what is — in comparison with the culture we have in fact historically created — the paltriest of levels.[36] Or, in other words, human beings have in fact been able to "distinguish themselves from animals,"[37] to go beyond the repertoire inherent in their physical constitution, only because they have been able to *augment* their 'physical constitution' by literally appropriating and molding *matter* into virtual extensions of their body. Only, that is, by overcoming the purely 'objective existence' of nature, by transforming it into physical apparatus directly subservient to our will, have we been able to subdue and to comprehend an initially hostile and mysterious environment. A pointed stick becomes a hoe: thus, and only thus, is the way opened to agriculture — and hence finally, on a more theoretical level, to botany. We learn to control fire: not merely to warm ourselves, but also, in time, to transform substances into more useful substances — this being the practical presupposition of chemistry. And so on.

The Marxian analysis of the human powers is for this reason an essentially *materialistic* one. The human 'creative' abilities cannot exist, according to Marx, unless they are *realized* through the appropriation of the natural world. They cannot be realized except as *modes of organization, and hence of activity*, of the natural substances which comprise (in the first instance) the human body and (in the further, truly *historical*, instances) that body as extended by appropriated nature. If I may be permitted a heuristic metaphor: human intelligence is regarded by Marx as the ensemble of talents and abilities residing in our natural and in our *produced* 'limbs'.

The occurrence of the word 'produced' at this juncture is, moreover, no accident. For it is the Marxist's understanding of the materiality of culture which enables him to appreciate the privileged theoretical role that must be accorded to the concept of *production*. Only when we understand that specifically human culture can exist only insofar as it is literally created, worked up from the raw material environment into expressions of our own agency — only then will we be able to give due appreciation to the real prerequisites of such culture. The artificial world, without which our complex sensory-motor repertoire would have no medium of existence, must constantly be produced and reproduced if our 'distinction from the animal' is to *remain* a distinction. It is for this reason that *the history of the development of the human powers in general must be integrated with the study of the development of the human productive powers in particular*. And it is also for this reason that any study of the history of culture which remains unenlightened by the study of the development of the technological prerequisites of the various stages in the development of culture must be considered a correspondingly abstract culturological view.

Nor is this all. The above conception of the materiality of human culture also provides, Marx and Engels believe, a vantage point from which the specific problem of the character and development of *social formations* can meaningfully be posed. They state:

[A] certain mode of production, or industrial stage, is always combined with a certain mode of co-operation, or social stage, and this mode of co-operation is itself a 'productive force'.[38]

If we are to understand the various stages of social development, that is to say, we must understand each such stage as a particular *productive deployment of human beings*, a deployment which must be functionally responsive to the particular character of the technology then available, if human existence at the assumed level of historical development is to be *produced*. The 'logic' of social structure is grasped only when the various social formations are understood to be various stages of response to the requirement that we organize our *labor* in correspondence with the technological means that have themselves thus far been produced. We can, on this basis, understand Marx's famous insistence that each stage in the development of human culture presupposes both a particular development of technology – of 'forces of production' – and, correspondingly, the emergence of social forms – 'relations of production' – without which that technology could not be utilized.[39,40]

It is no objection to this materialist thesis to observe that, after all, there are people who engage in 'theoretical activity' properly so-called. The thesis of the materiality of culture is not meant to deny this. Anyone familiar with the argument of Part I of *The German Ideology* will realize that, to the contrary, the social division between 'mental and material labor' is one of the most fundamental of all Marxian sociological conceptions. Marx and Engels indeed maintain that, in all but the most and the least productive societies, the existence of a class whose primary function is that of intellectual labor, and whose function is accordingly to manage – that is, *to rule* – society, is a functional prerequisite of the production of material culture at that level. As contrasted with the 'ruling class', each such society will also have a class whose function is directly 'material' in the peculiar sense that it is done in the fields, in the factory, etc. – i.e., in the sense that it may be characterized as 'manual'.[41] Moreover, Marx and Engels do not deny, but rather insist, that future (communist) society will also involve both sorts of function. Communism is conceived by them as that stage of society at which class distinctions will be abolished, in the sense that no persons will be *restricted* to merely manual, or indeed to merely intellectual, functions. Both sorts of function will continue to exist, but they will be unified in the life of each individual.[42]

The point is not to deny the existence – and the importance – of the *reflective function*. It is rather to stress that the development of material culture cannot properly be understood unless these functions are considered as a phase, an aspect, of a much richer process. It is of course true that cultural advance would be inconceivable without the intellectual functions of (a) reflecting upon previous practice and (b) formulating, in light of such reflections, plans and projects for the future. Any brand of 'Marxism' which lapses into denying this obvious truth may correctly be rejected as 'crude' and 'reductionist'. What must rather be insisted is that each intellectual discipline's possible accomplishments are stringently delimited by the character of the instrumentation, relevant to that discipline, which has historically been developed — this instrumentation's degree of development being itself a function of the general technical level of advance achieved by the culture in question. Advancement in *theory* therefore depends upon its 'descent' into the *practical* arena — i.e., upon its contribution to the development of those material forms without which future *theorization*, on a higher plane, would be impossible. Theory must inevitably contribute to the production of human existence on a higher *material* plane, if it itself is not to be stultified.[43]

Some paragraphs back I promised a 'sketch' of historical materialism; and that, of course, is all that I have given. I have not defended this conception against the various objections that will be advanced; nor have I provided anything but the most minimal of explications of its basic concepts. These tasks must be reserved for another occasion. I am acutely aware that these brief observations are far from adequate to provide a decisive refutation of rival, idealist and dualist, doctrines; but these observations are not intended to accomplish that task. Rather, I have at most provided what might be termed the *programmatic and intuitive basis* of the anti-reductionist, culturological materialism which lies at the heart of Marxian social theory. An entirely compelling refutation of idealism and dualism, from a Marxian standpoint, would require the enlargement of this intuitive basis into a fully articulated theory of the nature and (correlatively) the manner of development of the human powers. This task exceeds the bounds of this essay by hundreds, or perhaps even thousands, of pages.

The preceding discussion should nonetheless make it intuitively clear why we should take Marx's insistence that he is a 'materialist' at face value. The general program is indeed a materialistic one; it constitutes an attempt to conceive of the human powers as particular modes of activity — 'forms of motion' is an expression favored by Engels[44] — of matter that has come to be organized in a particular way. We have seen that the Marxian view rests, indeed, upon the basic contention that human culture in general is extended only insofar as human existence is literally extended by the *materialization* of new techniques; of apparatus that is so molded that new powers can have their medium of real existence. We have seen that without this fundamental assumption, the entire Marxian emphasis upon the historical fundamentality of the 'forces' and 'relations' of *production* would seem an arbitrary or even unintelligible claim. It should therefore be clear that Marxism rests upon a fundamental resistance to any conception which maintains that the human powers reside in some special substance distinct from the natural substances of which the body, and appropriated 'external' nature, are composed. Any philosophical conception which posits a special 'soulish' substance distinct from these natural substances, and wherein human intelligence is asserted to reside — any such *dualistic* or *idealistic* conception is quite foreign to the Marxian perspective. To such a non-materialist view, the fundamental Marxian assertion that the existence and the furtherance of the human powers depends upon the corresponding production of material culture, i.e., that the human 'Spirit' develops only insofar as it is quite literally materialized, must remain a closed book. The distinctively Marxian conception of human culture and history will, in other words, be foreign to all but those who maintain that *intelligence exists only insofar as the material world has come to be intelligently organized*.

It is also important to understand what this does not imply — namely, that the human powers, 'since they reside in the natural substances of which everything is composed', can be scientifically treated in the same way as these substances may be treated when they comprise the 'external world'. At no point have I depicted Marxian materialism in such a way that it appears to involve a reduction of our science of the human to the categories appropriate to the natural sciences. To the contrary, I have assumed throughout that human beings are distinguished from other entities in virtue of their 'intelligence', and I have explicitly argued in this section that Marxism attributes fundamental importance to the reflective function. It insists, however, that this be treated *as* a 'function' — i.e., as a peculiar mode of activity of what is nonetheless quite *material*.

It is for this reason that, though Marx and Engels do not deny the irreducible importance of the fact that human beings 'think', are 'conscious', yet it is true that they insist upon *not* using these terms in order to characterize the peculiar character of the human subject matter. The trouble with such terms is not that they are inapplicable, but rather that they are much too *abstract*, capturing at best a *feature* of human life. Marx and Engels *avoid* these abstractions because the *bête noire* against which they were attempting to do battle is precisely the tendency to hypostatize them, to consider them to be substances, entities that might possibly exist in their own right, and thus having (as Marx and Engels accuse the Young Hegelians of believing) their own autonomous historical development. To the contrary, "from the start the 'Spirit' is afflicted with the curse of being 'burdened' with matter."[45] The quotation marks are their own; the passage veritably drips with sarcasm. The point is that matter is a 'burden' — an 'affliction', a 'curse' — only from the point of view of those who have not yet understood that it is only through the actual *production* of human existence, the *working-up of matter into media of human intercourse with the world*, that we have in fact been enabled to 'know' what we do know. Man's mental and creative functions must be conceived as nothing other than modes of activity which a particular organization of matter makes possible. To put this point in terms that will be familiar to the contemporary, linguistically-oriented philosopher: these functions should not themselves be described in the substantive mode, but rather adjectivally ('human beings *are conscious*') or verbially ('human beings *think*'). We may thus avoid the hypostatization of these functions which is responsible for an entirely mistaken conception of their development.

Those interpreters of Marx that have found him *not* to be a reductionist, are thus not at all mistaken. Indeed, such interpreters are, in a certain respect, worthy of our applause; for their impact, *versus* certain apparently reductionist tendencies which have paraded under Marxism's banner, has (insofar) been salutary. But it should, by now, also be clear that to attack *Engels' particular variant of materialism* in the name of anti-reductionism is to make what is, from the point of view of the Marxist, as serious a blunder as any that is committed by the reductionist. For not only is materialism a fundamental aspect of the modern scientific worldview, as I argued in the preceding section; it is, just as significantly, the foundation upon which the entire Marxian edifice is built. Marxism is indeed an anti-reductionism; but the conclusion to which the present argument points is that it is an *anti-reductionism which must nonetheless be conceived in materialist terms*. To the Marxist, the correct form of anti-reductionism must maintain that when the natural substances that constitute the world are 'organized' in a particular manner, i.e., when they have become assimilated into the human repertoire, they have a qualitatively different character from that which they had when they belonged to what we (accordingly) call 'objective nature'. Not only must it be conceded, it must be *insisted*, that the categories appropriate to the human mode of material organization — the categories of 'production', of 'culture', of 'intelligence', and so on — are irreducible to the categories appropriate to the lower forms of material organization. But it must also be insisted that the Marxian view of 'production', of 'culture', of 'intelligence', and so on, cannot be understood unless the human form of existence which these concepts describe is nonetheless understood to be quite literally material. And, bearing this twofold insistence in mind, it is, I think, impossible to avoid the conclusion that the theoretical pivot on which the totality of Marxian theory turns is the anti-reductionist materialism which — by no coincidence whatsoever — Friedrich Engels did so much to elucidate and develop.

4. THE PRACTICAL IMPORT OF THE FOREGOING

I have accused Engels' critics of certain theoretical distortions concerning the material character of human existence and hence of historical change. But it must now be added that the issue is not merely of *theoretical* interest. For it is clear that any theoretical distortions concerning the manner in which the present developed out of the past will imply correlative *practical* distortions concerning the manner in which the present is most effectively to be transformed into the future. This proposition is axiomatic for anyone who has learned anything at all from the study of Marx and Engels, and I will assume that it does not require argumentative buttressing here. What I now propose to do is to provide a brief critical characterization of the Marxian tendency to argue that those who are guilty of opposition to materialist theory may also be expected to be guilty of specifically *bourgeois practice*.

For, indeed, Marx and Engels do not merely *criticize* those who reflect (or distort, or underplay) the thesis of the materiality of culture. All the more radically, they utilize this thesis in order to explain why this thesis has itself been neglected. *Historical materialism provides us with a means of understanding why historical materialism has itself not, heretofore, been the dominant method of understanding sociocultural phenomena.*

The first basic consideration lying behind this Marxian understanding concerns the distinction between intellectual and manual labor, already discussed briefly.[46] Marx and Engels maintain that the social differentiation between these two forms of activity characterizes all but the *least* and the *most* productive societies. Except in the case of either of these two extremes, the rational[47] organization of society requires the establishment of a particular class which, because it is freed from activity which *immediately* and *directly* pertains to the reproduction[48] of material culture, may engage in those more abstract, reflective tasks which are less obviously — but just as truly — promotive of such material reproduction. We find, in short, a division of society into two classes: one essentially confined to *manual*, and the other to essentially *intellectual*, functions. Now in a society of extreme primitiveness, such a division is not yet possible: society has not yet generated enough of a surplus (above subsistence) to support a stratum freed from the exigencies of direct toil. In a society of extreme productive potency, on the other hand, it becomes possible, or indeed essential, for the narrowness of function implied in this material/mental division to be *overcome*: not only can society now *afford* the 'all-round'[49] education of all of its members; such universal all-round education becomes, moreover, essential to the rational administration of a highly advanced industrial society. *Class divisions* accordingly characterize the enormous epoch lying between extreme social poverty, on the one hand, and extreme social plenty, i.e., communism, on the other.

The second basic Marxian consideration is this: the consciousness of people is an expression of those actual skills which they have mastered — and that is to say: of those particular functions they have been trained to fulfil. Each particular function will — just because it is a *particular* function — therefore imply the development of certain talents for dealing with reality, at the expense of the development of certain other talents. The differences in people's thoughts and values are, accordingly, conceived to be a function of the *practical* differences which characterize their productive lives. A botanist and a butcher may look at the same flower, but it is clear that their experiences of it — the actual features of it which they are, respectively, able to discriminate and appreciate — will be radically different.[50]

Once these two basic considerations are taken together, it would appear impossible to avoid the conclusion that *the systematic inability to fully understand the materiality of society's reproduction cycle must be traced to the underdevelopment of directly material talents — and hence ideas and values — in those that are characterized by that inability*. And this is the conclusion that Marx and Engels explicitly draw. They maintain that once the material/mental division of labor comes upon the historical scene,

> consciousness can really flatter itself that it is something other than consciousness of existing practice ... ; from now on consciousness is in a position to emancipate itself from the world and proceed to formation of 'pure' theory, theology, philosophy, ethics, etc.[51]

They thus suggest that the tendency to view intellectual activity as having a development independent of the development of material productive forces is the occupational malady of those whose primary task within the social production process is to concentrate on the intellectual phase of this process. The failure to perceive the actual character of material culture is thus due to the *Einseitigkeit* of the ruling class's participation in the production of that very culture. The tendency to attribute an independent existence to the human intellect is an aspect of the ideology of the privileged stratum of a society.

Now, if the development of a non-materialist bias is to be attributed to the particular function of society's dominant class, it is difficult to escape the extremely ironical conclusion that those who are unable to consider Marx a materialist — i.e., who, to varying degrees, depict him as a defender of the 'Spirit' against the crude encroachments of Engelsian matter — are reading Marx with a distinctively ruling-class bias. It would be very tempting to say that, insofar as this interpretative tendency has become predominant, Marx himself has fallen into the hands of the ideological representatives of the very class he wished to overthrow. For indeed, the values embodied in this outlook would seem to be the values of those who stand socially counterposed to the working class; and whose aspirations — hence practical politics — may therefore be counted upon to be effectively bourgeois.

Is this what we are to conclude? Not precisely. It is not that this analysis is entirely wrong; rather, as it has so far been stated, it is itself tainted by those 'crude and mechanical' tendencies I myself, at the outset of this paper, wished to reject. To maintain that a non-materialist interpretation of Marx must be a bourgeois interpretation is enticing to the polemicist — but it is not very subtle. There are at least two things wrong with such an entirely categorical formulation.

(A) In the first place, it ignores the possibility that the confusions lying behind the criticism of Engels are the innocent result of conceptual-scientific *error*, rather than being traceable to more profound — ideological — roots. This is an important distinction. Whatever Marx and Engels may have seemed to imply in their more programmatic pronouncements, they plainly did not believe that every time an error is made it is due to the *Einseitigkeit* of a person's life's-activity within the economic metabolism of society.[52] Some subjects are simply very difficult: and this judgment applies not only to such fields as quantum mechanics and theoretical economics, but also, of course, to philosophy. Failure to advocate any notion, Marxian materialism included, may be due to such difficulty; and since this is the sort of failure that may befall even those who have in fact been specifically trained to practice a given discipline, we are clearly not entitled to attribute each and every theoretical error to ideological roots — i.e., to the limitations inherent in one's particular social function. It is thus possible that, in the case of some of Engels' critics, the problem is indeed only that of innocent theoretical error; and so we are certainly not automatically entitled to dismiss such critics as being, in effect, bourgeois ideologists.

(B) There is yet another deficiency in the formula according to which opposition to materialism *must* be flatly termed 'bourgeois'. Thus, let us suppose that it is *true* that a given theorist's underplaying of the materialist content of Marxism is indeed due to his having been socialized as a practitioner of the more reflective functions. Even if this is so, it does not follow that he should be unqualifiedly characterized as a bourgeois ideologist. The basis of the problem here was explicitly and sensitively acknowledged by Marx and Engels; the point is that, as the crises afflicting capitalism become ever more acute, "a small section of the ruling class cuts itself adrift and joins the revolutionary class."[53] Such people, Marx and Engels maintain, "supply the proletariat with fresh elements of enlightenment and progress."[54] To this one may add the following: that insofar as members of the intelligentsia do participate in the socialist movement, we may expect that they will be characterized *both* by the ruling-class ideas which played such a significant part in the formation of their character, *and* by the anti-ruling-class ideas they are in the process of adopting. Insofar as Marxism finds recruits among the intelligentsia, we must expect to find hybrid ideological formulations which reflect the movement of such people *from* those tendencies which are bourgeois, *to* those which are authentically revolutionary.

If this latter analysis is accepted, two conclusions are inevitable. The first is that a flat characterization of Engels' opponents as 'bourgeois' is in violation of one of Engels' own most basic principles — namely, that a movement from one qualitative state to another must not be described in terms appropriate to either state taken in isolation, but rather in dialectical terms, i.e., terms which capture both aspects of this *process of becoming*. And if this principle is granted, we must accordingly avoid one-dimensional vilifications of the critics of Engelsian materialism, even if it is true that such critics approach Marxism inevitably trailing bourgeois clouds behind them. Owing to the presence of other, authentically Marxian tendencies in their thought and action, the commitment of such people to the project of *becoming Marxists* may in many cases be considered quite genuine.

But the second inevitable implication of viewing matters in the present light is that there is, nonetheless, a distinct sense in which these critics of Engels are yet to be regarded as captives of the ruling-class ideology. It may be correct, from a Marxian point of view, to tolerate Engels' anti-materialist critics because of the presence of 'other, authentically Marxian tendencies' in their work; but it must equally be insisted that such anti-materialism (insofar as it is not merely attributable to theoretical error) is *per se* a bourgeois tendency. At the very least we may note that, *to the extent* to which one interprets Marx as being some sort of dualistic 'synthesizer'[55] of materialism and idealism, one has not fully taken to heart one of the most significant claims of Marx himself: that the failure to grasp human history as being, most fundamentally, the development of the material creativity of our species, is a failure to be expected of precisely those whose active lives involve only indirect exposure to the day-to-day process of production and reproduction of the all-too-material human world.[56] It is thus a failure to be expected of these who have not themselves directly experienced the fundamentality of what we ordinarily call *labor*. For this reason, the serious Marxist must take Marx and Engels at face value when they assert that *the materialist conception of history is the theoretical bias peculiarly appropriate to the laboring class of society*. And it is in this sense that self-critical Marxists must beware of compromising the thesis of the materiality of culture, which is the basic point of distinction of that bias. The revolutionary character of Marxism would (despite all good intentions to the contrary) otherwise be at least seriously undermined; for this revolutionary character is itself rooted in the experiences and interests of those who know only too well how dependent upon the manipulation of 'matter' our civilization is.

Thus, just as a one-dimensional dismissal of the compromisers of materialism would, as I suggested two paragraphs back, be 'undialectical', so also would it be undialectical not to face up to the persistent problem of *the infusion of bourgeois ideas into Marxian theory itself*. If this problem is not addressed squarely, the danger exists that we will remain under the mystifying influence of the class-stratified social system that we professedly are attempting to transcend. A firm insistence upon the materialist character of Marxism, on the other hand, constitutes an essential aspect of the overcoming of such mystification, and thus inevitably contributes to practical self-understanding on the part of those who participate in the Marxian historical movement.[57] It is for this reason that the vindication of the thoroughgoing, yet subtle, materialism of Friedrich Engels is of more than merely theoretical importance.

NOTES

* I was aided in writing this paper by awards from the Louis M. Rabinowitz Foundation and the Research Foundation of the State University of New York.
[1] These terms themselves occur with an almost predictable frequency in the writings of the critics of Engels to be criticized here.
[2] The terms 'base', 'material foundation', etc., were of course used by Marx himself. See e.g. the celebrated Preface to *Contribution to the Critique of Political Economy*, (International Publishers, New York, 1970), pp. 20–22.

³ *The Betrayal of Marx* is indeed the title of a book published (Harper & Row, New York, 1975), edited by F. L. Bender, which makes all of the anti-Engelsian errors which this essay is intended to dissolve. (See also note 15.)
⁴ The latter term was actually coined by Plekhanov in his article 'Zu Hegels sechzigsten Todestag', published in *Neue Zeit* in 1891. Neither Marx nor Engels appears ever to have used this exact form of words.
⁵ Lukacs, Georg, *History and Class Consciousness* (M.I.T. Press, Cambridge, 1971), p. 3. (Originally published in 1923 as *Geschichte und Klassenbewusstsein*.)
⁶ Marx and Engels, *Selected Works* (Progress Publishers, Moscow, 1962), Vol. II., p. 373.
⁷ See e.g. *Karl Marx: Early Writings*, ed. by T. Bottomore (McGraw-Hill, New York, 1963), p. 155.
⁸ Lichtheim, George, *Marxism* (Praeger, New York, 1961), p. 247.
⁹ *Ibid.*, e.g., p. 258; and *passim*.
¹⁰ Avineri, Shlomo, *The Social and Political Thought of Karl Marx* (Cambridge University Press, Cambridge, England, 1970), p. 66.
¹¹ *Ibid.*, pp. 69–70.
¹² Dupré, Louis, *The Philosophical Foundations of Marxism* (Harcourt, Brace & World, New York, 1966), pp. 142–143.
¹³ Tucker, Robert, *Philosophy and Myth in Karl Marx* (Cambridge University Press, Cambridge, England, 1964), p. 184.
¹⁴ Jordan, Z. A., *The Evolution of Dialectical Materialism* (St. Martins Press, New York, 1967), *passim*.
¹⁵ Bender, ed., *op. cit.*, pp. 41–42.
¹⁶ Strictly speaking, 'physicalism' refers in particular to the view according to which *physical science* is, in principle, a wholly adequate science. There are other, less radical, possible reductionist positions. One might, for example, entertain the more moderate belief that, though biology is *not* reducible to physical science, psychology *is* in principle reducible to the science comprised by physical science *plus* biology. Physicalism is nonetheless the reductionist variant most often discussed, and the one (see Bender, ed., *op. cit.*, pp. 41–42) which Engels is typically accused of maintaining.
¹⁷ Engels, F., *Dialectics of Nature* (hereafter *DN*), (International Publishers, New York, 1940), pp. 15–16.
¹⁸ *Ibid.*, p. 17. (Were Engels (and Darwin) alive today, he (they) would say 'australopithecus' instead of 'monkey'.)
¹⁹ Saying that a 'materialist' believes that everything is made up of 'matter' is approximately as instructive as saying that a 'pessimist' is one who subscribes to 'pessimism'.
²⁰ Gould, S. J., 'Darwin's Delay', *Natural History* (December, 1974), pp. 68–70.
²¹ Quoted in *ibid.*, p. 69.
²² *Ibid.*, p. 69. The choice of the concept of 'heresy' is made by Gould himself.
²³ See note 39, below.
²⁴ Engels, F., *Anti-Dühring* (Foreign Languages Publishing House, Moscow, 1954), Part I, Ch. 12.
²⁵ Mill, J. S., *System of Logic* (J. W. Parker, London, 1843), Bk. III, Ch. VI, Section 2.
²⁶ The term 'emergent' was in fact originally coined by George Henry Lewes in his *Problems of Life and Mind*, 2nd Series (Trübner, London, 1875), Prob. V, p. 412, and has been used in most English-language discussions of this issue ever since.

²⁷ Lewes lived with his famous common-law wife, George Eliot, at 21 North Bank, from 1863 to 1880. Engels' home from 1870 to 1894 was at 112 Regent's Park Road. It was during the 1870s that both did most of the philosophical work in question here. I I have studied a map of London dating from that era, and would judge their residences to be roughly half a mile apart.
²⁸ Marx and Engels, *Selected Works* (Foreign Languages Pub. House, Moscow, 1962), Vol. II, p. 373.
²⁹ Marx and Engels, *Werke*, Vol. 20 (Dietz, Berlin, 1968), p. 513.
³⁰ *DN*, pp. 174–175. Emphasis mine. I may here refer the reader to other passages which make clear not only Engels' anti-reductionism, but also his specific belief in the irreducibility of *human* phenomena. See *DN*, 203 ff. and 209; and also the admirably lucid fragment, 'The Part Played by Labor etc.', referred to in note 29. It should be said in this connection that the young Lukacs had an excuse for his unkind interpretation of Engels that the latter's more recent critics do not have: *DN*, which is particularly rich in anti-reductionist passages, and hence which makes it clear that Engels' 'dialectic' involves no compromise of the role of human conscious activity, had not yet been published when *History and Class Consciousness* appeared.
³¹ Marx's doctoral dissertation, finished in 1841, might also be considered evidence of this attitude – Marx favors Epicurus's atomic theory to Democritus's because the former allows for human freedom – except that it is certain that Marx himself was not yet a materialist at that time.
³² Easton, L. D. and Guddat, K. H. (eds.), *Writings of the Young Marx on Philosophy and Society* (Doubleday, Garden City, 1967), p. 391. Emphasis Marx's.
³³ *Ibid.*, pp. 440–402.
³⁴ See note 2 for source.
³⁵ Marx and Engels, *The German Ideology*, Part One (International Publishers, New York, 1970). (Hereafter referred to as *GI*)
³⁶ *GI*, p. 51.
³⁷ *Ibid.*, p. 42.
³⁸ *Ibid.*, p. 50.
³⁹ Says Marx: "Darwin has interested us in the history of Nature's Technology, i.e., in the formation of the organs of plants and animals, which organs serve as instruments of production for sustaining life. Does not the history of the productive organs of man, of organs that are the material basis of all social organization, deserve equal attention?" It is, says Marx, "technology" which "discloses man's mode of dealing with nature," and which thus underlies the totality of culture – even, indeed, the "misty creations of religion." *Capital*, Vol. I (International Publishers, New York, 1967), p. 372. This passage proves, incidentally, Marx's awareness of his own connection with the Darwinian Revolution, certain aspects of which were discussed in the preceding section. Marx had already remarked to Engels, in a letter of Dec. 19, 1860 (see the *Selected Correspondence* of Marx and Engels (International Publishers, New York, 1942)): "Although it is developed in the crude English style, this [i.e., *The Origin of Species*] is the book which contains the basis in natural history for our view."
⁴⁰ The *locus classicus* of the concepts 'forces' and 'relations' of production is the source cited in note 2.
⁴¹ See e.g. *GI*, pp. 51–52.
⁴² *Ibid.*, p. 53.

⁴³ A complete Marxian account of the human powers would involve showing that the basis of the human reflective function – i.e., language – can itself be accounted for in materialist terms. Marx himself endorses this general position when he notes that even (spoken) language is "agitated layers of air." (*Ibid.*, p. 51.) A compelling account of how language can be treated as a material phenomenon (in Engels' sense of 'material') is the outcome of much recent philosophy – American philosophy in particular. Various papers by Wilfrid Sellars, Hilary Putnam and others are noteworthy in this regard. This philosophical work, whether or not it is carried out with awareness of affinity with Marxism, is clearly congenial to the latter.

⁴⁴ *DN*, passim.

⁴⁵ *GI*, p. 50, Marx and Engels are here speaking of language in particular, but their point is a perfectly general one.

⁴⁶ See note 41, above.

⁴⁷ This term refers here not to any 'ethical' rationality, but rather to the *functionality* of social relations with respect to technology.

⁴⁸ 'Reproduction' is the concept which captures the process of *production* when the latter is most adequately conceived – i.e., as an *ongoing* process which must *eo ipso* regenerate its own basis.

⁴⁹ Marx uses this and cognate formulations throughout his work, and not just in the early writings, to describe the situation in which the mental/manual division of labor has been abolished. The *Grundrisse* of 1857–58 (Penguin Books, Middlesex, England, 1973), by no means an early work, is particularly rich in such formulations.

⁵⁰ The example is my own, but the flavor is distinctly Marxian. See the extremely important manuscript, 'Private Property and Communism', in T. B. Bottomore (ed.), *Karl Marx: Early Writings*, (C. A. Watts & Co., Ltd., London, 1963), pp. 152–167, for Marx's discussion of the relationship between perception and social practice.

⁵¹ *GI*, pp. 51–52.

⁵² Marx's and Engels' discussions of ideology do not treat this issue explicitly, but it would be perverse to construe them in any other fashion.

⁵³ Marx and Engels, *Selected Works*, (Foreign Languages Publishing House, Moscow, 1962), p. 43.

⁵⁴ *Ibid.*, p. 43.

⁵⁵ By now it should be clear that the only 'synthesis' carried out by Marx in the *Theses on Feuerbach* involves no compromising of materialism, but rather consists in emphasizing (as only previous idealists had emphasized) the irreducible importance of human *intentional activity*: Marx's accomplishment is actually the recognition of the centrality of such activity – within the scope of what is nonetheless a thoroughly materialist theory. And if I may say so: This is just what Marx *says*.

⁵⁶ See above, pp. 31–33.

⁵⁷ As with other themes of this paper, so also here: my remarks are to be considered as programmatic, rather than as constituting a fully developed theory. A full development of the present point would require a more detailed exploration than can here be provided of the praxiological implications of historical idealist tendencies under discussion. Some of the best work in this respect is available in one volume: *Studies and Further Studies in a Dying Culture* by C. Caudwell, (Monthly Review, New York, 1971), *passim*.

Prof. Donald D. Weiss
Dept. of Philosophy
State University of New York at Binghamton
Binghamton, New York 13901, U.S.A.

BIBLIOGRAPHY OF THE WRITINGS OF BENJAMIN NELSON*

1933	'Robert de Curzon's Campaign Against Usury'. Unpublished M. A. Thesis, Columbia University.
1939–44	With Joshua Starr, 'The Legend of the Divine Surety and the Jewish Moneylender'. *Annuaire de l'Institut de philologie et d'histoire orientales et slaves* 7 289–338.
1944	'The Restitution of Usury in Later Medieval Ecclesiastical Law'. Unpublished Dissertation. Columbia University.
1947	'The Usurer and the Merchant Prince: Italian Businessmen and the Ecclesiastical Law of Restitution, 1100–1550'. *The Tasks of Economic History* (Supplemental Issue of *The Journal of Economic History* 7 104–122.
1949a	*The Idea of Usury: From Tribal Brotherhood to Universal Otherhood.* Princeton: Princeton University Press. [See also 1969a.]
1949b	'Blancard (the Jew?) of Genoa and the Restitution of Usury in Medieval Italy'. *Studi in Onore di Gino Luzzatto* I 96–116. Milan: A. Giuffrè.
1951	'The Modalities of Thought and the Logics of Action'. In *Value Conflicts, Moral Judgments, and Contemporary Philosophies of Education*. University of Minnesota. Minneo.
1953	Co-editor with Arthur Naftalin, Mulford Q. Sibley, and Donald C. Calhoun. *An Introduction to Social Science: Personality, Work, Community*. New York: Lippincott. Second edition, 1957; third edition, 1961.
1954	'The Future of Illusions'. *Psychoanalysis* 2 16–37. [See also 1956, 1959b, 1967b.]
1955	Co-editor with John Mundy and Richard E. Emery. *Essays in Medieval Life and Thought, Presented in Honor of Austin Patterson Evans*. New York: Columbia University Press.
1955–56	'Adventure of Ideas'. *Psychoanalysis* 4 44–46.
1956	Reprint of (1954). In *Man in Contemporary Society* II, pp. 958–979. New York: Columbia University Press.

* Reprinted, with corrections, from *On the Roads to Modernity: Conscience, Science, and Civilizations. Selected Writings by Benjamin Nelson*, edited by Toby E. Huff. Totowa, New Jersey: Rowman and Littlefield, 1980. With permission of the publisher.

1957

(a) With Marie L. Coleman. 'Paradigmatic Psychotherapy in Borderline Treatment'. *Psychoanalysis* 5 28–44.

(b) Preface to *Freud and the Twentieth-Century*. Edited and selected by Benjamin Nelson, pp. 5–8. New York: Meridian. [See also 1959c and 1962h.]

(c) Foreword to *Psychoanalysis and the Future. A Centenary Commemoration of the Birth of Sigmund Freud*. Edited by B. Nelson and the board of editors of *Psychoanalysis*, pp. v–x. New York: National Psychological Association for Psychoanalysis.

(d) 'On Dr. Walker's "Five Theories" '. *Psychoanalysis* 5 26–27.

1958

(a) Introduction to *Sigmund Freud on Creativity and the Unconscious: Papers on the Psychology of Art, Literature, Love, Religion*. Selected and edited by Benjamin Nelson, pp. vii–x. New York: Harper Torchbook.

(b) 'Social Science, Utopian Myth and the Oedipus Complex'. *Psychoanalysis and the Psychoanalytic Review* 45 120–126.

(c) 'Questions on Existential Psychotherapy'. *Psychoanalysis and the Psychoanalytic Review* 45 77–78.

(d) With Charles Trinkaus. Introduction to Jakob Burckhardt's *The Civilization of the Renaissance in Italy*. 2 volumes. New York: Harper Torchbooks, pp. 3–19.

1959

(a) 'The Great Divide'. *Psychoanalysis and the Psychoanalytic Review* 46 66–68.

(b) 'Communities – Dreams and Realities'. In *Community. Nomos II*, edited by Carl Friedrich, pp. 135–151. New York: Liberal Arts Press. [A revision of (1954).]

(c) *O século de Freud*; [Portuguese edition of *Freud and the Twentieth Century*]. São Paulo, Brazil.

(d) Review of *The Scholastic Analysis of Usury*, by J. T. Noonan (Cambridge, Mass.: Harvard University Press, 1957) *American Historical Review* 64 618–619.

1960 'Psychological Systems and Philosophical Paradoxes'. *Psychoanalysis and the Psychoanalytic Review* 47 43–51.

1961

(a) 'Contemporary Politics and the Shadow of the Sade'. *Psychoanalysis and the Psychoanalytic Review* 48 30–32.

(b) Review of *Consciousness and Society: The Reorientation of European Thought, 1890–1930*, by H. Stuart Hughes (New York: Alfred Knopf, 1958). *American Sociological Review* 26 473–474.

(c) Introductory Comment to 'Apocalypse: The Place of Mystery in the Life of the Mind', by Norman O. Brown. *Harper's Magazine* (May) 46–47.

1962

(a) 'Comments' (on Grant's 'Hypotheses in Late Medieval and Early Modern Physics'). *Daedalus* 91 613–616.

(b) Preface to *The Point of View for My Work as an Author* by Søren Kierkegaard, pp. vii–xxi. New York: Harper Torchbooks.
(c) 'Sociology and Psychoanalysis on Trial: An Epilogue'. *Psychoanalysis and the Psychoanalytic Review* **49** 144–160.
(d) 'Phenomenological Psychiatry, *Daseinanalyse* and American Existential Analysis: A "Progress" Report'. *Psychoanalysis and the Psychoanalytic Review* **48** 3–23.
(e) 'Faces of Twentieth Century Analysis'. *The American Behavioral Scientist* (February) 16–18. [See also 1965h.]
(f) Review of *Religion and Economic Action* by Kurt Samuelsson (New York: Basic Books, 1961). *American Sociological Review* **27** 856.
(g) Editor. *Psychoanalysis and the Social-Cultural Sciences: Contemporary Perspectives*. Special issue of *Psychoanalysis and the Psychoanalytic Review* **49**, 2.
(h) *Freud e il XX Secolo*. (Italian translation of *Freud and the Twentieth Century*.) Verona: Arnoldo Mondadori.

1963

(a) '*The Balcony* and Parisian Existentialism'. *Tulane Drama Review* **7** 66–79.
(b) 'Casuistry'. *Encyclopaedia Britannica* **5**, Reprinted Vol. 5, pp. 51–52 in the 1968 edition. Chicago.
(c) 'Sartre, Genet, Freud'. *Psychoanalysis and the Psychoanalytic Review* **50** 156–171.
(d) 'Über den Wucher.' *Kölner Zeitschrift für Soziologie und Sozialpsychologie.* **15**, Sonderheft 7 (1963): *Max Weber Zum Gedächtnis*, pp. 407–447. [A translation of passages from *The Idea of Usury*.]
(e) 'Hesse and Freud: Two Newly Recovered Letters'. *Psychoanalysis and the Psychoanalytic Review* **50** 11–16.
(f) Editor. *Psychoanalysis and Literature*. Special issue of *Psychoanalysis and the Psychoanalytic Review* **50**, 3.
(g) 'A Professor's Avowal of Personal Faith'. *Newsday* (March 26); 37.

1964

(a) 'Actors, Directors, Roles, Cues, Meanings, Identities: Further Thoughts on "Anomie"'. *Psychoanalytic Review* **51** 135–160.
(b) 'Religion and Development'. *Proceedings of the Sixth World Conference, Society for International Development*. Edited by T. Geiger and L. Solomon, pp. 67–68. Washington, D.C.
(c) 'In Defense of Max Weber. A Reply to Herbert Luethy'. *Encounter* **23** 94–95.
(d) 'Max Weber's *The Protestant Ethic: 1904–1964*'. Abstracts of Papers presented at the 59th Annual Meeting of the American Sociological Association, pp. 22–23. Montreal, Canada.
(e) '"Probabilists", "Anti-Probabilists" and the Quest for Certitude in the 16th and 17th Centuries'. *Actes du Xme congrès internationale d'histoire des sciences* (*Proceedings of the Tenth International Congress of the History of Science*), Ithaca, 1962. **1** 269–273.

1965
- (a) 'Self-Images and Systems of Spiritual Direction in the History of European Civilization'. In *The Quest for Self-Control: Classical Philosophies and Scientific Research*, edited by S. Z. Klausner, pp. 49–103. New York: Free Press.
- (b) 'The Psychoanalyst as Mediator and Double-Agent'. *The Psychoanalytic Review* **52** 45–60.
- (c) A reply to Herbert Marcuse. In *Max Weber und die Soziologie heute (Proceedings of the Fifteenth Annual Meeting of the German Sociological Society)*, edited by Otto Stammer, pp. 192–201. Tübingen: J. C. B. Mohr. [See also 1971b.]
- (d) 'Dialogues across the Centuries: Weber, Marx, Hegel, Luther'. In *The Origins of Modern Consciousness*, edited by John Weiss, pp. 149–165. Detroit: Wayne State University Press.
- (e) 'On Life's Way – Reflections on *Herzog*'. *Soundings* (Spring) 148–154.
- (f) 'Max Weber and Talcott Parsons as Interpreters of Western Religious and Social Development'. Abstracts of Papers presented to the *Annual Conference of the Society for the Scientific Study of Religion*. New York.
- (g) 'Storm Over Weber'. *The New York Times Book Review*, January 3, p. 23 and February 28, pp. 34ff.
- (h) 'Mental Healers and Their Philosophies'. (Reprint of 1962e). *Literature, Religion, Psychiatry*. Special issue of *The Psychoanalytic Review* **52** 131–136.
- (i) Review of *The Sociology of Religion* by Max Weber (Boston: Beacon, 1963). *American Sociological Review* **30** 595–599.
- (j) Editor. *Literature, Religion, Psychiatry*. Special issue of *The Psychoanalytic Review* **52**, 2.

1967
- (a) 'Reflections on Michaelson's Scholarly Study of Religion'. In *The Study of Religion on the Campus Today*, edited by K. D. Hartzell and H. Sasscer, pp. 26–32. Washington, D.C.: Association of American Colleges.
- (b) Reprint of (1954). In *Personality and Social Life* edited by Robert Endleman, pp. 563–576. New York: Random House.
- (c) *Usura e cristianesimo. Per una storia della genesi dell'etica moderna.* [Italian translation of *The Idea of Usury*, 1949a]. Florence: Biblioteca Sansoni.

1968
- (a) 'The Early Modern Revolution in Science and Philosophy: Fictionalism, Probabilism, Fideism, and Catholic "Prophetism"'. In *Boston Studies in the Philosophy of Science* **3**, edited by R. S. Cohen and Marx Wartofsky, pp. 1–40. Dordrecht, Holland: Reidel.
- (b) 'Scholastic *Rationales* of "Conscience", Early Modern Crises of Credibility, and the Scientific-Technocultural Revolutions of the 17th and 20th Centuries'. *Journal for the Scientific Study of Religion* **7** 157–177.
- (c) 'The Avant-Garde Dramatist from Ibsen to Ionesco'. *The Psychoanalytic Review* **55** 505–512.
- (d) Introduction to *The Bourgeois: Catholicism versus Capitalism in Eighteenth-Century France*, by Bernard Groethuysen, pp. vii–xiii. (New York: Holt, Rinehart and Winston).

(e) Editor. *Histories, Symbolic Logics, Cultural Maps.* Special Issue of *The Psychoanalytic Review* **55**, 3.
(f) Co-editor with Marie Coleman Nelson. *Roles and Paradigms in Modern Psychotherapy.* New York: Grune and Stratton.

1969
(a) *The Idea of Usury: From Tribal Brotherhood to Universal Otherhood.* Second edition, enlarged. Chicago: University of Chicago Press.
(b) 'Conscience and the Making of Early Modern Cultures: *The Protestant Ethic* Beyond Max Weber'. *Social Research* **36** 4–21.
(c) Introduction to *Madness and Society*, by George Rosen, pp. vii–ix. New York: Harper Torchbooks.
(d) 'Metaphor in Sociology'. Review of *Social Change and History*, by Robert Nisbet (New York: Oxford). *Science* **116** (19 December); 1498–1500.

1970
(a) 'Psychiatry and Its Histories: From Tradition to Take-Off'. In *Psychiatry and Its History: Methodological Problems in Research*, edited by George Mora and J. Brand, pp. 229–259. Springfield, Illinois: Charles C. Thomas.
(b) 'The Omnipresence of the Grotesque'. *The Psychoanalytic Review* **57** 506–518. (Reprinted in *The Discontinuous Universe*, edited by Sallie Sears and Georgiana Lord, pp. 172–185. New York: Basic Books, 1972.)
(c) 'Is the Sociology of Religion Possible? A Reply to Robert Bellah'. *Journal for the Scientific Study of Religion* **9** 107–111.
(d) Editor. *Philosophy, Technology and the Arts in the Early Modern Era*, by Paolo Rossi. New York: Harper Torchbooks.
(e) Editor. *Cultural Revolutions and Generational Conflicts.* Special Issue of *The Psychoanalytic Review* **57**, 3.
(f) Review of *The Sociology of Max Weber* by Julien Freund (New York: Pantheon, 1968). *American Sociological Review* **35** 549–50.

1971
(a) 'The Medieval Canon Law of Contracts, Renaissance "Spirit of Capitalism", and the Reformation "Conscience": A Vote for Max Weber'. In *Philomanthes: Studies and Essays in the Humanities in Memory of Philip Merlan*, edited by Robert B. Palmer and Robert Hamerton-Kelly, pp. 525–548. The Hague: Martinus Nijhoff. [See also 1972c.]
(b) Comment on Herbert Marcuse's Paper. In *Max Weber and Sociology Today*, edited by Otto Stammer, pp. 161–171. New York: Harper Torchbooks. [English edition of 1965c.]
(c) 'Afterword: A Medium with a Message: R. D. Laing'. In *R. D. Laing and Anti-Psychiatry*, edited by Robert Boyers and Robert Orrill. New York: Harper and Row, pp. 297–301. [Originally published in *Salmagundi* **16** (Spring) 199–201.]
(d) Introduction to 'Max Weber on Race and Society'. (Translated by Jerome Gittleman). *Social Research* **38** 30–32.
(e) Introduction to Emile Durkheim and Marcel Mauss, 'Note on the Notion of Civilization' (translated by B. Nelson). *Social Research* **38** 808–813.

(f) Review of *The Study of Literate Civilizations*, by F. L. K. Hsu (New York: Holt, Rinehart and Winston, 1969). *American Anthropologist* **73** 319–320.

1972
(a) 'Systems of Spiritual Direction' (An Autobiographical Essay). *Criterion* (A Publication of The Divinity School of The University of Chicago) **11** 13–17.
(b) 'Consciences, Sciences, Structures of Consciousness'. *Main Currents in Modern Thought* **29** 50–53.
(c) 'Droit Canon, Protestantisme et "Esprit du Capitalisme". A propos de Max Weber'. (French translation of 1971a) *Archives de sociologie des religions* **34** 3–23.
(d) 'Communities, Societies, Civilizations: Post-Millennial Views of the Masks and Faces of Change'. In *Social Development: Critical Perspectives*, edited by Manfred Stanley, pp. 105–133. New York: Basic Books.
(e) Review of *Science, Technology and Society in Seventeenth-Century England*, by Robert K. Merton (New York: Howard Fertig, 1970). *American Journal of Sociology* **78** 223–231. (Reprinted in *Varieties of Political Expression in Sociology* (An American Journal of Sociology Publication), pp. 202–210. Chicago: University of Chicago Press.)

1973
(a) 'Civilizational Complexes and Intercivilizational Encounters'. *Sociological Analysis* **34** 79–105.
(b) 'Weber's *Protestant Ethic*: Its Origins, Wanderings and Foreseeable Futures'. In *Beyond the Classics?* edited by Charles Y. Glock and P. Hammond, pp. 71–103. New York: Harper and Row.
(c) 'Priests, Prophets, Machines, Futures: 1202, 1848, 1984, 2001'. In *Religion and the Humanizing of Man*, revised second edition, edited by J. M. Robinson, pp. 37–57. Waterloo, Ontario: Council on the Study of Religion. [See also 1976e and 1977c.]
(d) Introduction to 'Max Weber on Church, Sect, and Mysticism' (translated by Jerome L. Gittleman). *Sociological Analysis* **43** 140.
(e) With Vytautas Kavolis. 'Comparative and Civilizational Perspectives in the Social Sciences and Humanities'. *ISCSC Newsletter*. Geneseo, New York. Excerpt in *Comparative Civilizations Bulletin* **5** (Spring) 13–14.
(f) 'An Overview'. Communication to the authors of 'Treatment of Psychosocial Masochism'. *The Psychoanalytic Review* **60** 365–372.
(g) 'The Games of Life and the Dances of Death'. In *The Phenomenon of Death: Faces of Mortality*, edited by Edith Wyschogrod, pp. 113–131. New York: Harper and Row.
(h) With Dennis Wrong. 'Perspectives on the Therapeutic in the Context of Contemporary Sociology: A Dialogue Between Benjamin Nelson and Dennis Wrong'. *Salmagundi* **20** (Summer–Fall) 160–195.
(i) With Donald Nielson. 'Civilizational Patterns and Intercivilizational Encounters'. (A selected and annotated bibliography.) *Bulletin of the International Society for the Comparative Study of Civilizations* **6** (Summer) 3–15.

(j)   With Jerome Gittleman. Introduction to 'Max Weber, Dr. Alfred Ploetz, and W. E. B. DuBois (Max Weber on Race and Society II)'. *Sociological Analysis* **34** 308ff.

1974
(a)   '*Eros, Logos, Nomos, Polis*: Their Shifting Balances and the Vicissitudes of Civilizations'. In *Changing Perspectives in the Scientific Study of Religion*, edited by Allan Eister, pp. 85–111. New York: Wiley-Interscience.
(b)   'Science and Civilizations, "East" and "West": Joseph Needham and Max Weber'. In *Boston Studies in the Philosophy of Science* **11**, edited by R. S. Cohen and M. Wartofsky, pp. 445–493. Dordrecht, Holland: Reidel.
(c)   'Max Weber's "Author's Introduction" (1920): A Master Clue to His Main Aims'. *Sociological Inquiry* **44** 269–278.
(d)   'De Profundis ... Responses to Friends and Critics'. *Sociological Analysis* **35** 129–141.
(e)   'On the Shoulders of the Giants of the *Comparative* Historical Sociology of "Science" – In *Civilizational Perspective*'. In *Social Processes of Scientific Development*, edited by Richard D. Whitley, pp. 13–20. London: Routledge.
(f)   With Harold Rosenberg. 'Art and Technology: A Dialogue Between Harold Rosenberg and Benjamin Nelson'. *Salmagundi* **27** (Summer–Fall) 40–56.
(g)   'Psychoanalysis and Psychohistory – An Analytic Perspective'. *Book Forum: An International Transdisciplinary Journal* **1** 254–262.

1975
(a)   'The Quest for Certitude and the Books of Scripture, Nature, and Conscience'. In *The Nature of Scientific Discovery*, edited by Owen Gingerich, pp. 355–371. Washington, D.C.: Smithsonian Institution Press.
(b)   'Copernicus and the Quest for Certitude: "East" and "West"'. In *Vistas in Astronomy* **17**, edited by A. Beer and K. Aa. Strand, pp. 39–46. Oxford: Pergamon Press.
(c)   'Max Weber, Ernst Troeltsch, Georg Jellinek as Comparative Historical Sociologists'. *Sociological Analysis* **36** 229–240.
(d)   'Quality of Life: Existence, Experience, Expression'. In *Systems Thinking and the Quality of Life*, edited by Clair K. Blong, pp. 19–20. Washington, D.C.: Society for General Systems Research.

1976
(a)   'On Orient and Occident in Max Weber'. *Social Research* **43** 114–129.
(b)   'Max Weber as a Pioneer of Civilizational Analysis'. *Comparative Civilizations Bulletin* **16** (Winter) 4–6.
(c)   'Vico and Comparative Historical Civilizational Sociology'. *Social Research* **43** 874–881.
(d)   Foreword to *The Narcissistic Condition: A Fact of Our Lives and Times*, edited by Marie Coleman Nelson, pp. 13–23. New York: Human Sciences Press.
(e)   'Prêtres, prophètes, machines, futurs: 1202, 1848, 1984, 2001'. [French translation of 1973c]. In *Les Terreurs de l'an 2000*, pp. 227–246. Paris: Hachette.

(f) With Niklas Luhmann. 'A Conversation on Selected Theoretical Questions: Systems Theory and Comparative Civilizational Sociology'. *The Graduate Faculty Journal of Sociology* 1 1–17.

(g) Editor. *Freud and Neo-Marxisms – Erikson and Psychohistories*. Special issue of *The Psychoanalytic Review* 63, 2.

1977

(a) *Der Ursprung der Moderne: Vergleichende Studien zum Zivilisationsprozess.* (*The Origins of Modernity: Comparative Studies of Civilizational Process*). Frankfurt am Main: Suhrkamp.

(b) Review of *Business, Banking, and Economic Thought in Late Medieval and Early Modern Europe: Selected Studies*, by R. de Roover, edited by Julius Kirshner. (Chicago: The University of Chicago Press, 1974). *Journal of European Economic History* 6 487–491.

(c) 'Priester, Propheten, Maschinen, Zukünftiges: 1202, 1848, 1984, 2001'. (German translation of 1973c). In *Die Schrecken des Jahres Zweitausend*, edited by Henry Gacanna, pp. 244–267. Stuttgart: Klett.

(d) 'Tradition and Innovation in Law and Society: Comparative Historical and Civilizational Perspectives. A Paper in Progress'. Prepared for the Conference at Freiburg (September 18–25, 1977), sponsored by the Institut für Historische Anthropologie. Unpublished.

NAME INDEX

Achinstein, P.   27, 29, 46, 48, 134, 181
Adams, F. D.   203
Adorno, T.   282
Agassi, J.   2
Agrippa of Nettesheim   142
Akabori, S.   209, 211
Albritton, C. C.   203
Amira, K. v.   311, 318
Aquinas, T.   315
Arendt, H.   305
Aristotle   147, 157–158, 184, 186, 197, 202, 217, 294
Ashby, W. R.   103, 133, 134
Auden, W. H.   173
Augustine   309
Avineri, S.   176, 181, 334, 335, 337, 356
Ayala, F. J.   47, 134

Babb, M.   262
Bachelard, G.   318
Bacon, F.   41, 142, 317, 326, 344
Baer, K. E. v.   184–185
Baeyer, A. J. F. v.   152
Barash, D. P.   219, 223, 224, 230, 233
Baring-Gould, S.   306, 314, 319
Barker, S.   181
Barrett, W.   260, 262
Bartley, W. W.   282
Bartolus of Sassoferrato   305
Baumgardt, C.   316, 319
Beach, F. A.   219, 233
Becker, G. S.   222, 233
Beckner, M.   43, 45, 46, 48, 49, 52, 53, 55, 56–61, 63, 72, 76, 85, 100, 102, 133, 134
Beckwith, J.   233
Beer, A.   365
Bell, A.   152
Ben-David, J.   151, 170

Bender, F. L.   335, 356
Berger, J.   206, 211
Berkeley, G.   243
Bernal, J. D.   152, 170, 198, 203, 206, 210, 211
Bernard, Saint   312
Bertalanffy, L. v.   250, 261
Bigelow, J.   47, 48, 50, 135
Blong, C. K.   365
Bohr, N.   29, 146
Boltzmann, L.   140
Boorse, C.   46, 134
Bottomore, T. B.   356, 358
Boyd, W. K.   319
Boyers, R.   363
Brand, J.   363
Brodbeck, M.   41
Bromberger, S.   2
Bronowski, J.   256, 257, 262
Brooks, C.   256, 262
Brown, G. E.   196
Brown, N. O.   360
Brown, Peter   319
Buchler, J.   203
Buck, R. C.   135
Bullock, T. H.   250
Bunge, M.   29, 30
Burckhardt, J.   360
Burian, R.   29, 33–40, 41
Burnet, F. M.   197
Butts, R. E.   27, 29

Cairns-Smith, A. G.   201, 202, 203, 206, 207, 211
Calhoun, D. C.   359
Campbell, B.   234
Campbell, D. T.   46, 90, 104, 112, 113, 131–132, 134
Camus, A.   216, 231
Cantor, G.   186

367

NAME INDEX

Carnap, R.   2, 5, 6, 25, 29, 133, 134, 177, 265
Carus, P.   305, 319
Caspar, M.   316, 317, 319
Caudwell, C.   358
Cavanna, H.   366
Chargaff, E.   211
Chasseneux, B.   312–314
Chomsky, N.   223
Chorover, S.   233
Clausius, R.   188
Cohen, R. S.   25, 29, 30, 31, 41, 135, 170, 233, 262, 280, 325, 330, 362, 365
Cohn, N.   319
Colbert, J. B.   318
Cole, L. J.   296, 297, 319, 321
Coleman, M. L. [Benjamin Nelson's wife] see Nelson, M. C.
Colletti, L.   171
Collingwood, R. G.   180, 316, 319
Colodny, R. G.   29, 30, 31
Conway, J.   192
Copernicus, N.   143, 147, 198, 328
Coulton, G. G.   303, 304, 305, 319
Crick, F. H. C.   191, 198, 208, 209, 210, 211
Cummins, R.   46, 86, 134
Curry, W. C.   319

Dahrendorf, R.   176, 181
Darwin, C.   179, 188–189, 195, 197, 198, 203, 212, 340–341, 356, 357
Davidson, J.   211
De Roover, R.   366
Dee, J.   142
Degens, E. T.   209, 211
Deloria, V.   301
Democritus   357
Denom, L. E.   213
Descartes, R.   155, 198, 237, 239, 241, 339, 342
Deschamps, P.   307, 319
Dewey, J.   195, 212–213
Di Angelis, M.   181
Dicke, R. H.   24
Dijksterhuis, E. J.   316, 319

Dingler, H.   272
Dilthey, W.   177
Dobzhansky, T.   134
Dombrowski, H. D.   166, 171
Douglas, M.   319
Dupré, L.   335, 356
Durant, W.   202
Durkheim, E.   300, 363
Dusek, V.   218, 234
Dyson, F.   190, 203

Easton, L. D.   357
Eberhard, W.   326, 330
Eccles, J. C.   258–259, 262
Edelman, G. M.   46, 134
Edison, T. A.   152
Edwards, R. P.   134
Eglinton, G.   211
Egner, R. E.   213
Engels, F.   137, 235, 237, 242, 332–339, 341–349, 351–355, 356, 357, 358
Eichenbaum, H.   232, 233
Eichmann, A.   305
Einstein, A.   21, 24, 146, 187, 198, 316
Eiseley, L.   181
Eisenberg, L.   232, 233
Eister, A.   365
Eliade, M.   183, 185, 202
Eliot, G.   357
Emery, R. E.   359
Enc, B.   46, 48, 134
Endleman, R.   362
Epicurus   357
Erikson, E. H.   260
Euclid   283, 284, 285, 287–291, 292, 293
Eutychius, patriarch of Alexandria, (877–940) 306
Evans, E. P.   295, 296, 297, 309, 311, 312, 313, 314, 315, 319, 321

Fairbank, J. K.   330
Farnell, L. R.   319
Fauconnet, P.   319
Feigl, H.   2, 29, 30, 41
Feuerbach, L.   164

NAME INDEX

Feyerabend, P.   1–2, 4, 5, 6, 8–24, 25, 26, 27, 28, 29, 30, 33–41, 165, 170, 262
Fischer, K.   238
Fisher, R. A.   189, 203
Flanagan, O.   181
Florkin, M.   203, 211
Foster, M.   42
Frank, P.   2
Frankl, V.   247, 262
Frege, G.   186, 187
Freud, A.   260
Freud, S.   260, 261
Freund, J.   363
Friedlander, J.   234
Friedrich, C.   360
Frisch, C.   320
Fryer, A. C.   306, 307, 319, 321

Galileo   28, 198, 323
Gallop, P. M.   209, 211
Gardner, M.   203
Gebhardt, C.   245
Geiger, T.   361
Gerth, H. H.   319
Ghiselin, M. T.   203
Giedymin, J.   27, 30
Giles, E.   234
Gingerich, O.   365
Ginsberg, M.   319
Gittleman, J.   365
Glansdorff, P.   55, 56, 112, 122, 134
Glock, C. Y.   364
Gödel, K.   187–188
Goodman, N.   2
Gould, S. J.   203, 233, 356
Grant, E.   360
Greenberg, D. S.   152, 171
Gregory, M. S.   233
Grillot de Givry, E. A.   319, 321
Groethuysen, B.   362
Grünbaum, A.   280
Guddat, K. H.   357
Gutting, G.   28, 30

Habermas, J.   164, 171, 278, 280, 281, 282

Haeckel, E.   199
Haken, H.   196, 203
Halfmann, J.   153, 170, 171, 173–180
Hallen, B.   42
Hamerton-Kelly, R.   363
Hammond, P.   364
Han Yü   328
Hanson, N. R.   2, 27
Harris, E. E.   202
Harris, M.   228, 233
Hartman, H.   203
Hartzell, K. D.   362
Hegel, G. W. F.   139, 238, 243, 334, 336, 340, 341
Heisenberg, W.   1, 34, 35, 301, 319
Hempel, C. G.   2, 5, 6–8, 25, 26, 30, 46, 107, 134, 284, 294
Hering, E.   199
Hesse, M.   2
Hessen, B.   155, 171
Hilbert, D.   187
Hirschleifer, J.   222, 233
Ho Peng Yoke   326, 330
Hoagland, M.   211
Hobbes, T.   239, 339, 344
Hodgkin, A. L.   248
Hollingshead, L.   231
Horkheimer, M.   280
Howard, D.   280
Hrosvitha (Hrotsuit)   306
Hsu, F. L. K.   364
Hubbard, R.   233
Huff, T. E.   359
Hughes, H. S.   360
Hull, D.   46
Hume, D.   141, 158, 159, 162
Hutchins, R. M.   320
Hutton, J.   200, 203
Huxley, A. F.   248
Hyde, W. W.   311, 314, 319

Ibsen, H.   137
Inouye, H.   233

Jacobsen, E.   226, 233
Jakobson, R.   193, 203
James, W.   176, 193, 195, 203, 301, 320

NAME INDEX

Jammer, M. 316, 320
Janik, A. 181
Jay, M. 280
Jeffress, L. A. 202
Jerne, N. K. 197
Jesus 309
Johanssen, W. 179
Jordan, Z. A. 335, 356

Kant, I. 150, 154, 155–163, 165, 166, 167, 171, 174, 178–179
Kaplan, N. 170
Katchalsky, A. 206, 211
Kavolis, V. 364
Kepler, J. 158, 198, 299, 315–317, 320, 328
Keynes, J. M. 318, 320
Kierkegaard, S. 361
Kilman, R. 174, 181
Kirshner, J. 366
Klare, C. 280
Klausner, S. Z. 362
Klein, M. 28, 30
Koch, R. 189
Koehler, W. R. W. 321
Koestler, A. 250, 261, 262, 320
Kolata, G. B. 126, 134
Kordig, C. R. 27, 28, 30
Koyré, A. 317, 320
Krämer, H. [Henricus Institoris] 309, 321
Kropotkin, P. A. 137
Kuhn, T. S. 1–2, 8–12, 14–15, 20, 21, 24–25, 26, 28, 30, 34–35, 141–142, 149, 150, 153, 165, 171, 175, 255, 324

Lakatos, I. 2, 26, 27, 30, 141–142, 153, 165, 170, 181, 294
Lancre, P. de 305
Lang, M. 157, 162, 163, 171
Lange, R. 233
Langer, S. K. 223, 233
Langton, E. 303, 320
Layton, E. 151, 171
Lea, H. C. 302, 309, 312, 314, 320
Lear, J. 320

Leavis, F. R. 261, 262
Lecky, W. E. H. 315, 320
Leeds, A. 218, 226, 233, 234
Leibniz, G. W. 158–159, 162, 236, 243
Lenin, V. I. 242
Lenormant, F. 320
Leplin, J. 27, 30
Lévi-Strauss, C. 137, 227
Lewes, G. H. 342, 356, 357
Lewontin, R. C. 203, 213, 233
Lichtheim, G. 334, 335, 336, 337, 356
Lieber, M. 231, 233
Long, C. H. 202
Lord, G. 363
Louis XIV 318
Luhmann, N. 168, 171, 366
Lukacs, G. 332–333, 334, 335, 336, 356, 357
Lysenko, T. D. 144, 189

McCarthy, T. 177
McCown, E. R. 234
Machamer, P. 26, 28, 30
McIntyre, D. B. 200, 203
Mackie, J. L. 126, 135
MacLeish, A. 256, 257, 262
Mandrou, R. 315, 320
Manier, E. 47, 48, 52, 53, 57, 71–72, 74, 84, 97, 100, 133, 135
Marcuse, H. 362, 363
Marett, R. R. 300
Martin, M. 27, 28, 30, 42
Marx, K. 137, 151, 161–162, 164, 171, 176–177, 263, 324, 331–336, 341, 343–355, 356, 357, 358
Matheja, J. 209, 211
Matson, F. W. 250, 262
Mattack, R. D. 203
Mauss, M. 363
Maxwell, G. 2, 29, 30
Maxwell, J. C. 188
Mead, G. H. 195
Mendel, G. 189, 198
Merton, R. 324, 364
Michelet, J. 318, 320

NAME INDEX

Michelson, A. A.   20, 27, 187
Michurin, I. V.   189
Mill, J. S.   41, 140, 141, 342, 356
Mills, C. W.   319
Miller, L.   233
Milne, E. A.   24
Mintz, S. L.   262
Mitroff, I.   174, 181
Monod, J.   119, 135, 194, 203
Monter, E. W.   320
Mora, G.   363
Morley, E. W.   20, 27, 187
Morton, J. A.   152, 171
Mundy, J.   359
Mungello, D. E.   325, 330
Murphy, M. T. J.   211
Musgrave, A.   30, 181, 294

Naftalin, A.   359
Nagel, E.   2, 5, 6–8, 25, 26, 31, 46, 49, 108, 133, 135
Nakayama, S.   302, 320
Needham, J.   295, 297, 299, 300, 301, 311, 314, 317, 320, 323, 324–325, 329
Nelson, B.   231, 232, 233, 234, 300, 359–366
Nelson, M. C.   360, 363, 365
Neumann, J. v.   188, 190–192, 198, 202, 203, 204
Neurath, H.   211
Neurath, O.   265
Newton, I.   24, 97, 113, 140, 154–155, 157–161, 164, 165, 166, 178, 179, 198, 318, 340
Nielson, D.   364
Nisbet, R.   363
Noble, D.   151, 171
Noonan, J. T.   360
Nord, F. F.   211
Nordenskiöld, E.   203

O'Connell, M.   280
Odum, E. P.   213
Onsager, L.   200
Orrill, R.   363
Overton, C. E.   210–211

Paecht-Horowitz, M.   206, 211
Palmer, R. B.   363
Palter, R.   31
Parmenides   292
Pasteur, L.   189
Pattee, H. H.   194, 203
Paul the Deacon (Paulus Diaconus, c. 720–800)   306
Pauling, L.   198
Peano, G.   186, 187
Peirce, C. S.   188, 195, 203
Perutz, M. F.   211
Philolaus   143, 147
Piaget, J.   248, 257
Pignot, J.-H.   312, 313, 320
Planck, M.   21, 28, 31, 187
Plato   186, 187, 217, 244, 285, 286, 289, 336
Plekhanov, G. V.   356
Pliny   316
Plutchik, R.   226, 234
Polanyi, M.   111–112, 113, 115–116, 135
Pólya, G.   283, 284, 294
Popper, K. R.   2, 33, 34, 39, 41, 42, 112, 113, 135, 140–141, 153, 167–168, 171, 173, 177, 179, 263–280, 281, 282, 324
Powers, W. T.   49, 135
Price, D. J. de Solla   171
Prigogine, I.   55, 56, 112, 113, 122, 134
Ptolemy   147
Putnam, H.   1, 25, 31, 358
Pythagoras   186, 187

Quine, W. V. O.   2, 179, 194, 203

Racine, J.   318
Radner, M.   30, 31
Raphael   199
Reichenbach, H.   2, 177
Rescher, N.   60, 135
Rich, A.   209, 211
Rieke, R.   181
Ripley, R. L.   297
Robin, P. A.   295, 320

Robinson, J. M.   364
Rosen, E.   317, 320
Rosen, G.   363
Rosenberg, H.   365
Rosenberg, N.   151, 171
Rosenblueth, A.   47, 48, 50, 84, 135
Ross, W. D.   202
Rossi, P.   363
Roux, W.   197
Ruse, M.   47, 84, 88, 90, 126, 127–128, 135
Rutherford, E.   29
Russell, B.   186, 187, 213, 243, 245
Russell, E. S.   202
Russell, J. B.   305, 321

Sahlins, M. D.   222, 234
Samuelson, P. A.   222, 234
Samuelsson, K.   361
Sasscer, H.   362
Schäfer, L.   281
Schaffner, K.   45, 135
Schapiro, M.   305, 307, 321
Scheffler, I.   26, 27, 31, 42
Schilpp, P. A.   30, 31, 134
Schreier, H.   233
Schrödinger, E.   113, 199, 203, 252
Schulman, L. S.   203
Scopes, J. T.   179
Scriven, M.   29, 30
Sears, S.   363
Seeger, R. J.   233
Sefter, S.   209, 211
Seiden, P. E.   203
Seitz, W. A.   251
Sellars, W.   1, 2, 8, 25, 26, 31, 358
Semon, R. W.   199
Shakespeare, W.   340
Shapere, D.   16, 27, 28, 31
Shimony, A.   55, 133, 135, 255–256, 262
Sibley, M. Q.   359
Simon, H. A.   60, 135
Silvers, A.   233
Sivin, N.   301, 302
Skeat, W. W.   321
Skinner, B. F.   195, 258, 259–260, 261

Smart, J. J. C.   1, 25, 31
Smith, W. Robertson   321
Smythies, J. R.   247, 250, 261, 262
Snow, C. P.   261
Socrates   186, 285–286, 289
Spencer, H.   140, 184–185, 202
Sperry, R. W.   258
Spiegel-Rösing, I.   171
Spinoza, B.   237–242, 245
Spiro, M. E.   321
Sprenger, J.   309, 321
Spurr, H. M. B.   296
Stachel, J.   29
Stacy, P.   225, 234
Stalin, J.   331
Stammer, O.   362, 363
Stanley, M.   364
Starr, J.   359
Stegmüller, W.   157, 158–160, 171
Stent, G. S.   211
Strand, K. Aa.   365
Sutch, D.   233
Swift, J.   187
Szabó, A.   294

Tarski, A.   6
Taylor, H. O.   304, 321
Taylor, R.   50, 133, 135
Tertullian   304, 321
Theophilus the Penitent   305–308
Thorndike, L.   318, 321
Topitsch, E.   30, 31
Toulmin, S.   2, 157–158, 165, 171, 181
Trinkaus, C.   360
Trivers, R. L.   216, 224, 230, 234
Tucker, R.   335, 356
Tullock, G.   222, 234
Turing, A. M.   185, 188, 192, 202, 204–206

Ulam, S. M.   192

Valois, N.   321

Waddell, H.   321
Wade, N.   232, 234